Transcription Regulation in Prokaryotes

Transcription Regulation in
Prokaryotes

Rolf Wagner

Heinrich-Heine University, Düsseldorf, Germany

OXFORD
UNIVERSITY PRESS

Great Clarendon Street, Oxford OX2 6DP

Oxford University Press is a department of the University of Oxford.
It furthers the University's objective of excellence in research, scholarship,
and education by publishing worldwide in

Oxford New York

Athens Auckland Bangkok Bogotá Buenos Aires Calcutta
Cape Town Chennai Dar es Salaam Delhi Florence Hong Kong Istanbul
Karachi Kuala Lumpur Madrid Melbourne Mexico City Mumbai
Nairobi Paris São Paulo Singapore Taipei Tokyo Toronto Warsaw

with associated companies in Berlin Ibadan

Oxford is a trade mark of Oxford University Press
in the UK and in certain other countries

Published in the United States
by Oxford University Press Inc., New York

First published 2000

British Library Cataloguing in Publication Data

A catalogue record for this book is available from the British Library

Library of Congress Cataloging in Publication Data

(Data applied for)

ISBN 0 19 850354 7

1 3 5 7 9 10 8 6 4 2

Typeset by RefineCatch Limited, Bungay, Suffolk
Printed in Great Britain
on acid-free paper by
Biddles Ltd., Guildford, Surrey

Foreword by Dr Peter Geiduschek

The dating of historical foundations is seldom complete as well as precise. (Indeed, an uncertainty principle of sorts may prevail: precision is won at the cost of completeness.) Nevertheless, the specific foundations of this book on Transcription Regulation in Prokaryotes can be dated fairly precisely to forty years ago by two documents, whose central significance to its subject matter can hardly be doubted: the discovery of RNA polymerase was announced tersely in 1959 by Weiss and Gladstone; the operon model of gene regulation was presented in detail in 1961 by Jacob and Monod (following brief partial presentations in 1959 and 1960 in the Comptes Rendus of the French Academy of Sciences).

The interweaving of genetic and biochemical analysis of transcription has been central to its profound development during the subsequent forty years. For approximately the first decade, the bacteria and bacteriophages formed the principal focus of this endeavor (in effect, the only real game in town). This was mostly a matter of technical necessity, but partly also a social phenomenon. The technical barriers to tackling the core questions of gene regulation in multicellular organisms were numerous and formidable. Recapitulating the brilliant success of the phage group with animal viruses seemed like a promising starting point, but even this effort required another decade of preparation. In the meantime, the excitement of new discoveries, the magnetism of brilliant expositors, and the rapid establishment of a school of especially talented researchers, all exerted a powerful pull in favor of the bacteria. Then recent social innovation also played an important part: the Phage Course, which served as a kind of master class for molecular biologists, drew students from a wide variety of backgrounds, including many from the physical sciences, whose natural tendency to reductionism was likely to be reinforced by a wide-ranging ignorance of Biology. But a bacteriocentric reductionism was also in the minds and the hearts of those who came to the work with a broad background in Biology or medical training. An important example of this thinking is represented by Monod and Jacob's summary article in the Symposium volume of the 1961 Cold Spring Harbor meeting on Cellular Regulatory Mechanisms. The article includes a section on 'regulation and differentiation in higher organisms' with closing comments on cancer, couched in terms of a negative regulation model of the operon. Of course, these were intended as speculative proposals offered as frameworks for thinking and discussion, as guides to the perplexed. It was made clear that 'eventually . . . differentiation will have to be studied in differentiat(ing) cells'. Cloning and the spectacular development and enrichment of techniques enabling molecular genetics in eukaryotes, and especially in metazoans, has progressively transformed these early perspectives. The early proposals are less important for their specific content than for signalling a determination to formulate fundamental problems of genetics, development (and, by implication, evolution) in concrete and generalizable, molecular and mechanistic terms.

Comparable technical developments have also enormously broadened our perspective on gene regulation in the bacteria in two ways: by enlarging the territory for exploring regulatory phenomena and circuits, and by greatly facilitating the deeper exploration of mechanism. This has generated a revised sense of the importance of studying gene regulation in the bacteria (or, speaking more generally, in the prokaryotes). The prokaryotes represent the vast preponderance of organisms by number, and the largest part of the earth's biomass. It is in the prokaryotes that the greatest proportion of evolution has taken place and they represent the largest part of currently existing biological diversity. Because we know so little about most of them, the prokaryotes must represent an enormous reservoir of unexplored science, and of yet to be discovered novelty. At the same time, a deeper, quantitative, and mechanistically sophisticated study of transcription and its regulation in the bacteria continues to offer outstanding opportunities to form a coherent understanding of cell function and genetic processes based on reaction mechanism and molecular structure. At the most basic level of structure and mechanism, current research on bacterial transcription is laying down foundations that will support the next new level of understanding of transcription in all cells and organisms.

This is the task to which this book is dedicated. It has evolved out of a course that Professor Wagner has taught to students in the third and fourth years of University studies in Germany; such a course might be offered also to students in the last year of an honors undergraduate degree elsewhere, or, in the US, in a senior level undergraduate or first-tear graduate course. The book combines the structure of a textbook with coverage of its topics at a level that is usually reserved for reviews and monographs. The ample list of references that it provides is weighted toward the secondary review literature; a full listing of its primary reserach sources might have nearly doubled its lenght. In our current setting of Medline and of journals widely available on line, this strikes me as a prudent choice. Assessing the research literature critically and in detail is, of course, an essential part of the training of scientists. Doing that with every paper to which reference might be made in the context of this book is neither a useful nor a practical option.

Preface

Our present day understanding of Molecular Biology has been largely influenced by the rapid progress within the field of transcription. Transcription represents a fundamental step in the flow of genetic information, and studying this process has inspired many scientists all over the world. The rapidly increasing amount of data collected in recent years has contributed enormously to our present perception of gene regulation. This is true for transcription in both prokaryotes and eukaryotes, although in prokaryotes transcription is a more central process and thus has a much larger impact on the overall regulation of gene expression. This is simply due to the fact that in prokaryotes the macromolecular synthesis reactions of replication, transcription and translation are not compartmentalized, but are tightly coupled processes which directly influence each other. As a consequence we find a number of specific regulatory features that are unique to prokaryotes, whereas others are found in both prokaryotes and eukaryotes. These distinctions make studies of prokaryotic transcription particularly interesting. Moreover, in recent years we have seen a tremendous increase in information in the transcription field, with many molecular details of the transcription apparatus and mechanistic details of many regulatory events emerging. Hence, our understanding of the process of transcription and its regulation has evolved in the last years from a coarse view of the components and processes to a more detailed picture at higher resolution. Over the years, I have tried to pass on this interesting development to the students that attended my lectures in prokaryotic transcription regulation. One recurring question with which I have been faced many times, and which has been often asked by the students was whether I could recommend a textbook that covers all the recent details—a book that represents a comprehensive description of prokaryotic transcription and its regulation. I always answered evasively and recommended a collection of original journal articles and reviews. Clearly, however, students prefer books to original journal articles or reviews. So gradually the idea was born to summarize, sort and digest all the growing information on transcription and its regulation which I taught in my lecture, and to present it in a textbook.

It took a long time from the moment when the idea was born to the point when the work was actually started. This moment was triggered by the fortuitous coincidence that I met Dr Lulu Stader, who at that time was a commissioning editor at Oxford University Press. Without her enthusiastic encouragement I certainly would have hesitated much longer, and may be not yet have even started to write this textbook. She helped me to overcome the initial activation threshold and competently guided the first stages of the preparation of the manuscript where I realized that I had a lot to learn. This is why I am especially grateful to her. Of course I was underestimating the efforts and time necessary for such a project. Thoughtlessly I decided to do everything by myself, including artwork for the figures and all the typing. Two consequences arose from this. Firstly, the project was slowed down considerably, and secondly, I am solely responsible for all the errors or inconsistencies in the text or the figures. I hope that above all this book contributes towards helping students successfully tackle the complex subject of transcription regulation. The book is also intended as a guide for the more advanced reader who seeks a comprehensive description of prokaryotic transcription regulation. I would be very happy if this book also serves to raise or extend the reader's general scientific interest for Molecular Biology, and if it passes on some of the joy and affection I always felt for Molecular Biology then I will be exceptionally pleased.

R. W.

Düsseldorf
September 1999

Acknowledgements

This book is dedicated to *E. coli* which has provided most of the information presented in this book and from which much more exciting new information can be expected in the future.

I also like to express my deep gratitude to the friendly staff at Oxford University Press for their kind and professional help. I am especially thankful to Dr. Cathy Kennedy, Esther Browning and John Grandidge. They really did a great job and it was a great pleasure for me to work with them! In addition, I wish to thank all those who have provided me with information and allowed me to adopt figures and data that had been published before. The list of their names is too long to be written out here, but credit is given to each of them in the figure legends whenever a special allowance has been granted.

Certainly I have neglected numerous duties while I was deeply involved in writing this manuscript. I apologize to all those who had to suffer from my lack of attention during this time. In particular I like to thank all the members of my lab who have patiently supported my efforts in getting this manuscript ready for publication. I am indebted to Peter Geiduschek for his readiness to write the Foreword for this book. Last but not least, I would especially like to thank Sabine Kaul for her encouragement, support and patience. Although she will probably never read the book, she is the happiest person to see it finished.

R. W.

Contents

1

General introduction

In a very simplified way living organisms may be characterized by their capacity to undergo self-reproduction. Generally, self-reproduction requires a complicated system of biomolecular reactions. Those reactions are normally performed by macromolecules and carried out in a compartment, for example, a cell. Living cells contain all the necessary molecules required for reproduction. For a complicated organism the number of such macromolecules may be immense. To simplify matters, the macromolecules may be grouped into a few categories. They could be divided operationally into groups of **structural**, **functional** or **informational molecules**. They can, however, also be classified according to their chemical nature. If we do not consider small molecules and storage compounds to be essential for reproduction, such a division leaves us with only two classes of molecules, namely **nucleic acids** and **proteins**. Proteins fulfil structural (e.g. keratin or collagen) and functional tasks (e.g. as enzymes, antibodies, or regulatory and transport functions). Nucleic acids are similarly versatile. They serve as structural, functional *and* information-providing molecules. There are two types of nucleic acids, **DNA** and **RNA** (see Box 1.1), both of which are suitable for the transfer of information. DNA is chemically more inert, and thus serves for permanent storage.

In most cases DNA represents the hereditary material, and a (complete) copy is passed from one cell to the next or from one generation to the next. The process of DNA duplication is described as replication. In contrast to DNA, RNA can be regarded as a short-lived 'blueprint' from the DNA. Hence, RNA represents a transient copy of the DNA. The copying process from DNA to RNA is termed **transcription**, and generally only one or several functional units or segments of the DNA sequence are transcribed at a time. Such (functional or structural) units are termed genes. The RNA products or transcripts may serve different functions. They can either act directly as structural or functional molecules (e.g. transfer RNAs (tRNAs) or ribozymes; Table 1.1), or they are used as construction plans (messenger RNAs (mRNAs)) for the synthesis of proteins.

The protein biosynthesis reaction, which converts the information stored in a mRNA sequence into the amino acid sequence of a protein, is termed **translation**. Hence, in a strongly simplified way, living cells can be characterized by

Box 1.1 Nucleic acid building blocks and transcription scheme

The monomeric building blocks of nucleic acids are termed **nucleotides**. Nucleotides are composed of three molecular constituents:

(1) aromatic purine bases **adenine** and **guanine** or pyrimidine bases **uracil**, **cytosine**, and **thymine** in DNA;
(2) either one of the C5 sugars **ribose** or **deoxyribose**;
(3) **phosphate** residues.

As a convention, atomic positions are numbered; to distinguish the atomic positions within the sugars and the bases the positions within the sugar molecules are indicated by a prime (C1′ to C5′). For example, the aromatic bases are linked through a glycosidic bond between the C1′ of the sugars and either the N1 position of pyrimidines or the N9 position of purine bases (Fig. B1.1a).

In polynucleotides the individual building blocks (nucleotides) are linked by phosphodiester bonds between the 5′ and 3′ OH groups of the sugars. By convention, a polynucleotide sequence is written from the 5′ end to the 3′ end usually in a left to rightward direction.

In *DNA* the C5 sugar is **deoxyribose** and the pyrimidine bases are **thymine** and cytosine. DNA generally exists as a long helical molecule of two complementary strands of polynucleotide chains with opposite polarity (Box 6.1). *RNA* differs from DNA as the C5 sugar is **ribose** instead of deoxyribose. Moreover, the pyrimidine thymine does not normally occur in RNA but is replaced by **uracil.**

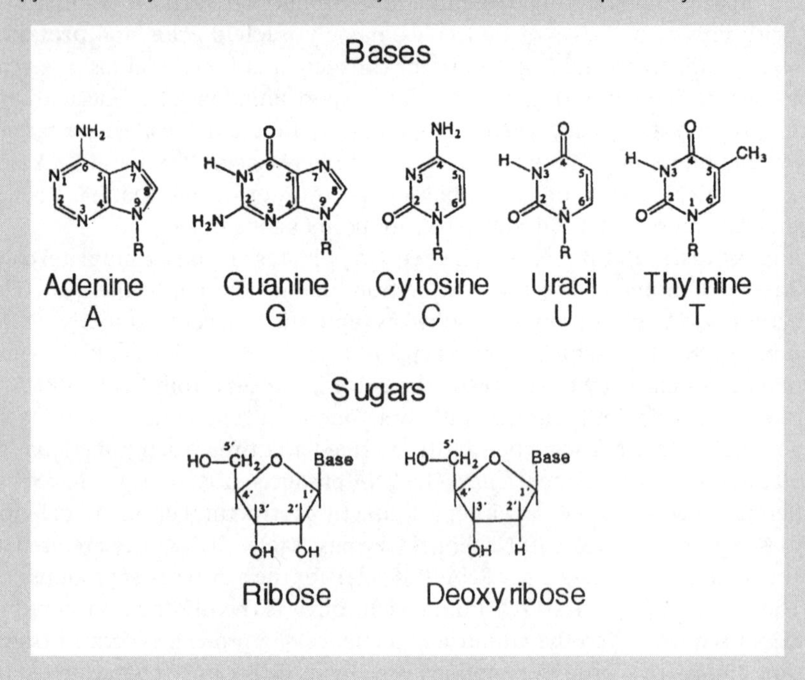

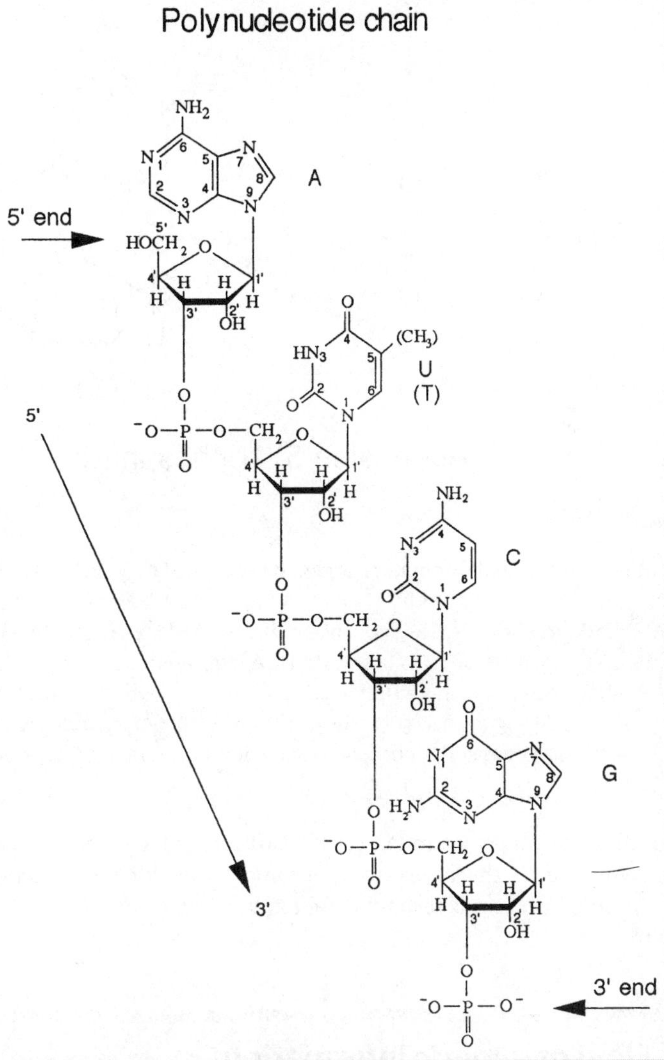

Polynucleotide chain

Fig. B1.1a Structure of nucleic acids and their building blocks

During *transcription* the sequence of one *DNA* strand is converted into a sequence of *RNA*. The substrates for this reaction are ribonucleotide 5′ triphosphates (NTPs). ATP is shown as an example (Fig. B1.1b).

The NTPs are linked in a Mg^{2+}-dependent reaction catalysed by DNA-dependent RNA polymerase under release of pyrophosphate (PPi)

$$NTP + (NMP)_n \longrightarrow (NMP)_{n+1} + PPi$$

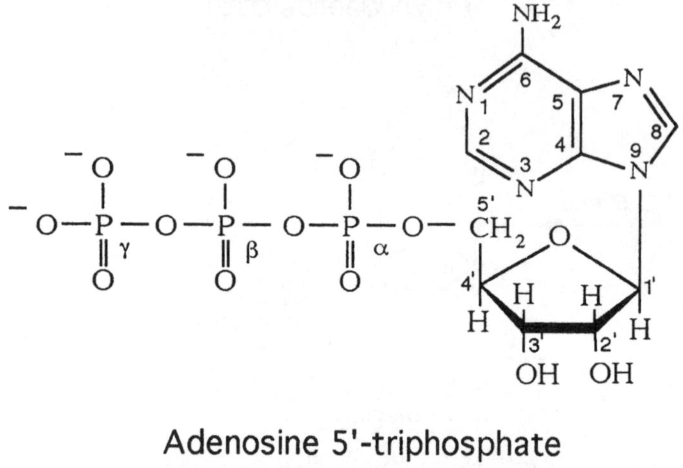

Adenosine 5'-triphosphate

Fig. B1.1b Chemical structure of ATP

RNA polymerase adds mononucleotide units to the 3′ hydroxyl ends of the growing polymer chain $(NMP)_n$. The resulting product (RNA) thus has 3′–5′ phosphodiester bridges. The sequence of the formed RNA is complementary to the DNA template strand. In contrast to DNA replication, which can only elongate an existing chain, RNA transcription can start *de novo*, which means that the 5′ end is formed by a normal NTP instead of a primer polynucleotide. Natural RNA therefore always has a 5′ triphosphate end if it is not modified after transcription.

three fundamental reactions, **replication**, **transcription** and **translation**, which produce all the necessary macromolecules for reproduction (Box 1.2). The combined process is termed **gene expression**.

1.1 Flow of genetic information

During the biomolecular reactions of replication, transcription and translation information is passed from one class of molecule to the next. There is an inherent hierarchy in this flow of information. It can not usually be reversed. This means that a protein sequence is never directly transformed into a RNA or DNA sequence. Moreover, proteins are never made directly from a DNA sequence. This flow is not always completely unidirectional, however, since we know that RNA can be reversibly transcribed into DNA, at least in higher organisms or during retrovirus replication (Fig. 1.1).

Box 1.2 Principal cellular macromolecular reactions

The flow of genetic information in all living cells entails three major biosynthetic reactions for the production of macromolecules. The products resulting from these reactions are classified as **DNA**, **RNA**, and **protein**.

The first process, the duplication of DNA, is termed **replication**. During replication an identical copy of the double-stranded DNA is synthesized by a **semi-conservative** copying mechanism. That means, the newly synthesized DNA consists of one 'old' strand originating from the parental DNA molecule and one 'new' strand, the **daughter strand**, which is synthesized according to the rules of base pair complementarity (adenine pairs with thymine (A:T), guanine pairs with cytosine (G:C). Replication is carried out by enzymes termed **DNA polymerases**.

The second step, **transcription**, denotes the copying process of the permanent information stored in DNA to an RNA molecule. Only one strand (the coding strand) of DNA is transcribed into a complementary sequence. For instance, the DNA sequence of the coding strand 3′AGTC . . . is converted to the RNA transcript 5′UCAG . . . (adenine → uracil; guanine → cytosine; thymine → adenine; cytosine → guanine). The reaction is carried out by enzymes termed DNA-dependent RNA polymerases and involves the formation of phosphodiester linkages between ribo NTP substrates (see Box 1.1). The reaction propagates in 5′ to 3′ direction with respect to the growing RNA chain. The principal chemical reactions are outlined in Chapter 1.3, and the scheme is depicted in Fig. 1.3.

During the third step, **protein synthesis**, nucleic acid sequences are converted into amino acid sequences. The process during which the sequence information inherent to an RNA molecule is transferred to an amino acid sequence of a protein is termed **translation**. The rules for translation of a nucleic acid sequence into the amino acid sequence of a protein are defined by the **genetic code**. With few exceptions the genetic code is almost universal for all living species. Four different aromatic bases (two purines, adenine and guanine, and two pyrimidines, uracil, and cytosine: see Box 1.1) are used to encode a total of 20 different amino acids. Each amino acid is defined by a sequence of three nucleotides (triplet). Since the number of possible combination of 4 bases in a three-letter code is larger than the number of different amino acids (64 *versus* 20) the genetic code is degenerated. That means, one specific amino acid is encoded by more than one triplet (see table below).

In some cases one step of the directed genetic flow can be reverted. In those cases RNA sequences can also be used as templates for the synthesis of (a single stranded) DNA. This reaction is termed **reverse transcription**, and carried out by enzymes designated **reverse transcriptases**. Reverse transcription is found in many higher organisms and represents an essential mechanism for the replication of retroviral genetic material.

RNA can also be directly synthesized from RNA templates in a process termed **RNA replication**. This reaction is catalysed by **RNA replicases**, which are encoded by many viruses or RNA phages.

First position (5′ end)	Second position U	C	A	G	Third position (3′ end)
U	UUU Phe	UCU Ser	UAU Tyr	UGU Cys	U
	UUC Phe	UCC Ser	UAC Tyr	UGC Cys	C
	UUA Leu	UCA Ser	UAA Stop	UGA Stop	A
	UUG Leu	UCG Ser	UAG Stop	UGG Trp	G
C	CUU Leu	CCU Pro	CAU His	CGU Arg	U
	CUC Leu	CCC Pro	CAC His	CGC Arg	C
	CUA Leu	CCA Pro	CAA Gln	CGA Arg	A
	CUG Leu	CCG Pro	CAG Gln	CGG Arg	G
A	AUU Ile	ACU Thr	AAU Asn	AGU Ser	U
	AUC Ile	ACC Thr	AAC Asn	AGC Ser	C
	AUA Ile	ACA Thr	AAA Lys	AGA Arg	A
	AUG Met	ACG Thr	AAG Lys	AGG Arg	G
G	GUU Val	GCU Ala	GAU Asp	GGU Gly	U
	GUC Val	GCC Ala	GAC Asp	GGC Gly	C
	GUA Val	GCA Ala	GAA Glu	GGA Gly	A
	GUG Val	GCG Ala	GAG Glu	GGG Gly	G

The standard genetic code

Table 1.1 Transcription products

Type	**Species**	**Amount of total RNA (%)**	**Molecular weights (kDa)**
tRNAs *	60†	17	25
rRNAs *	5S§ rRNA		35
	16S rRNA	80	500
	23S rRNA		1000
mRNAs	~2000†	3	Heterogeneous

* Stable RNAs. † Different RNAs. § RNA molecules are often characterized by their sedimentation velocity coefficients as a means of their size. Sedimentation coefficients are defined by the equations: $s = \frac{dx/dt}{\omega^2 x}$, where x is the distance of the boundary from the centre of rotation in centimetres, t is the time in seconds, and ω is the angular velocity in radians per second. For sedimentation in water at 20°C the coefficients are given as $s_{20,w}$. A sedimentation coefficient of 1×10^{-13} seconds is called a **Svedberg unit** or simply a **Svedberg (S)**. 5S RNA thus means that the RNA has a sedimentation coefficient of 5×10^{-13} seconds.

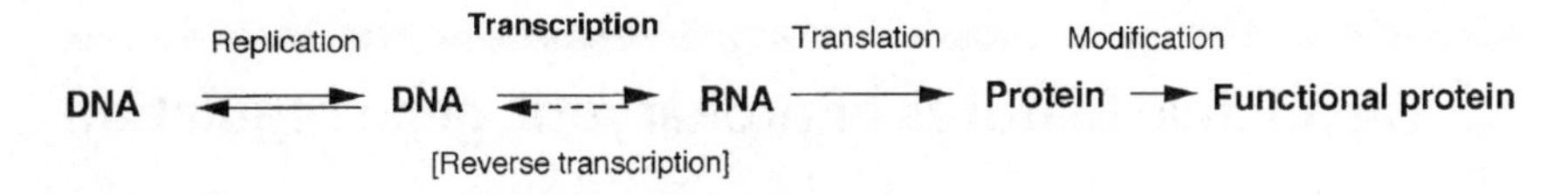

Figure 1.1 The flow of genetic information. The formal flow of genetic information goes from DNA to functional proteins. Transcription (synthesis of RNA) takes a central position between replication (DNA synthesis) and translation (protein synthesis). Sometimes functional proteins are only obtained after post-translational modification. Reverse transcription (dashed arrow) is not normally found in prokaryotes.

The hierarchical order of reactions does not mean that the individual synthesis steps are independent. We will see below that, particularly for bacteria, replication, transcription and translation are intricately coupled, and do not take place separately.

Now consider that there is a really huge number of genes stored in the DNA. Even a bacterial cell, which may be considered as a simple or 'primitive' organism, contains more than 4000 genes. To avoid 'mass confusion' the synthesis reactions of replication, transcription and translation have to be correctly timed (e.g. only defined genes have to be expressed at a certain time, and they must be correctly tuned (e.g. some products are required in a higher stoichiometry or concentration than others). In other words, qualitative and quantitative controls have to be exerted during the synthesis of macromolecules in a cell. The ability of a cell to change the synthesis pattern of macromolecular products in a qualitative or quantitative way is generally described as **regulation of gene expression**. Regulation can take place at any synthesis stage, namely during replication, transcription or translation. In addition, regulation is observed at stages after the macromolecular products have been synthesized. DNA, RNA and protein can be modified after their synthesis, which consequently affects their functionality or stability. Often functional molecules are only generated after post-replicative, post-transcriptional or post-translational processes.

Regulation does not only entail timely ordered reactions according to fixed programs. Living organisms, especially bacteria, often encounter rapidly changing environmental conditions. To adapt to such changes or to escape from a harmful environment cells have to express different sets of functional molecules employing flexible regulatory systems. Such flexible adaptations require complex systems of sensing and signal transduction. It is clear that regulation has to encompass complex networks to allow orchestrated adaptation and cell growth. We will see below that the central step for regulation within the cascade of biochemical reactions is *transcription*.

1.2 Distinctive features of prokaryotic gene regulation

Among the battery of regulatory steps listed above, regulation of transcription plays a central and predominant role in bacteria. There are several reasons why in prokaryotic cells *transcription* is the central platform for the control of transfer of information. A comparison of the different organization of prokaryotic and eukaryotic cells, and of the differences in the mechanism and structure of the transcription machinery will help in understanding the central role of prokaryotic transcription and its regulation. In eukaryotes there are three distinct classes of enzymes (**RNA polymerases**) which are specialized for the transcription of RNA molecules with separate functions (Table 1.1). These different enzymes are designated as RNA polymerase I, II and III. Ribosomal RNAs (rRNAs) are transcribed by RNA polymerase I; the heterogeneous nuclear RNA (hnRNA) pool, which comprises the precursors of the mRNA fraction, is transcribed by RNA polymerase II; RNA polymerase III is responsible for the transcription of the small RNAs, including 5S RNA and tRNAs (Table 1.1). In bacteria, however, there is only *one* type of RNA polymerase which has to transcribe *all* the different RNA species. This means that, in addition to the control of prokaryotic RNA polymerase activity, it is also necessary that the specificity of the enzyme has to be regulated such that it is able to carry out transcription of fundamentally different RNA molecules. The existence of specialized RNA polymerases constitutes a basic difference between prokaryotes and eukaryotes but it is certainly not the major reason for the special importance of transcriptional regulation in bacteria. The reason for the most important implications on the regulation of bacterial gene expression is given by another fundamental difference: in bacteria there are no compartments, like a nucleus or cytoplasm, into which different cellular activities are separated as in eukaryotes. Hence, in bacteria all the macromolecular synthesis reactions, like those depicted schematically in Fig. 1.1, occur simultaneously in the same compartment. As a consequence, each of the reactions, e. g. replication, transcription or translation, has a direct influence on the respective other process. This means that, while a part of the genome is replicated, transcription begins at the new (and the old) DNA strand long before replication is finished. At the same time ribosomes will start to translate the nascent RNA concurrent with all the other synthesis reactions. Hence, growing polypeptides are synthesized on nascent RNA chains while the genome is being replicated. For all the reactions to proceed in parallel fashion it is a prerequisite that the progression of the synthesis steps follow the same polarity. In fact, propagation of all the macromolecular synthesis reactions is in accordance with a 5′ to 3′ direction. This means that during protein synthesis mRNA templates are translated into a sequence of amino acids from their 5′ to the 3′ ends. During translation the polypeptides grow from the amino to the carboxyl terminus. Thus transcribing RNA polymerases and ribosomes may all be bound to the same RNA molecule while marching in the same direction. Often several ribosomes translate in the wake of RNA polymerase, forming a translational battery (**polysomes**), where all ribosomes are engaged in the synthesis of identical proteins (Fig. 1.2). It is important that the rate of propagation of the two processes is synchronized. The underlying phenomenon is called **transcriptional-translational coupling**. Transcriptional-translational coupling is a specific feature of prokaryotic gene expression. It plays an essential role in many aspects of regulation (see Section 5.3)

At high transcription initiation frequencies (a parameter which, as will be seen later, is related to promoter strength) several RNA polymerases may

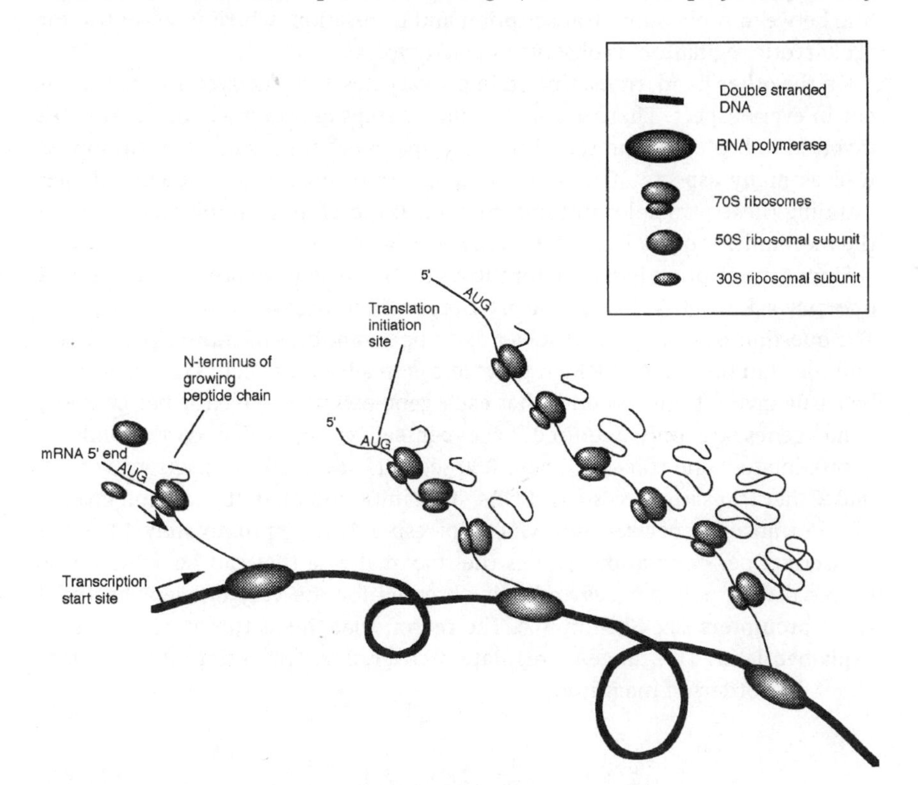

Figure 1.2 Transcriptional–translational coupling. Transcription and translation are simultaneous reactions in prokaryotic cells which lack compartimentation into nucleus and cytoplasm. Ribosomes start to synthesize proteins before transcripts are completed. Usually more than one ribosome is engaged with protein synthesis on the same mRNA transcript (polysomes). Depending on the transcription initiation frequency several RNA polymerases may transcribe the same gene simultaneously. The transcription start site on the DNA is indicated by an arrow. Translating ribosomes start at the AUG initiation codon on the growing mRNA chain. The 5′ ends of the growing mRNA chains are marked.

march one behind the other on the same DNA template. The situation will become even more complicated because of topological constraints of the DNA template. During the progress of transcription the DNA template and RNA polymerase have to rotate relative to each other. Given the size and the complexity of the components involved this is a formidable problem which has to be solved by the cell. These topological questions, briefly addressed here, will be discussed in more detail in later chapters of this book. Because of compartmentalization of eukaryotic cells the major macromolecular synthesis reactions such as replication, transcription and translation are all separated. The different molecules have to be transported or processed (spliced) before the next step of information transfer can be carried out. Hence, no intricate coupling between replication, transcription and translation, which is so central for prokaryotic regulation, is observed in eukaryotes.

On the other hand, transcription in prokaryotes and eukaryotes is not different in every aspect. There are many similar steps and homologous structures involved in both kingdoms, and the principal mechanisms of transcription, as well as many aspects of their regulation, are of universal character. Understanding these principles in prokaryotes will therefore certainly help to solve the mechanisms of transcription of higher organisms.

Since transcription is central for the regulation of gene expression in bacteria one may ask what is the **regulatory range** which is covered by transcription. The question is, what are the lower and upper numbers of transcripts per second that can be formed by RNA polymerase in a bacterial cell. The answer can be easily given if one assumes that each gene exists as one copy per genome. Some genes are only required once per cell cycle, which corresponds to approximately one transcript per 2000 seconds or 5×10^{-4} transcripts per second. Other transcripts, such as rRNAs, have initiation rates at a fast cell growth of one transcript per second, which corresponds to approximately 1.5×10^3 transcripts per generation. This is the maximal rate that can be achieved by bacterial RNA polymerases and it is observed for the very efficient bacterial rRNA promoters (see Section 8.7). The reason that this is the maximal rate is explained later. The range of regulation covered by transcription thus spans about four orders of magnitude.

1.3 Biochemical nature of the transcription reaction

Like any macromolecular synthesis cycle the process of transcription may be broken down into the substeps of **initiation**, **elongation** and **termination**. Each of these steps is complex by itself, and all of the steps are subject to specific regulation.

The actual transcription reaction is known to be a copying mechanism from the information contained in the sequence of a DNA molecule into an RNA chain, or more precisely, transcription can be defined as the sequence-dependent synthesis of RNA according to one strand of double helical DNA. What is the mechanism and the chemical nature of this biochemical reaction? The process of transcription involves the stepwise addition of nucleoside monophosphates to a growing polyribonucleotide chain according to the principle of base complementarity (Fig. 1.3).

The substrates are ribonucleoside 5′ triphosphates (NTPs) which are linked in an enzymatic phosphodiester reaction to the free 3′ hydroxyl group of the growing RNA chain. The enzyme catalysing the reaction is termed DNA-dependent RNA polymerase. The reaction involves many repetitive steps of nucleotide addition until the sequence of the respective RNA product molecule is completed. At this point, the transcription complex decomposes and the RNA is released. As explained previously, the progress of RNA synthesis proceeds in a

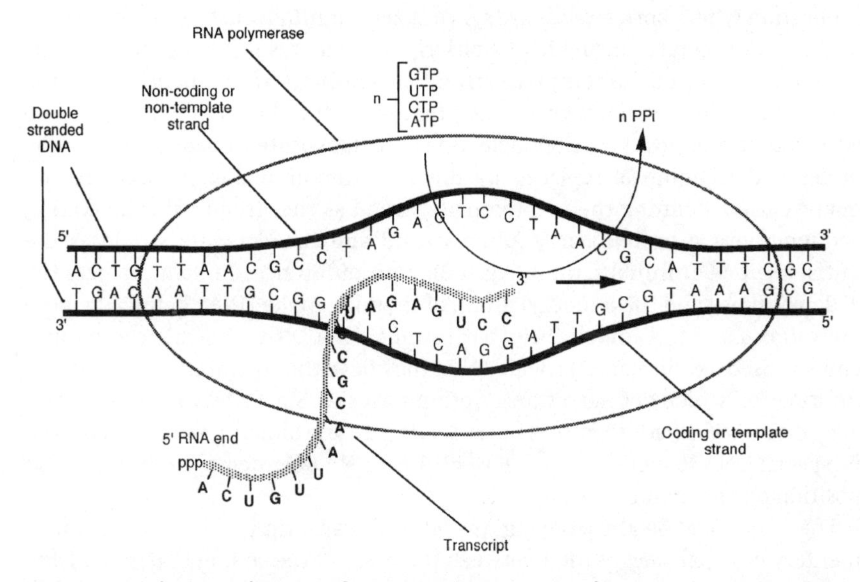

Figure 1.3 Schematic diagram of a transcription process. The transcription reaction occurs at a strand-opened DNA duplex indicated by two dark lines. Bases are indicated by captial letters. The space of an elongating RNA polymerase is shown by a grey ellipse. The transcription complex proceeds in the direction of the horizontal arrow (in the 5′ direction of the coding strand). Substrate NTPs are added stepwise to the growing RNA chain via phosphodiester bonds and pyrophosphate (PPi) is released during each addition step of this reaction. The growing RNA chain (thick grey line) has a sequence complementary to the coding or template strand. The RNA sequence shown in bold type is identical to the corresponding sequence of the non-coding or non-template strand except that thymine is substituted by uracil. The 5′ and 3′ ends of the nucleic acids are labelled. The 5′ end of the RNA transcript carries a triphosphate from the initiating first NTP.

5′ to 3′ direction. Note that the polynucleotide synthesis in nature always proceeds in a 5′ to 3′ direction. The reaction involves α-β-phosphodiester hydrolysis of the incoming NTPs and is Mg^{2+}-dependent (Box 1.1).

1.4 What makes RNA transcription different from DNA replication?

Although the enzymatic reaction leading to RNA has much in common with the synthesis of DNA, there are notable differences in the biosynthesis of both polynucleotides. DNA in its natural form consists of two complementary strands which are antiparallel and interwound, forming a right-handed double helix (see Chapter 6). In contrast to the **semi-conservative** process of DNA replication where both strands are synthesized simultaneously, only one strand of the DNA is copied during RNA transcription. The resulting product (RNA) is the complement of the **template strand** (or **coding strand**), hence it has the exact nucleotide sequence of the non-template strand (or non-coding strand) with the exceptions that ribonucleotides are substituted for deoxyribonucleotides and thymine is replaced by uracil. Different nomenclatures are frequently used, defining the non-template strand as the strand with the **coding sequence** (for a protein) or + **(plus) strand** and the template strand as the **antisense** or – **(minus) strand** (Fig. 1.3). Both differences to RNA, namely the 2′ deoxyribose and the 5 methyl group of thymine make DNA chemically more inert (Box 1.1). Most importantly, both strands of a DNA molecule can in principle be used as templates, provided that they have the corresponding sequence information which defines a transcription start site. Start sites for transcription are normally termed **promoters** (see Section 2.1). Promoters are characterized by special DNA sequences and precisely direct RNA polymerase to the initiation position on the DNA template.

The sequence-specific copying process of transcription involves complementary base pair recognition between the bases of the coding DNA strand and the incoming substrate ribonucleoside triphosphates. This recognition process requires that a stretch of nucleotides of the double helical DNA template is exposed transiently to allow base pair formation. DNA strand separation occurs when RNA polymerase binds to the promoter during initiation (**promoter melting**). The strand separated complex is termed the **open complex**. The DNA in open complexes forms a **transcription bubble** which moves along with the elongating RNA polymerase.

One of the characteristics of DNA replication is that it cannot be initiated *via* a *de novo* strand synthesis. It requires a primer molecule with a free 3′ hydroxyl group from which to start. This primer can in principle be DNA or RNA. In contrast, RNA transcription can be initiated at any start site (promoter) without the necessity of a primer sequence. (This observation is frequently used as one of the arguments for the notion that RNA evolved earlier than DNA, implying that the latter has developed at later stages of evolution.) There is an additional difference between RNA and DNA polymerases with respect to chain elongation. Although DNA polymerase is able to synthesize long DNA chains replication is actually a **distributive process**. This means that DNA polymerase which has left the template can rebind to the 3′ hydroxyl terminus of any priming strand and extend released 3′ ends. In contrast, the reaction catalysed by RNA polymerase is essentially **processive** and, once RNA is released, the same RNA chain cannot be further elongated. During initiation at many promoters very short RNA products are repetitively released before processive synthesis of a long RNA chain occurs. During this **abortive reaction** RNA polymerase stays bound to the promoter. It will be seen later that the switch between abortive cycling and productive elongation is an important step to regulate the efficiency of transcription (see Chapter 3).

Finally, DNA represents the conserved genetic information and has more or less to be synthesized only once per cell cycle. In contrast, the requirements for different RNA molecules in a cell are very complex, and are under constant change throughout the lifetime of a cell. The synthesis of RNA molecules has to be switched on and off in a precisely tuned way. It must be coordinated with the production of related compounds and adjusted to the demands of cell growth or to changes in the environment. Hence, it is subject to multiple facets of regulation. Within this framework of complex regulatory circuits *transcription* accounts predominantly for the adaptation to the individual synthesis efficiencies of the cell by an intricately coupled network of different control mechanisms.

1.5 The products of transcription

Before the details of the transcription apparatus and the mechanisms of transcription are discussed something about the products of the process should be known, namely RNA. RNA molecules present in living cells are usually classified according to their function or life time (Table 1.1).

RNA molecules are commonly divided in the fraction of **mRNAs** and **stable RNAs**. Stable RNAs consist of the rRNAs which are essential components of the translational machinery (ribosomes), and tRNAs, which fulfil quite a number of different cellular functions. As their main task, however, tRNAs serve as adapter molecules, providing the protein biosynthesis machinery (ribosomes) with activated amino acids for the process of translation. The mRNA fraction, on the other hand, represents the pool of RNA molecules which is used as templates for translation into the different cellular proteins. This RNA fraction is therefore heterogeneous, consisting of a large number of different molecules (approximately 2000 for *Escherichia coli*).

In recent years an additional set of RNA molecules has been characterized with specific functions: small nuclear RNAs (snRNAs), as well as small nucleolar RNAs (snoRNAs) which function in RNA splicing and rRNA maturation. In addition, guide RNAs (gRNAs) which are involved in a post-transcriptional process called RNA editing have been characterized, as have small stable RNAs (7S, 10S tmRNA or 4.5S) which are part of the signal recognition particles or which have special functions during translation. While snRNAs and gRNAs are specific to eukaryotes, the remaining RNA species are also found in prokaryotes or have at least prokaryotic analogues.

Moreover, several RNA molecules are found as catalytic subunits of enzymes (e.g. M1 RNA in the processing enzyme RNaseP). Owing to the capacity of RNA to form a large variety of different structures, which are dynamic and interconvertible, special RNA molecules have evolved that carry out enzymatic reactions. No proteins are required for the function of these molecules. They can act as real enzymes and, by analogy to protein enzymes, are called ribozymes. Many different functions can be catalysed by such ribozymes. Some of the ribozymes appear to be of older evolutionary origin as protein enzymes. Hence, the existence of ribozymes again supports a very early role of RNA in evolution.

Summary

In all living organisms there is transfer of genetic information from one generation to the next. This transfer of information is accomplished by a process termed gene expression, which is characterized by a sequence of different forms of macromolecular biosynthetic reactions, namely DNA synthesis (*replication*), RNA synthesis (*transcription*) and protein synthesis (*translation*). Generally, in higher organisms, gene expression occurs within separated cellular compartments. In prokaryotic cells, however, there are no subcellular compartments and the different reactions occur simultaneously without local separation. Hence, in bacteria the synthesis reactions of DNA, RNA and proteins are intricately coupled and their regulations are mutually affected. Regulation of gene expression takes place at every single step of the biosynthetic reactions leading from DNA to protein. In bacteria the major step at which regulation occurs is transcription. Changes in the transcriptional efficiency account for a difference of four orders of magnitude in the frequency at which different RNA molecules are formed. The transcription reaction is catalysed by enzymes (termed DNA-dependent RNA polymerases) and involves the stepwise addition of ribonucleoside triphosphates into a growing RNA chain. The RNA product, which is complementary in sequence to one DNA strand, the coding strand, and corresponds in sequence to the non-coding strand, is synthesized in a 5′ to 3′ direction. Different classes of RNA molecules are synthesized by the same bacterial enzyme. These RNA products are either short-lived mRNA molecules, which serve as blueprints for translation, or stable RNAs, which are constituents of the translation machinery and which can exhibit structural and catalytic functions by themselves.

Further reading

Losick, R. and Chamberlin, M. J. (eds) (1976) *RNA Polymerase*. New York: Cold Spring Harbor Laboratory.

Stryer, I. (1991) *Biochemie* Heidelberg: Spektrum Akademischer Verlag.

Watson, J. D., Hopkins, N., Roberts, J., Steitz, J. A. and Weiner, A. (1987) *Molecular Biology of the Gene*, 4th edn. California: Benjamin-Cummings.

The Genetic Code (1966) Cold Spring Harbor Symposium Quantitative Biology, Vol. 31. New York: Cold Spring Harbor Laboratory.

2

The 'players' or cellular components necessary for transcription

This chapter gives a summary of the components necessary to constitute a transcription complex. It begins with a brief explanation of the DNA structures known as specific start sites of transcription, the promoters. Then follows a description of the different subunits of RNA polymerase. In this section some recent information is provided about the conserved structures and functional domains that have been identified and localized within the amino acid sequence of bacterial RNA polymerase subunits. It is followed by several paragraphs that deal with the structure and function of alternative σ factors, which are the subunits providing the specificity for the RNA polymerase to initiate transcription for special sets of genes. In addition, some information is presented on the function of anti-sigma factors in the control of σ factor activity. The chapter is closed with two short sections on the three-dimensional structure of *E. coli* RNA polymerase, and some observations indicating that the composition or structure of RNA polymerase may be modified during different stages of the bacterial growth cycle.

2.1 Transcription start sites: the promoter

Transcription involves binding of DNA-dependent RNA polymerase to a double helical DNA template. Initiation of the transcription reaction is restricted to defined sites within such a DNA template. These sites are called **promoters**, and they are defined by a conserved DNA sequence. As will be seen below, several subsets of genes with common regulatory features, or **regulons**, frequently require special promoter structures (see Chapter 8). They will be discriminated by alternative **specificity factors**, which join RNA polymerase to make an initiation-competent enzyme. The general structure of promoters responsible for the transcription of standard genes during exponential growth ('housekeeping genes') can be identified on the basis of their conserved primary sequence. These conserved elements were first recognized as a result of sequence comparison of known bacterial and phage transcription start site sequences. At this point only a minimal description of a standard promoter will be given. More details of the rather complex and diverse topology of promoters are found in Chapters 3 and 6.

2.1.1 Core promoter elements

A minimal description of a standard promoter entails three structural elements:

1. A highly conserved sequence region, which is characterized by the conserved hexameric sequence of the non-coding strand 5′-TATAAT-3′. This sequence usually centres 10 base pairs upstream of the first nucleotide of the transcript to be formed (RNA 5′ end). Hence, this element is termed the **–10 region**. Occasionally it is also called Pribnow box.
2. A second recognition element, centred about 35 base pairs upstream of the first transcribed nucleotide, which has the conserved sequence 5′-TTGACA-3′ (again, this is the sequence of the non-coding strand). This sequence element is called the **–35 region**.
3. The sequence in between the –10 and –35 elements is apparently not conserved. However, the distance between the two elements has a characteristic length of 17±1 base pairs. This sequence is called the **spacer region** and defines the third feature of a standard core promoter. Sequence changes within the spacer region have only little direct influence on the promoter activity, while a different spacer length drastically affects the performance of a promoter. A standard *E. coli* consensus promoter indicating the above three structural features is presented in Fig. 2.1.

Whereas direct contacts between the –35 and the –10 hexamer sequences and RNA polymerase can be shown, the available evidence suggests that nucleotides

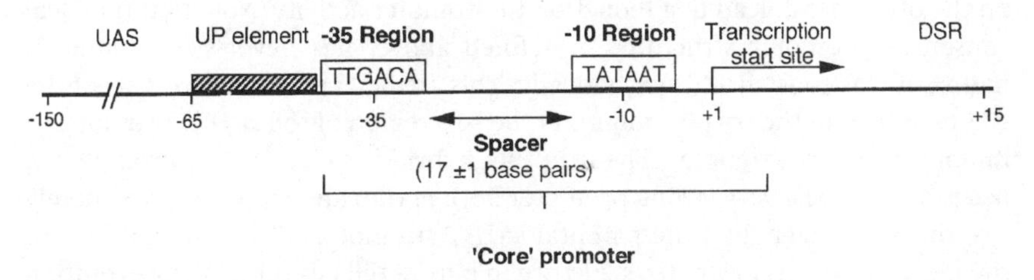

Figure 2.1 *E. coli* consensus promoter structure. The core and flanking structural elements of a σ^{70} consensus promoter are shown. The core promoter consists of the –35 and –10 regions separated by a spacer sequence of 17 ± 1 base pairs. The promoter is further characterized by the upstream and downstream flanking elements; DSR, downstream sequence region; UP element and UAS, upstream activating sequence. The consensus sequences given in the –35 and –10 regions correspond to the non-template strand sequence. Numbers indicate nucleotide positions relative to the transcription start site which is defined as + 1.

within the spacer are not directly bound. However, the spacer is certainly important in positioning the two consensus hexamers at the correct distance *and* angle for optimal contacts with the surface of the RNA polymerase. Even if the same number of spacer nucleotides are present, this angular orientation can be disturbed when the spacer sequence contains a different purine-pyrimidine composition, which is known to affect the helical **twist** of the DNA (see Box 6.1). Although there is no fixed primary sequence requirement for the spacer region it is plausible that sequence changes within the spacer change the efficiency of the promoter because such sequences may affect the DNA twist. Insertion or deletion of one or more base pairs, however, can partly be accommodated by different spacer sequence compositions without significantly affecting the promoter efficiency. Note that in helical DNA a difference of a single base pair between the –35 and the –10 regions will not only alter the relative distance between the two recognition elements; it will also change their angular orientation by roughly 35° (10.5 base pairs correspond approximately to one helical turn (360°) of B-DNA). In conclusion, both *twist* and *spacer length* contribute to the optimal orientation of the –10 and –35 promoter recognition elements relative to the binding surface of RNA polymerase. Since the **superhelical density** of a template directly affects DNA twist it is clear from the above considerations that the superhelicity will also affect the efficiency of many promoters (see Chapter 6).

The efficiency of a transcription initiation process is dependent on the degree of correspondence with the conserved structural elements which define a promoter according to the above description. Single base change mutations confirm the importance of the three promoter elements. Most of the mutations affecting the activity of promoters can be found in the conserved elements. Substitutions that cause a better match with the consensus sequence almost always improve the function of the promoter. Deviations from the consensus, on the other hand, lead to a reduction in promoter activity. Note that the ideal consensus promoter structure, as defined above, has never been found in nature. It was constructed synthetically. This artificial promoter consists of the –35 region from the *trp* promoter and the –10 region of the *lac*UV5 promoter, a mutant of the *lac* promoter. The resulting hybrid is called the tac promoter. It has proven to be a very strong promoter both *in vitro* and *in vivo* (approximately five times stronger than the parental *lac*UV5 promoter). The apparent lack of the ideal consensus promoter structure in nature tells us that during evolution promoters have obviously not been optimized for *strength* but more likely for their capacity to be regulated. The definition of promoter strength and the way in which it can be determined are discussed in Chapter 3.

A special group of promoters, which has no or only weak homology to the –35 region is characterized by the DNA sequence motif TG, one base pair upstream of the –10 element. These promoters are classified as **extended –10 promoters**. Apparently, the TG motif provides a supplementary recognition sequence which can substitute for the lack of the –35 region. In other cases, where there is no or only weak homology to the –35 region, and no TG sequence flanking the –10 region is found, it is known that transcription factors (activators) are responsible for the activity of such promoters.

A fascinating question that still remains to be answered is which signals in the higher order structure of DNA determine where the RNA polymerase finds the correct sites for binding, unwinding the DNA and forming a transcriptionally competent complex amongst the thousands of base pairs (see Chapter 3 for details of these reactions). There have been many attempts to answer this question by means of computational analysis of sequence parameters and their biophysical consequences on the structure and stability of DNA. It has been shown for several promoters, for instance, that complex formation with RNA polymerase causes the DNA to bend. Bending, in turn, may lower the activation energy for strand separation and may thus directly facilitate strand opening, which is necessary for the formation of a transcriptional competent complex. To determine whether promoter structures are associated with DNA curvature computer programs developed to predict the DNA conformation were tested. Using such programs to localize curvature among a defined DNA sequence more than 100 different *E. coli* promoters were compared. Interestingly, the majority of the analysed promoters appear to be curved, with the centre of the bend localized between the –10 and –35 regions. In many cases biochemical analyses have confirmed the presence of DNA curvature within the promoter core regions. Today it is clear that, apart from the primary DNA sequence, there is a higher order structural correspondence among promoters which is recognized by RNA polymerase.

One important step during transcription initiation is base pair opening within the promoter region (**melting**). This step is catalysed by RNA polymerase and leads from the **closed** to the **open promoter complex** (see Section 3.3). The capacity for melting is determined by the stability of the double helical DNA, and inversely correlates with the free energy for strand separation. The stability of the DNA, in turn, depends greatly on the percentage of AT base pairs, which are known to be thermodynamically less stable than GC base pairs. The relationship between double helix stability and promoter activity was compared by free energy computations based on the dinucleotide free energies for strand separation. For a large number of promoter sequences it was found that within 500 nucleotides around the transcription initiation site the –10 region is the least stable part of the sequence. Although no clear correlation could be established between the rates of RNA polymerase–promoter open complex formation and the free energies, the investigation strongly suggests that the instability or melting properties of the –10 region play a significant role in promoter function.

2.1.2 The importance of flanking sequences

The three elements described above, the –10 region, spacer and –35 region, characterize the **core promoter structure**. It is known that sequence elements outside of this classical core often contribute much to the activity of a given promoter. For instance, at many but not all promoters, sequences upstream of the promoter core (–60 to –40 relative to the transcription start site) have a profound effect on activity. These sequences are usually AT-rich, and their presence can increase promoter efficiency by as much as 30-fold. The corresponding elements have been termed **upstream sequence regions (USR)** or **UP elements** (see Section 6.2.1). In addition, in the case of many strong promoters, like the early promoters of phage T5, **downstream sequence regions (DSR)**, ranging from +1 to +20, have been identified, which also contribute strongly to promoter activity. The arrangement of such flanking promoter elements is shown in Fig. 2.2.

2.1.3 The start site position

The position of the first nucleotide of the transcript is defined as +1. For most of the transcripts analysed the starting nucleotide is A (47%) or G. CTP or UTP are also occasionally found as the first nucleotide substrates, however. There is no fixed distance between position +1 and the last nucleotide of the –10 region. In most of the transcripts the initiation point is somewhere between five and nine base pairs downstream of the 3′ nucleotide of the –10 sequence. It is clear that the information for the structure of the initiation complex which positions

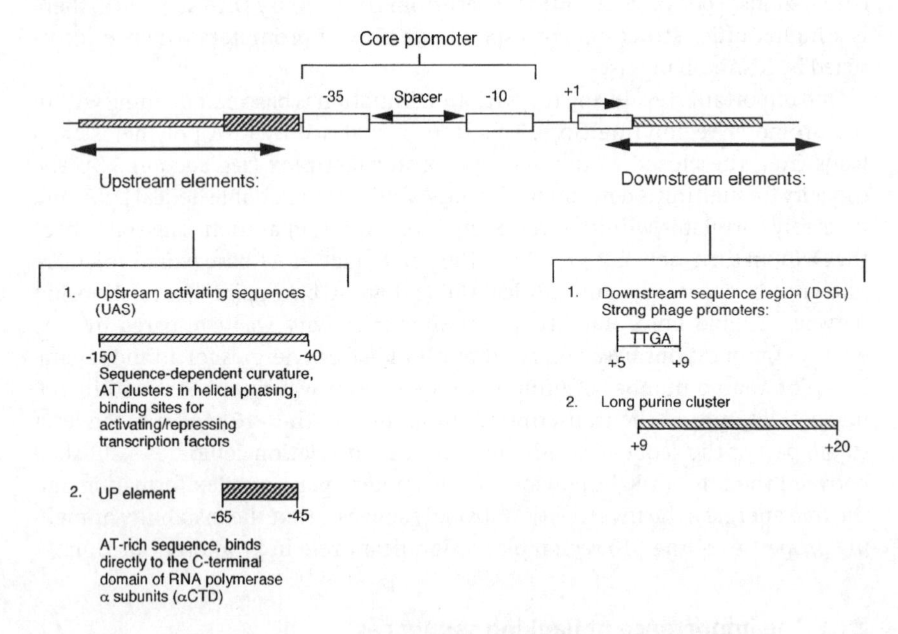

Figure 2.2 Flanking elements of bacterial promoters. Different upstream and downstream flanking elements of typical prokaryotic promoters are listed. Numbers indicate sequence positions relative to the transcription start site.

the transcription start site cannot be easily deduced from the primary structure of the promoter sequence. The higher order structural organization of the DNA and the RNA polymerase determines in sterical consequence where the transcription start site is located. This can be inferred, for example, by the observation that there is no preference for a certain start sequence at promoters with similar or identical sequences around the initiation region; e. g. the two promoters for the *araBAD* operon and *gal*P1 have the same start site sequence, 5′-CATAC-3′. While the *gal*P1 promoter initiates from the first A, the *araBAD* operon is initiated at the second A (Fig. 2.3).

There is frequently not only a single position from which initiation starts but several consecutive nucleotides can be accepted as start sites. A typical example is promoter P2 from the ribosomal rRNA operon (*rrnB*). In this case, transcription can start at any one of a row of Cs within the sequence 5′-ACCCCG-3′ (Fig. 2.4). Consequently, the corresponding transcripts found within the cell differ in length by one nucleotide.

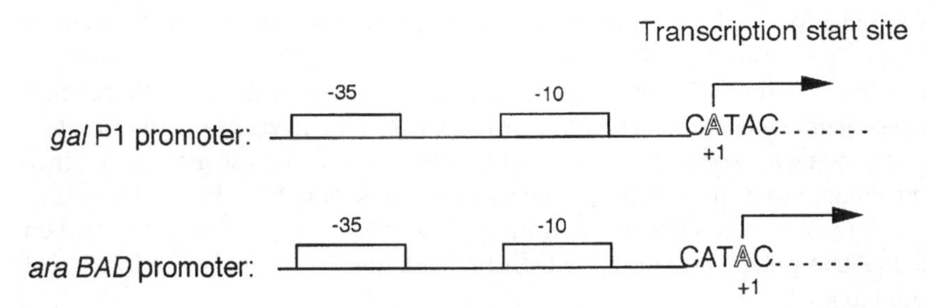

Figure 2.3 Variable start sites for identical start sequences. The *gal*P1 and the *araBAD* promoters are schematically depicted. Identical sequences surrounding the transcription start region (arrows) are shown in capital letters. The different start nucleotides are indicated with outlined letters and labelled +1.

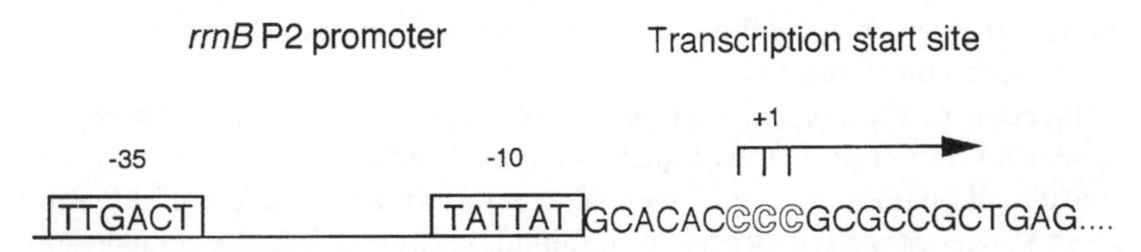

Figure 2.4 *rrnB*P2 transcription start sites. The sequence of the transcription initiation region and the –10 and –35 sequences of the *E. coli* ribosomal RNA operon P2 promoter is shown. Transcription is initiated from either one of three consecutive Cs shown outlined and marked with the arrow.

2.1.4 Complex promoters

In many cases genes are controlled not by a single promoter structure but by a complex arrangement of several promoters. Depending on the direction of transcription such composed promoter arrangements are defined as **tandem** (adjacent promoters transcribing in the same direction), **convergent** (adjacent promoters transcribing towards each other) or **divergent** (adjacent promoters transcribing away from each other). The distance of such complex promoters can vary considerably between only a few and several hundred base pairs. Frequently, composite promoters may also have overlapping consensus elements. Notable examples are the P*lac* and the P*gal* promoters (Fig. 2.5).

2.2 RNA polymerase

The machinery which catalyses DNA-dependent RNA synthesis is the enzyme RNA polymerase (EC 2.7.7.6). It was first described in 1959. This section summarizes findings on RNA polymerase of *E. coli*; this has been the most thoroughly investigated and can be taken as a prototype for all bacterial polymerases. In a fast growing *E. coli* cell there are roughly 3000 RNA polymerase molecules. Since the volume of a single cell can be estimated to be about 3×10^{-16} l this corresponds to an intracellular concentration of about 10 to 20 μM. The concentration of *free* RNA polymerase in the cell is estimated, however, to be only a fraction of this (about 1%).

Bacterial DNA-dependent RNA polymerase is a complex enzyme, composed of four different subunits: α, β, β' and σ. The catalytic activity to transcribe a DNA template resides in the so-called **core enzyme (E)**. The core enzyme consists of two α, one β and one β' subunits, hence it has the subunit structure $\alpha_2\beta\beta'$. The RNA polymerase core is capable of carrying out normal transcription elongation. Specific initiation of transcription, however, requires the **holoenzyme (Eσ)** which, in addition to the core subunits, contains the σ subunit or specificity factor (Table 2.1).

Direction of the precise start point and formation of the initial complex capable for transcription thus requires a higher complexity of the RNA polymerase subunit composition. As is seen later, transcription initiation of different subsets of genes is often directed by different σ factors. While the composition of the core enzyme of bacterial RNA polymerases, $\alpha_2\beta\beta'$, is necessary and sufficient for the basic transcription reaction, the σ subunits provide specificity for the correct initiation of transcription of different sets of genes under different cellular growth conditions or the expression of different **regulons** (see Chapter 8).

After the initiation cycle has been completed, the σ subunit leaves the

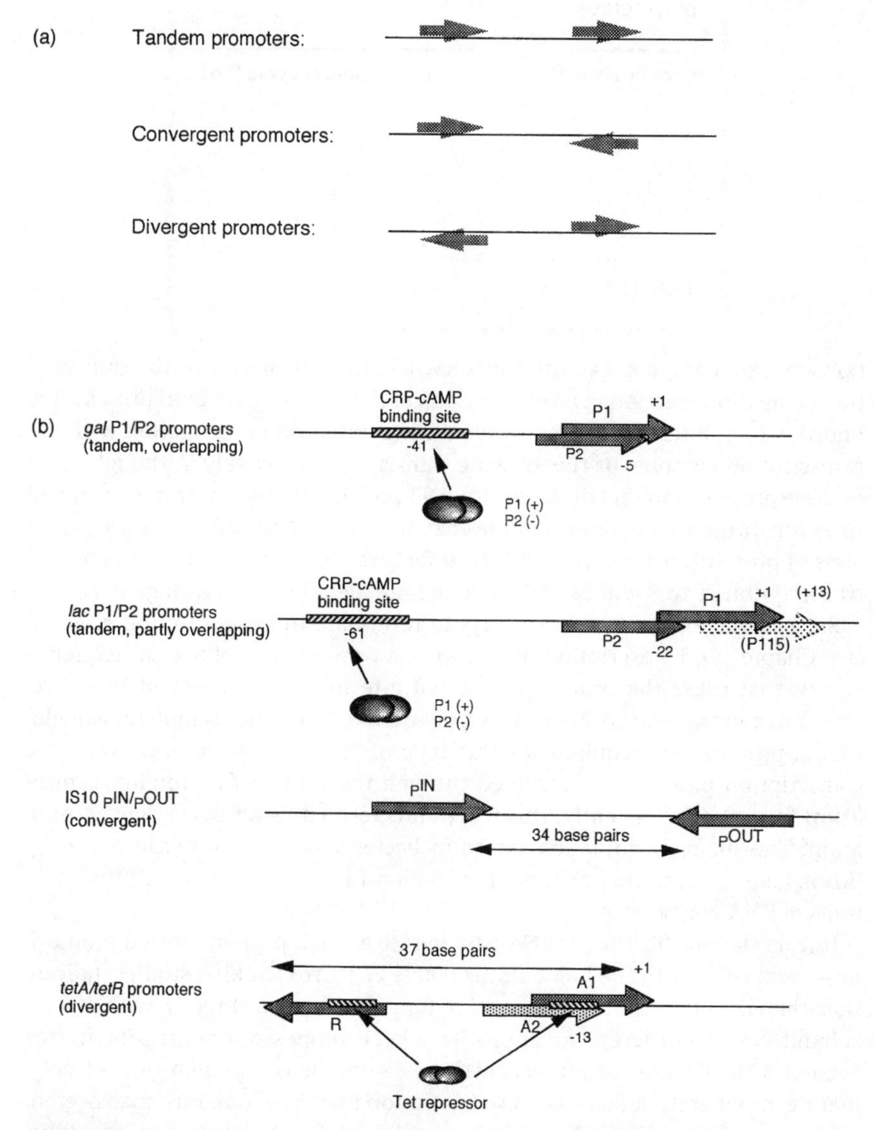

Figure 2.5 Complex promoters. (a) Composite promoters. Arrows point in the direction of transcription. (b) Complex promoters: examples shown are the overlapping tandem *gal*P1/P2 and *lac*P1/P2 systems (including the minor P115 promoter), the convergent IS element pIN/pOUT system, and the divergent *tetA/tetR* promoters. Numbers indicate nucleotide positions relative to the start of the major rightward promoter. The catabolite regulator protein CRP and the Tet repressor are shown and their respective binding sites are indicated by shaded rectangles.

Table 2.1 Subunit composition of bacterial RNA polymerases

Core enzyme [E]	Holoenzyme [Eσ]
$2 \times \alpha$	$2 \times \alpha$
$1 \times \beta$	$1 \times \beta$
$1 \times \beta'$	$1 \times \beta'$
	$1 \times \sigma$
Total (kDa) 378.8	449.0

transcription complex, and the core enzyme faithfully performs the transcription elongation reaction in an entirely processive way. A number of proteins are known which interact at some points during initiation or elongation with the transcription complex or the growing transcript, respectively. Although some of these proteins interact directly with RNA polymerase they are not considered to be constitutive components of the enzyme. They are considered as a separate class of proteins, termed **transcription factors**. The structure and function of transcription factors will be addressed in more detail later (see Chapter 7).

Bacteria have two alternative ways to terminate the transcription reaction (see Chapter 5). Transcription may stop as a consequence of special sequence signals that cause the transcript to adopt a termination proficient structure. The transcript is released and RNA polymerase leaves the template. No additional proteins are required for this type of termination. Alternatively, the transcription process is interrupted through the action of a protein (**termination factor**). Consequently, this reaction is termed factor-dependent termination. The major termination factor in bacteria is the hexameric protein **ρ (Rho)**. Like transcription factors, termination factors are not constitutive subunits of RNA polymerase.

During the purification of RNA polymerase a small protein termed ω notoriously copurifies with the holoenzyme (Table 2.2). Cross-linking studies indicate that the ω protein is in close contact to, or associated directly with, the β′ subunit. Several different functions have been proposed for this protein (see Section 8.5). Whether or not ω constitutes a unique component of RNA polymerase, or what its actual role in transcription might be, remains unanswered.

The assembly of the different subunits *in vitro* for the formation of a functional RNA polymerase holoenzyme is not an arbitrary process but occurs sequentially in the following ordered scheme:

$$2\alpha \rightarrow \alpha_2;\ \alpha_2 + \beta \rightarrow \alpha_2\beta;\ \alpha_2\beta + \beta' \rightarrow \alpha_2\beta\beta' \equiv \text{core enzyme [E]};$$
$$\alpha_2\beta\beta' + \sigma \rightarrow \alpha_2\beta\beta'\sigma \equiv \text{holoenzyme [E}\sigma\text{]}$$

Not all RNA polymerases are multisubunit enzymes. Some phages, like T7 or

Table 2.2 *Escherichia coli* RNA polymerase subunits

Subunit	Number of amino acids	Molecular weight (kDa)	Stoichiometry
α	392	36.5	2
β	1342	150.6	1
β'	1407	155.2	1
σ^{70} *	613	70.2	1
ω†	91	10.1	0.5–2

* σ^{70} is the specificity factor necessary for transcription of the housekeeping genes.
† ω copurifies with RNA polymerase. It is generally not considered to be a constitutive subunit.

SP6, encode special RNA polymerases which consist of only a single polypeptide with molecular weight of 90–100 kDa. These polymerases recognize phage-specific promoters with high affinity and do not require additional specificity or σ factors for initiation. They are very efficient enzymes and their transcription elongation rate is about fivefold faster than that of their bacterial counterparts (see Section 4.2). T7 and SP6 RNA polymerases have, therefore, gained widespread use for preparatory *in vitro* transcription reactions in the laboratory, when large amounts of RNA are required.

In contrast to the phage polymerases, RNA polymerases from higher organisms have a much more complex subunit structure than the bacterial enzymes. Eukaryotic RNA polymerases are composed of more than 10 different subunits. Interestingly, RNA polymerases from archaea show a comparably high complexity as eukaryotic enzymes (Table 2.3). As will be discussed below, there is a significant homology between the largest subunits of eukaryotic and prokaryotic RNA polymerases, however.

An intensively studied question is next discussed; namely, what is known about the structure and function of the different subunits of bacterial RNA polymerase, and can individual functional domains be assigned within the isolated RNA polymerase subunits? Large parts of the amino acid sequences of the different subunits are highly conserved within different bacterial species. Much valuable information on the function of the individual subunit proteins has been obtained in the past from the study of mutant subunit genes or from direct biochemical analysis. From such studies sites in the protein subunit can be correlated with a particular function of the RNA polymerase. The following sections detail the present knowledge on the structural and functional domains of the different bacterial RNA polymerase subunit proteins.

Table 2.3 Compositions of eukaryotic and archaeal RNA polymerases

Eukaryotes (*Saccharomyces cerevisiae*)			**Archea (*Sulfolobus acidocaldarius*)**	**Bacteria (homologous subunits)**
RNA polymerase I	**RNA polymerase II**	**RNA polymerase II**		
190	220 (185)	160	A′ (101)	β′ (155.2)
135	150	128	B (122)	β (150.6)
			A″ (44)	β′ (155.2)
49		82		
43		53		
40	44.5	40	D (30)	α (36.5)
		37	E (27)	
34.5	32	34	F (12)	
		31	G (13.8)	
27	27	27	H (11.8)	
		25		
23	23	23	K (9.7)	
19	16	19	L (10)	
14.5	14.5	14.5	I (9.7)	
12.2	12.6		M (5.5)	
10α/10β	10α/10β	10α/10β	N (7.5)	

Numbers indicate molecular weights in kilodaltons. Archaeal subunits are indicated by Arabic letters: the molecular weight is given in parentheses.

2.2.1 The α subunit

The α subunit, the product of the *rpoA* gene, is a 36.5-kDa protein, which is 329 amino acids in length. The involvement of the α subunits in core assembly has long been known. More recent evidence, however, indicates that α plays an additional important role in direct DNA binding upstream to certain promoters and in the interaction with transcriptional regulators. The molecule has been shown by limited proteolysis and nuclear magnetic resonance (NMR) spectroscopy to be composed of two independently structured domains which are connected by a flexible linker. The amino-terminal domain (αNTD; amino acids 8–241) has proven to be sufficient for subunit assembly. It contains two conserved structural segments (amino acids 30–75 and 175–210) which are also found in eukaryotic, archaeal and plastid RNA polymerase subunits. The structure of the αNTD has been resolved by X-ray crystallography at a resolution of 2.5 Å. These domains are known to be involved in the assembly process that gives the core enzyme. This finding might suggest that the assembly process is conserved between prokaryotic and eukaryotic RNA polymerases. Based on hydroxyl radical protein footprinting amino acids 30–75 are proposed to be in direct contact with the β subunit, whereas amino acids 175–210 are believed to contact β′ (Box 2.1).

Studies with α subunit mutants which are partially deficient in the assembly of core polymerase suggest that binding of α to the β and β′ subunits during assembly occurs in an asymmetrical way, with one α subunit in contact with β and the other with the β′ subunit. No contact regions have been established between the α and σ subunits.

When truncated α subunits consisting of only the N-terminal two-thirds of the molecule (αNTD) are reconstituted into RNA polymerase, it turns out that the resulting molecules are active to initiate and transcribe from some but not all promoters. The transcription of genes which require the action of specific activator proteins is not supported by RNA polymerases that lack the C-terminal domain of the α subunit. Among the activator proteins affected are transcription factors like CRP, OxyR, OmpR, CysB, AraC, MelR, Ada, IHF or FNR, whose function will be characterized in a later section (**Class I transcription factors**, see Chapter 7). Subsequently it has been shown that the carboxy-terminal domain (αCTD), consisting of amino acids 249–329, is necessary for interaction with the corresponding activator proteins. This has been corroborated by mutational analyses of the αCTD. It should be noted, however, that for certain transcription units, activators like CRP can also be grouped into **Class II transcription factors**, depending on the site of interaction with the upstream promoter region. At Class II CRP-dependent promoters a second site of interaction between CRP and α subunit has been identified by genetic and cross-linking experiments; this is located at the αNTD (amino acids 162–165, see Section 7.3.2).

The isolated αCTD was shown to be capable of forming dimers and binding to several promoters which contain AT-rich sequence regions upstream of the core promoter (a so-called UP element; see Section 6.2.1). Although segments responsible for upstream promoter DNA binding and interaction with Class I transcription factors are located within the same region of the α subunit, several different amino acid positions are involved in both reactions. The flexible linker between the αCTD and αNTD, comprising amino acids 240–249, is considered to facilitate the necessary motion between the two subunit domains to fulfil their function (see Figs 7.4 and 7.6).

The involvement of the α subunit in controlling RNA polymerase activity was already predicted by the early discovery that, upon infection of *E. coli* cells by T4 phages, the α subunits were modified by addition of an ADP-ribosyl group to a specific amino acid residue (Arg265). Since the αCTD has no effect on the T4 *late* transcription it is believed that modification at Arg265 alters the transcriptional properties of the polymerase in favour of the *early* transcription of the T4 phage genes. Interestingly, many amino acid substitution mutations

Box 2.1 Footprinting methods

The contact regions within protein–nucleic acid complexes can be analysed by comparing the accessibilities of nucleotide or amino acid positions in the complex with those in the free nucleic acid or the free protein. Biochemically the accessibilities are usually determined by modification with chemical reagents or through limited enzymatic hydrolysis. Although the method is generally applicable to both proteins and nucleic acids, the majority of techniques have been developed to study contact sites within DNA or RNA. In the case of protein–DNA complexes, DNAse I or specific exonucleases are generally used as cleaving enzymes for the DNA, while specific peptidases can be used to cleave the protein. To map sites within a nucleic acid molecule at higher resolution chemical reagents, which can be either base specific, single-strand-specific, or which are able to modify the sugar–phosphate backbone are employed (Table B2.1). Based on knowledge of the chemical attack it is often possible to assign the accessible position to the major or minor groove of the nucleic acid structure. For instance, the reagent **dimethylsulphate (DMS)** reacts with N7 of guanosine (major groove) or N3 of adenosine (minor groove). Some reagents have the advantage of allowing modification within living cells, thus giving structural information under *in vivo* conditions. It is important that the modification or cleavage reactions are performed at low frequency per molecule, conditions termed 'single hit'. After cleavage or modification the accessible sites are analysed by denaturing gel electrophoresis of the corresponding nucleic acid fragments. Each cleavage results in a nucleic acid fragment of characteristic

Table B2.1 DNA or RNA Footprinting reagents

Reagent	Site of attack	Information
Dimethylsulphate *	N7 of guanine	Binding to major groove
	N3 of adenine	Binding to minor groove
	N1 of cytosine	Single-stranded
Potassium permanganate *	5,6 double bond, thymine > cytosine	Single-stranded, changes in helical parameters
Diethyl pyrocarbonate	N7 of guanine or adenine	Changes in helical parameters, single-stranded
1,10-phenanthroline-Cu(I)	C1′ and C4′ of deoxyribose	Minor groove binding, change in DNA structure (B > A ≫ Z)
Hydroxyl radicals †	Ribose or deoxyribose	Backbone contacts

* Reagent can be applied *in vivo*
† Reagent can also be applied for protein footprinting.

length. Detection normally involves autoradiography of the nucleic acid, which has been labelled on one strand at one end before the cleavage. Alternatively, non radioactive nucleic acids can be analyzed and the modified sites detected by a primer extension reaction employing a reverse transcriptase system. Generally, reverse transcription does not read through certain modified nucleotides, thus the resulting cDNAs will terminate at a position where a chemical modification has occurred. The resulting nucleic acid fragments obtained after modification reactions performed with the free nucleic acid and the protein–nucleic acid

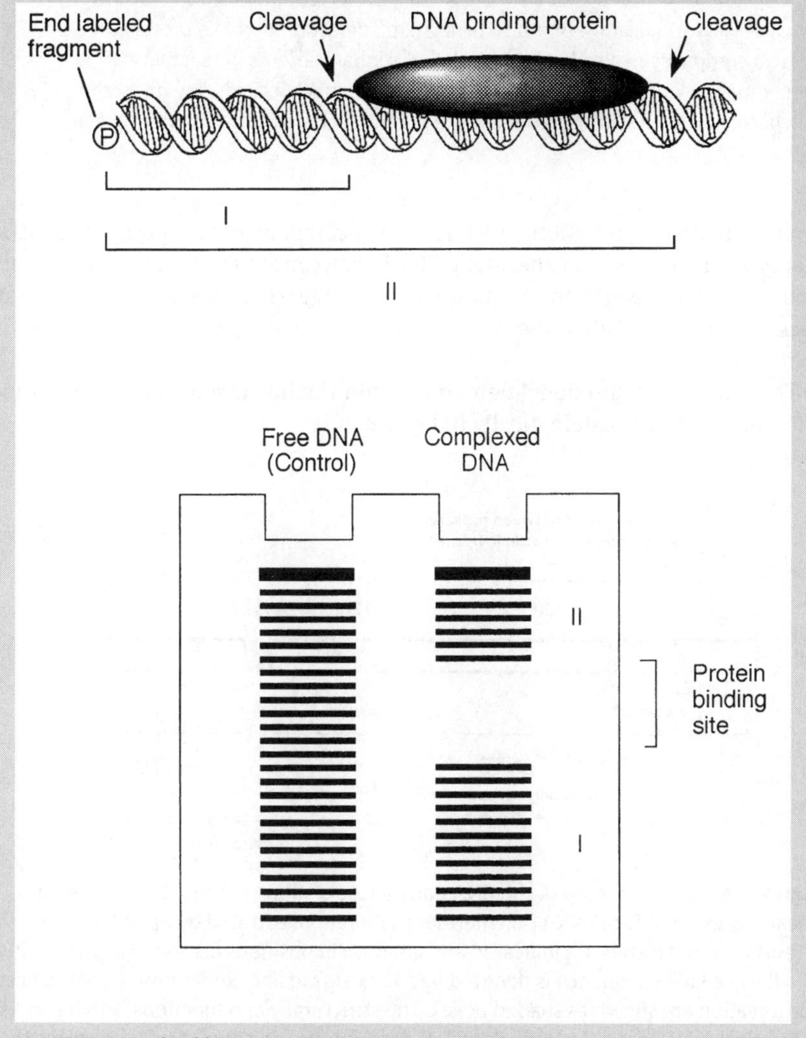

Fig. B2.1 Principle of footprinting experiments

complex, for instance, are separated site by site on a denaturing polyacrylamide gel. Identification of the positions of modification or cleavage is made possible by a sequencing reaction run in parallel with the unmodified nucleic acid. Positions of tight protein contact are characterized by a clear reduction in the reactivity of that position within the complex. Strong protection from chemical modification or reduction in the accessibility towards nucleases therefore results in a strong reduction or lack of the corresponding fragments on the sequencing gel. This lack of bands in the complexed samples has led to the expression 'footprints' (Fig. B2.1).

It should be noted that the accessibility of nucleotide positions towards chemical modification or limited enzymatic digestion within a protein–nucleic acid complex can also be enhanced. Such enhancements in reactivity of complexed nucleic acids, compared to the free nucleic acids, are generally characteristic of conformational changes that occur during complex formation.

which affect the function of αCTD are located near the position of ADP-ribosylation. The notion that the αCTD is not conserved between prokaryotic and eukaryotic polymerase subunits may suggest, however, that disparate mechanisms are likely to be involved in the activation of transcription in the two kingdoms.

The location of functional domains within the linear sequence map of the α subunit are shown schematically in Fig. 2.6.

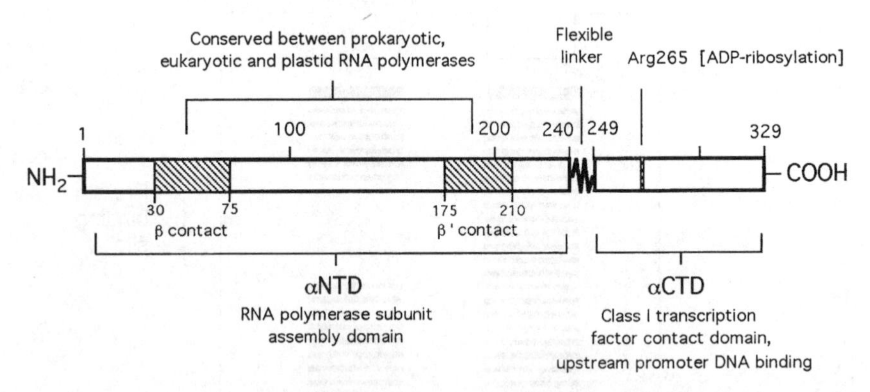

Figure 2.6 Structural map of the RNA polymerase α subunit. The 329-amino acid linear sequence of *E. coli* RNA polymerase α subunit is illustrated by open bars. The N- and C-terminal ends are indicated and amino acid positions are marked. The position of a flexible linker sequence is denoted by a dark zigzag line. Regions of high sequence conservation are shown as shaded boxes. The structurally and functionally independent N-terminal (αNTD) and C-terminal (αCTD) domains are indicated. Arg265 denotes the position of bacteriophage T4 modification.

2.2.2 The β subunit

The gene for the β subunit, *rpoB*, is encoded together with the β′ gene (*rpoC*) in one transcription unit. The β subunit is the second largest subunit of RNA polymerase. It is 1342 amino acids in length (150.6 kDa) and, together with the β′ subunit, constitutes the catalytic core of the enzyme. Bacterial β subunits show considerable homology to the second largest subunits of all eukaryotic multisubunit RNA polymerases, indicating a high conservation of the principal functions of these subunits among different organisms. The regions of sequence conservation are distributed in clusters over the whole molecule and have been grouped into nine domains, termed A–I.

Many mutants resistant to the antibiotic **rifampicin** have been isolated and characterized. The structure of this antibiotic, which inhibits RNA transcription initiation in bacteria, is shown in Fig. 2.7.

All the Rif^{r} conferring mutations are located within four distinct regions of the β subunit. Rifampicin inhibits the phosphodiester formation between the first nucleotides of a growing RNA chain (no inhibition is observed at the elongation phase). It is reasonable, therefore, to assume that the rifampicin binding site is close to the site of nucleotide addition, which forms the catalytic centre of the enzyme. This conjecture has been confirmed by cross-linking studies employing rifampicin derivatives containing nucleotides spaced at different lengths from the drug. Based on such studies the antibiotic is predicted to bind 15 Å away from the initiating nucleotide and within about 2 Å of positions -2 and -3 of the template. In the same way, cross-linking experiments have led to the identification of three distinct sites within the β subunit which are close to the 5′ end of the nascent transcript (Lys1065, His1237 in segments H and

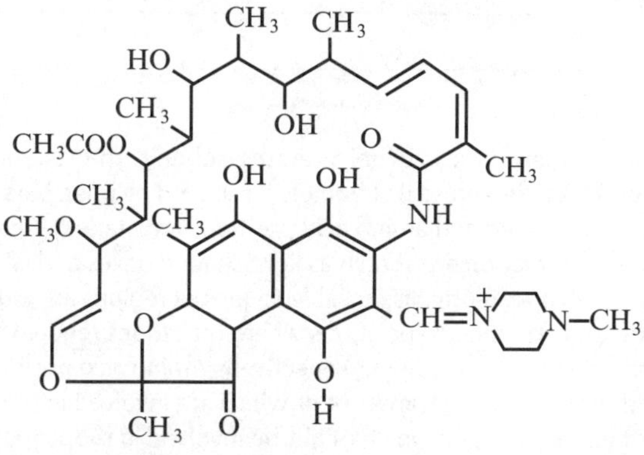

Figure 2.7 Chemical structure of rifampicin. The chemical structure of rifampicin (M_r 823), a semisynthetic derivative of rifamycin B, is shown. The antibiotic selectively inhibits bacterial transcription initiation at very low concentrations (0.01 μg/ml) by binding to the β subunit. Chain elongation beyond the third phosphodiester bond and eukaryotic transcription are not inhibited.

I, and a region between Asp516 and Arg540 in segment D). The C-terminal conserved elements H and I have been proposed to make subunit–subunit contacts to both the α and the β′ proteins. Furthermore, the existence of numerous mutants with altered termination and pausing properties documents that the β subunit is also involved in pausing and termination. These mutations map in the conserved regions C, D, F and I.

Several mutations in the β subunit have been described which affect one of the major bacterial regulatory mechanisms, namely the **stringent response**. This kind of control can be observed at amino acid starvation and involves the action of the small effector nucleotide **guanosine tetraphosphate (ppGpp)**. The β subunit of RNA polymerase is, therefore, considered to be directly involved in the stringent response mediated by ppGpp. Details of this response are described in Section 8.5.

Based on the identification of two 'dispensable' regions within the β subunit structure and from homology alignments of β subunit primary structures from different organisms it is obvious that the respective large subunits from both prokaryotes and eukaryotes are folded in three independent structural domains. A summary of the functional domains within the primary structure of the β subunit is presented in Fig. 2.8.

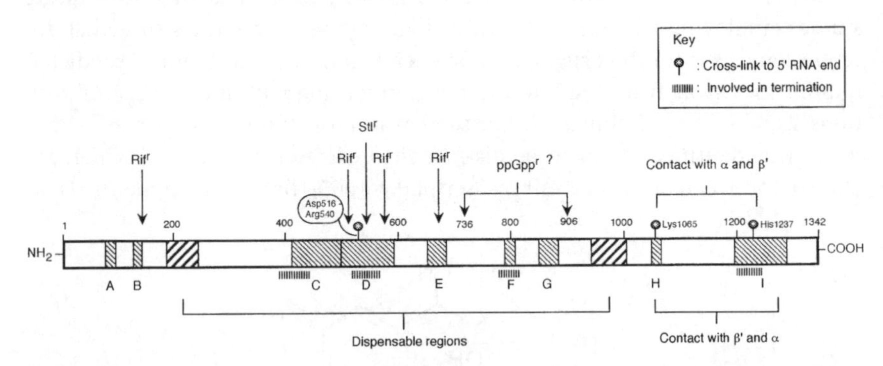

Figure 2.8 Structural map of the RNA polymerase β subunit. The 1342-amino acid linear sequence of *E. coli* RNA polymerase β subunit is illustrated by open bars. The N- and C-terminal ends are indicated and amino acid positions are marked. Regions of high primary structure conservation are shown as light shaded boxes and labelled with capital letters A–I. Two apparently dispensable sequence regions are indicated as dark shaded boxes. Rif^r and Stl^r point to positions where mutations render the β subunit resistant to rifampicin or streptolydigin, respectively. Amino acid positions which have been cross-linked to the growing transcript or which are involved in termination are marked. ppGpp^r denotes a region which might be involved in the stringent response affecting the interaction with the effector nucleotide ppGpp.

2.2.3 The β′ subunit

The largest subunit of RNA polymerase consists of 1407 amino acids (155.2 kDa). As described for the β subunit, considerable homology is found between β′ and the largest subunits of all eukaryotic polymerases. Overall, there are eight segments of sequence conservation with an average of 70% amino acid similarity. The conserved segments have been termed A–H. Note, however, that the C-terminal domain of the largest subunit of eukaryotic RNA polymerase II is not found in the β′ subunit. This specific region for eukaryotes is required for the assembly of pre-initiation complexes, which exhibit activator specificity. This can be taken as an indication that differences may exist between prokaryotes and eukaryotes for the transcription activation step. Within the conserved N-terminal region (segment A) of the β′ subunit there is a Cys_4 Zn finger-like motif, which is assumed to be involved in the binding of nucleic acids (more information on protein motifs involved in RNA or DNA binding can be found in Box 2.2). Within the conserved segment C a region of homology to the DNA binding cleft of the *E. coli* DNA polymerase I suggests the presence of a DNA binding motif.

Analysis of the sites involved in resistance to several antibiotics which affect elementary steps of transcription has significantly improved knowledge of the structure–function correlations of the β′ subunit. For instance, mutations conferring resistance to the antibiotic **streptolydigin**, which inhibits the rate of transcription elongation, have been mapped in segment F. Since similar mutations have also been mapped in segment D of the β subunit it is assumed that both subunits together constitute the binding site for the antibiotic. This

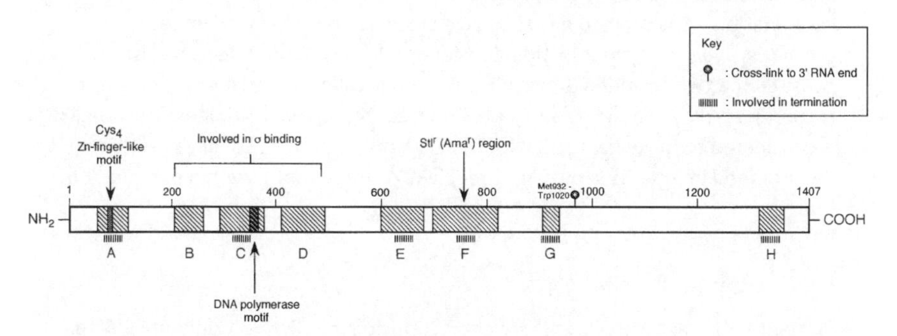

Figure 2.9 Structural map of the RNA polymerase β′ subunit. The 1407-amino acid linear sequence of the *E. coli* RNA polymerase β′ subunit is illustrated by open bars. The N- and C-terminal ends are indicated and amino acid positions are marked. Regions of high primary structure conservation are shown as light shaded boxes and labelled with capital letters A–H. Amino acid positions which have been cross-linked to the growing transcript or which are involved in termination are labelled. A putative DNA binding Zn-finger-like motif and a region of strong structural similarity to DNA polymerase are shown as dark shaded segments. Stl^r and Ama^r denote regions where resistance mutations towards streptolydigin or α-amanitin have been mapped.

Box 2.2 Protein structures involved in nucleic acid recognition

The structural analysis of many DNA and RNA binding proteins from bacteria and eukaryotes has allowed the assignment of a number of recurring primary and secondary structural motifs which function as binding or specificity domains within the amino acid sequences of the corresponding proteins. Frequently occurring motifs are shown in Fig. B2.2.

Helix-turn-helix motifs

This fold is found in many prokaryotic and phage regulatory proteins (see, for instance, the FIS–DNA complex, Fig. 7.14) but occurs also in several families of eukaryotic transcription factors, such as the homeodomain proteins or the POU-specific proteins. The helix-turn-helix motif (HTH) is characterized by two α helices linked by a β turn of four amino acid residues with an invariant glycine at the second position. The second helix makes contacts within the major groove of the target DNA in which it fits precisely. This helix is designated the recognition helix, therefore.

Zinc-finger motifs

Zn-finger domains are widespread in eukaryotic transcription factors (more than 1000 Zn-finger proteins have been characterized). They are not often found in prokaryotic transcription factors, however. This class of DNA binding proteins is structurally rather heterogeneous. The common characteristic element is a stabilizing coordination of the protein structure by a zinc atom. This coordination is maintained either by two cysteine and two histidine residues (Cys_2–His_2-type) or by four histidine residues (His_4-type). Zn-finger motifs consists of a two stranded β-hairpin and a single α helix. The central zinc atom is coordinated by the side chains of two cysteines and two histidines. The α helix is considered to make contacts with the DNA major groove. Usually Zn-fingers are arranged in repetition of two, three, five, or more motifs within one binding protein.

Leucine zipper motifs

Proteins with this type of motif are characterized by a DNA recognition helix that is linked directly to a dimerization domain. Leucine zipper proteins are mostly found in eukaryotes. The characteristic structure is represented by a region of basic amino acid residues (positively charged) at the N-terminal half followed by a C-terminal α helix, where leucines (or similar hydrophobic residues) are present at every seventh position (heptad repeat). Because of the structural parameters of the α helix (approximately 3.6 amino acids per turn) all the hydrophobic residues point in the same direction. If two such domains are aligned in parallel a self-complementary hydrophobic dimerization interface is formed.

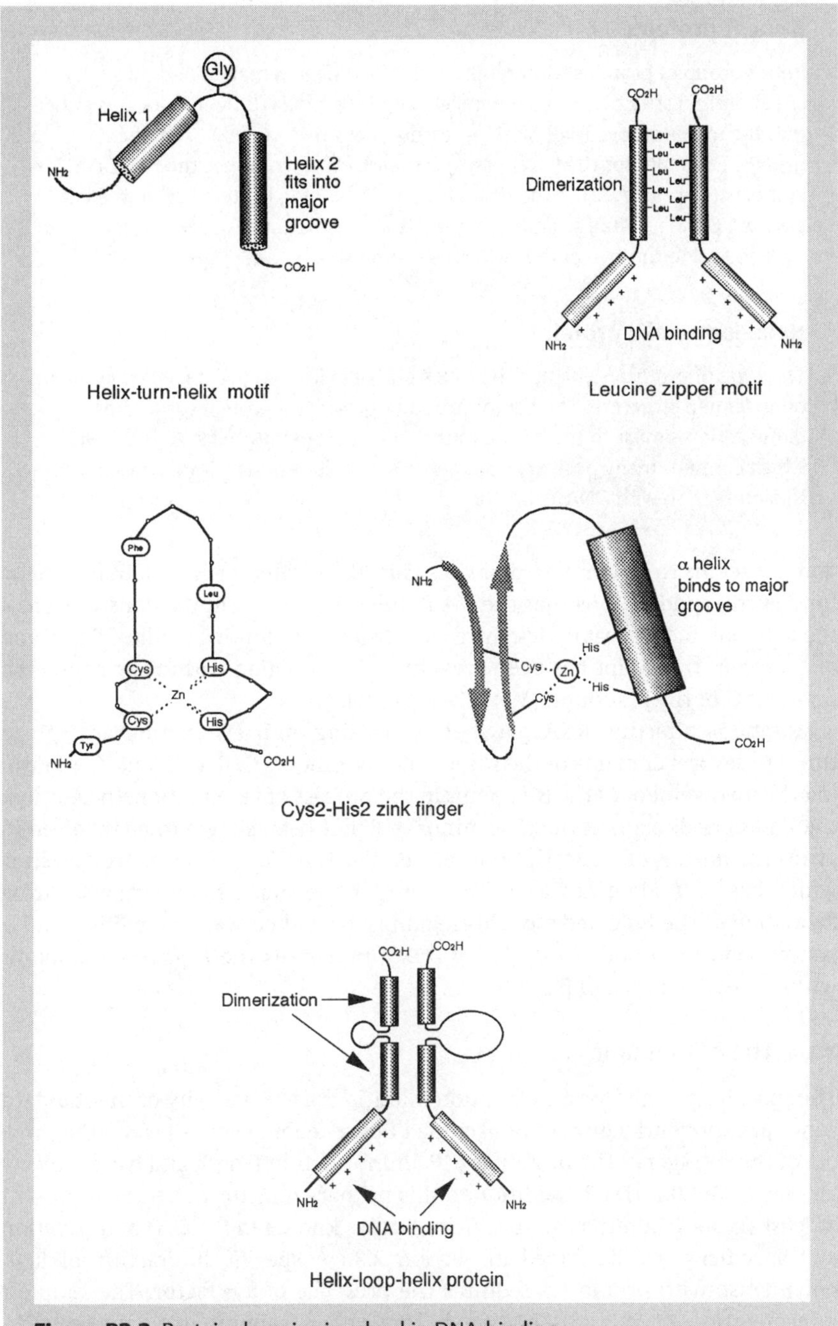

Figure B2.2 Protein domains involved in DNA-binding

β sheet proteins

In this group of proteins recognition of the DNA helix is maintained through a β sheet protein structure. An antiparallel pair of β strands (β ribbon) locates either into the major groove (e.g. MetJ repressor–operator complex) or the minor groove. Examples for the latter case are the HU–DNA complex, the IHF–DNA complex (see Fig. 7.12) or the eukaryotic TATA-binding protein TBP. In the latter case the β sheet forms a saddle-shaped structure that fits exactly into the widened minor groove of the DNA recognition site.

Helix-loop-helix proteins

This type of protein is characterized by a pair of helices which are linked by a loop of undefined structure. The helical structure provides a protein surface for dimerization similar to the helices in the leucine zipper proteins. A third helix, which contains many positively charged amino acid residues, is responsible for the interaction with DNA.

domain seems to be composed of sites partially located in each subunit, therefore. Further indications that the β′ subunit participates in the constitution of the RNA polymerase catalytic centre stem from cross-linking studies. The 3′ end of a nascent transcript has been cross-linked to a region within the conserved segment G of the β′ subunit (Met932–Trp1020).

Mutations affecting RNA polymerase pausing and termination are spread through several domains of the β′ sequence (segments C, E, F, G and H), indicating the involvement of the β′ subunit in the process of termination. In addition, the Zn-finger domain at the C-terminus of β′ has been shown to be involved in **antitermination** of transcription (see Section 5.4). There is also a region in β′ which has been identified as a site where σ^{70} appears to make contact during formation of the holoenzyme. This region is located between domains B and D (amino acid positions 201–477). Fig. 2.9 summarizes the functional domains within the linear map of β′.

2.2.4 The σ^{70} subunit

The specificity factor for transcription initiation of the majority of the standard genes at exponential growth of bacteria ('housekeeping genes') is σ^{70}, the product of the *rpoD* gene. The protein is 613 amino acids in length and has a molecular weight of 70.2 kDa. In addition to this primary σ factor there are at present at least six more alternative specificity factors known in *E. coli*. (Their function and specificity are described in Section 2.3.) A specific interaction of RNA polymerase with promoters requires the presence of a σ factor. The simplest interpretation of this observation is that the σ subunit itself makes DNA sequence-specific contacts. As shown below, this assumption has proven to be correct.

Comparison of different bacterial species has shown that homologous σ factors to the *E. coli* σ^{70} subunit exist. Based on their common amino acid sequences these proteins are considered to belong to the same family of σ^{70} proteins. Members of this family contain four regions with a high degree of sequence conservation. These four regions have been divided into subregions. Within the linear map of σ^{70} a number of different functions can be located. These functions include binding to the core RNA polymerase, DNA binding (recognition of the consensus –10 and –35 promoter elements), DNA melting and interaction with certain transcriptional activators (Class II activators).

It is known that σ^{70} does not bind to promoters in the absence of core RNA polymerase. Apparently, region 1 at the N-terminal end of the molecule plays a critical role in this inhibition. A disordered stretch of 22 acidic amino acid residues (188–209) within subregion 1.2, which is close to the DNA-binding region of the protein, sterically inhibits DNA interaction. In addition, the highly acidic amino acid residues repel the negatively charged DNA phosphate backbone electrostatically. (Note that σ^{70} is a very acidic protein, negatively charged at neutral pH.) It is clear that a substantial conformational change within the protein has to occur upon core RNA polymerase binding. A recent study has provided evidence that region 1 of σ^{70} is required for early steps in initiation complex formation, namely the **isomerization** from the **closed** to the strand-separated **open promoter complex** and subsequent formation of an initiating complex (see below).

Region 2 of σ^{70} contains the most highly conserved amino acids. Within this domain there is a set of nine residues which are absolutely invariant among the major σ factors of diverse eubacteria. A hydrophobic core within subregion 2.1 (amino acids 361–390) was shown by deletion analysis to be responsible for RNA polymerase core binding. A single conserved glutamine in region 2.2, corresponding to position 406, is also involved in core RNA polymerase interaction.

To allow sequence-specific incorporation of nucleotides into a growing transcript, the DNA in the coding region has to be converted into a stretch of single-stranded nucleotides. This strand opening is an important step and occurs during initiation after the first binding of RNA polymerase to the promoter. The process of converting closed into open complexes is termed isomerization or promoter melting. During this melting step approximately 12–17 base pairs, corresponding to one and a half helical turns of DNA, are broken. The single-stranded bases enable the base-specific copying steps of nucleotides from the coding strand within the active centre of the enzyme. Melting from the closed to the open complex requires the presence of σ factors. It can be concluded, therefore, that DNA melting during initiation is either directly or indirectly mediated by σ. In fact, it was found that σ^{70} subregions 2.3 together with 2.1 play an essential role in DNA melting. The two regions show considerable similarity to single-stranded nucleic acid binding proteins. Subregion 2.3, which is close to the DNA binding domain, contains a high proportion of solvent-exposed aromatic residues. The aromatic ring stacking interactions between nucleotide bases of the non-template strand and the aromatic amino acid side chains of subunit 2.3 are certainly important for the DNA melting process. This has been explicitly shown for Thr429 by a systematic mutagenesis study. The clusters of aromatic residues within subregion 2.3 are flanked by basic amino acids which are presumably involved in charge neutralization. The notion that subregion 2.3 participates in promoter melting is also confirmed by chemical probing and cross-linking data.

Recognition and binding of the –10 promoter hexamer involves amino acids in subregion 2.4 (amino acids 437–440). This region forms an **amphipathic α helix** with the hydrophobic residues facing the conserved hydrophobic core of the protein, while the residues important for binding the –10 promoter sequence are solvent-exposed on the opposite face of the helix. It is assumed that sequence-specific binding of the non-template strand stabilizes the transcription bubble in an open promoter complex. This assumption is supported by a series of cross-linking results. The template strand would thus be available for the catalytic activities of RNA polymerase. This kind of interaction between σ^{70} subunit and the –10 promoter element is confirmed by mutations within subregion 2.4 which suppress single base substitutions in the –10 promoter hexamer (e.g. a Gln to His change at position 437 suppresses down-mutations at –12 of the –10 hexamer, and a Thr to Ile change at position 440 alters the nucleotide requirement for optimal binding at the same position). The conserved amino acids involved in –10 element recognition have been termed the **rpoD box.** The rpoD box is found in primary σ factors which all recognize the same –10 promoter sequence. Consistent with the above notion is the observation that, among the alternative σ factors which recognize different –10 sequences, the corresponding amino acids are less conserved.

The established information about region 3 is scarce. Subregion 3.1 has a weak similarity with a **helix-turn-helix (HTH)** DNA-binding motif (see Box 2.2). Subregion 3.2, however, seems to be involved in core RNA polymerase binding. This can be inferred from a deletion analysis, where removal of 25 amino acids from subdomain 3.2 causes reduced affinity of the mutant σ subunit for core RNA polymerase binding. The assumption that region 3.2 is involved in core RNA polymerase binding is also confirmed by a recent mutational analysis. Base change mutants within subregion 3.2 were shown to revert the phenotype induced by the complete lack of the regulatory nucleotide ppGpp. It was shown for the mutant σ factors that their affinity for core RNA polymerase was apparently changed. Subregion 3.2 might, therefore, be involved in ppGpp-dependent regulation of transcription (see Sections 8.5 and 8.6).

Region 4 at the C-terminus of the σ^{70} subunit has been shown to participate in the recognition of the –35 promoter element. A DNA-binding motif (HTH motif) in the centre of a block of 30 amino acids within subregion 4.2 is highly conserved between primary σ factors. This structural element shows considerable similarity to many phage or bacterial DNA-binding proteins. Evidence for the direct involvement of the HTH element in recognition of the –35 region comes from mutational studies showing that base changes in that region suppress mutations in the –35 sequences of certain promoters.

Whether interaction of σ^{70} with the –10 and –35 promoter elements occurs simultaneously or in a sequential manner has been a focus for recent research. The two promoter elements are spaced by about two helical DNA turns. According to topographical considerations a simultaneous interaction at both sites of σ^{70} can probably only be achieved if the DNA within the complex is strongly bent. The observation of a pathway of discrete intermediate complexes from the initial promoter binding to the open strand-separated initiation complex, extending in the downstream direction, argues in favour of the sequential model. In addition, the kinetics of the σ subunit-directed promoter recognition have revealed that the –35 region is first bound by σ subregion 4.2 (see below). Direct contacts between subdomain 2.4 of σ and nucleotides within the –10 region of the non-template strand occur subsequent to this interaction.

Finally, a group of three or four basic amino acids near the C-terminus is possibly involved in interaction with **activator proteins** like AraC, which controls the expression of genes necessary for uptake and catabolism of arabinose (see Chapter 7) or the phage λ cI protein which regulates phage expression. Several additional examples for the binding of transcriptional activators to conserved regions of the σ^{70} subunit exist. For example, the transcription factors PhoB (involved in phosphate regulation), MalT (involved in the regulation of the maltose operon) and the catabolite regulator protein CRP are supposed to make contacts to a conserved element upstream of subregion 4.2. Mutants have been characterized which have base changes in the HTH motif of subregion 4.2 (Asp570, Glu575, Glu571, Arg596) known to be responsible for binding the –35 region. These mutants are defective in PhoB-, AraC-, MalT- or phage λ cI-protein-dependent activation, indicating that these activators bind to the HTH region. In addition, a T4 phage protein which functions as an **anti-sigma factor** (**AsiA**, see below) has been shown to associate with the C-terminal 63 amino acids of subdomain 4.2. The functional domains within the linear map of σ^{70} are summarized in Fig. 2.10.

Interestingly, transcription activators shown to interact with σ^{70} are known to affect promoters which do not have a strong –35 consensus structure. It is assumed, therefore, that transcription factor interaction with σ^{70} can compensate for the absence of the normal interaction between σ^{70} and the consensus –35 sequence.

The crystal structure of a σ^{70} subunit fragment from *E. coli* RNA polymerase has been resolved at 2.6 Å resolution. This fragment contains the residues 114–446. The derived structure allows explanation of many of the conclusions listed above, such as the interaction of σ^{70} with the core RNA polymerase, and binding of σ^{70} to the consensus –10 promoter elements. In addition, the derived structure suggests an immediate explanation for the inhibition of σ^{70} binding to DNA in the absence of RNA polymerase owing to sterical shielding of the binding domain by the acidic stretch of subregion 1.2.

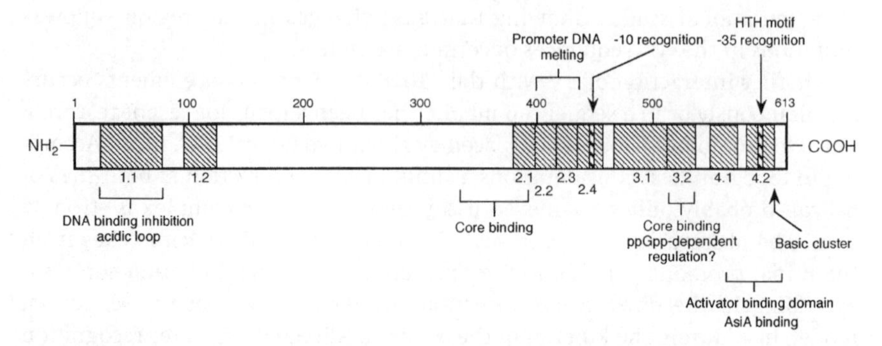

Figure 2.10 Structural map of the RNA polymerase σ^{70} subunit. The 613-amino acid linear sequence of the *E. coli* RNA polymerase σ^{70} subunit is illustrated by open bars. The N- and C-terminal ends are indicated and amino acid positions are marked above the structure. Structurally conserved domains 1-4 and subdomains (1.1, 1.2, 2.1, etc.) are shown as shaded areas and labelled below the structure. Subdomain 1.1 contains clusters of acidic amino acids and inhibits direct DNA-binding in the absence of the core polymerase. Amino acid sequences within subdomains 2.4 and 4.2, shown as separately shaded regions and marked by arrows, are known to interact directly with the –10 and –35 promoter elements, respectively. Subdomains 2.2 and 2.3 are known to be involved in promoter DNA melting. Regions 2.1 and 3.2 are likely to interact with core RNA polymerases. Mutations in subdomain 3.2 point to a function of σ^{70} in ppGpp-dependent regulation during the stringent response. A helix-turn-helix (HTH) DNA-binding motif and a cluster of basic amino acids involved in transcription factor recognition are marked within the C-terminal part of the molecule. Within subdomain 4.2 the position of the bacteriophage AsiA protein (anti-sigma factor) interaction, which inhibits the –35 recognition of bacterial promoters, is also indicated.

2.3 Alternative σ factors

Alternative σ factors direct RNA polymerase to initiate transcription at specific promoters to allow differential expression of a particular subset of genes. Alternative σ factors were first described for *Bacillus subtilis* and its phages SP01 and SP82, where they were shown to play a role in altering phage growth or in changing the program of transcription during developmental processes such as sporulation. Since then, 10 different σ factors have been characterized in *B. subtilis*. This number of characterized σ factors exceeds that of any other bacterial species (Table 2.4).

It is now clear, however, that alternative σ factors are a common feature in all bacteria, and at present there are seven different σ subunits known to exist in *E. coli* (Table 2.5).

The specificity of RNA polymerase for promoter sequences can be modified

Table 2.4 σ factors of *Bacillus subtilis*

σ factors *	Gene	Upstream recognition sequence (–35 region)	Number of spacer nucleotides	Downstream recognition sequence (–10 region)	Function
Vegetative factors					
σ^{A} (σ^{43}, σ^{55})	*sigA, rpoD*	TTGACA	17 ± 1	TATAAT	Housekeeping genes, early sporulation
σ^{B} (σ^{37})	*sigB*	$(^{A}/_{G})GG(^{A}/_{T})TT(^{A}/_{G})A$	14	GGGTAT	General stress response
σ^{C} (σ^{32})		AAATC	15	$TA(^{A}/_{T})TG(^{C}/_{T})TT(^{T}/_{G})TA$	Postexponential gene expression
σ^{D} (σ^{28})	*sigD, flaB*	TAAA	15	GCCGATAT	Chemotaxis, flagellar gene expression
σ^{H} (σ^{30})	*sigH, spoOH*	$(^{A}/_{G})(^{A}/_{G/C})AGGA(^{A}/_{T})(^{A}/_{T})T$	14	$(^{A}/_{C})GAAT$	Competence, early sporulation
σ^{L} (σ^{54})	*sigL*	TGGCAC	5	TTGCANNN	Expression of degradative enzymes
Sporulation factors					
σ^{E} (σ^{29})	*sigE, spoIIGB*	$(^{T}/_{G})(^{A}/_{C})ATA(^{A}/_{G})(^{A}/_{T})$	14	$CATACA(^{A}/_{C})T$	Early mother cell gene expression
σ^{F} ($\sigma^{spoIIAC}$)	*sigF, spoIIAC*	$GCAT(^{A}/_{G})$	15	$GG(^{A}/_{C})(^{A}/_{G})A(^{A}/_{G})(^{A}/_{C})T(^{A}/_{T})$	Early forespore gene expression
σ^{G}	*sigG, spoIIIG*	$G(^{A}/_{C})AT(^{A}/_{G})$	18	$CAT(^{A}/_{T})(^{A}/_{C})TA$	Late forespore gene expression
σ^{K} (σ^{27})	*sigK, spoIVCB:spoIIIC*	AC	17	CATANNNTA	Early mother cell gene expression

* Alternative designations are given in parentheses.

Table 2.5 σ factors of *Escherichia coli*

σ factors	Upstream recognition sequence (–35 region)	Number of spacer nucleotides	Downstream recognition sequence (–10 region)	Function
σ^{70}	TTGACA	17±1	TATAAT	Housekeeping genes at exponential growth
σ^{s} (σ^{38}) *	CCGGCG	17 ± 1	CTATACT	Stationary phase expression
σ^{H} (σ^{32})	TNtCNCCCTTGAA	13–17	CCCCATtTA	Heat shock genes
σ^{E} (σ^{24})	GAACTT	16 (ATAAA)	TCTGAT	Extreme heat shock, extracytoplasmic stress
σ^{F} (σ^{28})	TAAA	15	GCCGATAA	Flagella synthesis and chemotaxis
σ^{N} (σ^{54})	ttGGcaca	4	ttGCA	Nitrogen-regulated genes
FecI (σ^{19})§	AAGGAAAAT	17	TCCTTT	Ferric citrate transport genes, extracytoplasmic stimuli
T4 σ^{gp55}	None †		TATAAATA	Late T4 genes
SPO1 σ^{gp28}	TNAGGAGANNA	15–16	TTTNTTT	Middle phage genes
SPO1 $\sigma^{gp33,34}$	CGTTAGA	18 ± 1	GATATT	Late phage genes

* σ^{s} and σ^{70} will recognize the same promoters. The indicated sequence corresponds to the *fic* promoter, the only known promoter that is exclusively recognized by σ^{s}.
† σ^{gp55} promoters do not have a consensus –35 recognition element.
§ The promoter sequence of the *fecA* operon comprising five transport genes is given.

by different σ factors. Some of the alternative factors are absolutely specific for their own promoter sequence, while others have rather overlapping promoter specificities. Considerable flexibility in promoter use is thus possible through the cellular levels of alternative σ factors. It is not surprising, therefore, that the cellular levels of the different σ factors are subject to specific regulation. Regulation of σ factors can take place in different manners. Their *concentration* can either be controlled by synthesis or protein stability. On the other hand, their *activity* can be modulated by modification (e.g. cleavage of a pro sequence in the case of *B. subtilis* σ^{K}) or through the interaction with inhibitors (anti-sigma factors; see below).

2.3.1 Stationary phase-specific σ factor σ^S (σ^{38})

When *E. coli* cells enter the stationary phase and stop growing, many genes necessary under exponential growth conditions are no longer expressed. Instead, a set of genes important for survival at stationary phase is turned on (see Section 8.3). Many of these genes are transcribed by the holoenzyme $E\sigma^S$. The stationary phase-specific sigma factor, σ^S, is encoded by the *rpoS* gene and, based on its molecular weight (38 kDa), is also termed σ^{38}. σ^S is not only the alternative σ factor for the expression of stationary phase-specific genes but also controls a number of genes required under stress conditions, such as changes in the osmolarity or under nutritional deprivation. Interestingly, σ^S shares many of the promoter specificities with the primary σ factor σ^{70}, which means that many promoters can be recognized by both $E\sigma^S$ *and* $E\sigma^{70}$. This is unusual, since the other alternative sigma factors, described below, respond more or less to their own specific promoter sequences (see Table 2.5). There appears to be a certain conservation of C residues in the –35 region of many but not all, σ^S-dependent promoters, which suggests that nucleotides in the –35 region are partly responsible for the discrimination between holoenzymes containing σ^{70} or σ^S. In addition, the selectivity for transcription of stationary phase-specific genes by σ^S seems to require special conditions, such as auxiliary factors or a DNA template with lower superhelical density (see Box 6.4). At exponential growth the number of σ^S molecules in the cell is very small, at the limit of detection, while σ^{70}, the principal σ factor is present in approximately 1000 copies per cell. However, at the stationary phase the level of σ^S reaches about 30% that of σ^{70}. Control on the concentration of σ^s is maintained at several levels. Regulation occurs during transcription of the *rpoS* gene but also at post-transcriptional stages during translation. Furthermore, the concentration of σ^S is controlled by protein stability (see Section 8.3).

2.3.2 Heat shock σ factors σ^H (σ^{32}) and σ^E (σ^{24})

When the growth temperature of a bacterial culture increases above 40°C a number of genes become active. This reaction is called the **heat shock response** (see Section 8.2). The heat shock genes code for proteases and a family of proteins termed **chaperones**, which help to refold or degrade thermally denatured proteins. The temperature-dependent transcription of the heat shock genes requires RNA polymerase containing the σ^H subunit ($E\sigma^H$). The concentration of the σ^H subunit increases about 20-fold following transition to a higher growth temperature. Because of the short half-life of σ^H ($t_{1/2} \approx 1$ min) its amount in the cell, and concomitant the synthesis of the heat shock proteins, decreases rapidly. When cells are shifted to a temperature of 50°C the heat shock response is more extreme, and continuous synthesis of the heat shock proteins is essential for survival. Transcription of σ^H, the *rpoH* gene product, is now maintained by a second heat shock σ factor σ^E or σ^{24}, the *rpoE* gene product. Eσ^E will not recognize σ^H or σ^{70}-controlled promoters, and appears to be responsible for growth at very high temperatures and for thermotolerance. In addition, the Eσ^E polymerase transcribes genes in response to periplasmic stress. σ^E apparently belongs to a group of σ factors regulating extracytoplasmic functions, responding to extracytoplasmic stimuli. The two heat shock factors σ^H and σ^E are barely detectable under steady-state growth at normal temperature. They rapidly become abundant following a shift to temperatures above 40°C. As is the case in the regulation of σ^S, the concentration and activity of σ^H in the cell is regulated by transcription, translation and protein stability (see Section 8.2). In addition, the activity of σ^E is modulated by **anti-sigma factors** (see below).

2.3.3 σ^F (σ^{28}), factor for flagella gene expression and chemotaxis

The *rpoF* gene product σ^F, also termed FliA, is a minor σ factor which is involved in the synthesis of the proteins necessary for flagella formation and **chemotaxis**. This group of 18 genes is coordinately expressed and enables bacterial cells to move in response to gradients in the concentration of the chemical environment. The number of σ^F molecules per cell approaches about 50% that of σ^{70}. Its activity, however, is controlled by a second protein, FlgM, which functions as an anti-sigma factor, forming a 1:1 complex with σ^F, thereby rendering it inactive (see below). Expression of the flagella genes requires, in addition to the alternative σ^F, the presence of a Class I transcription factor (FlhDC).

2.3.4 σ^N (σ^{54}), factor for nitrogen-regulated genes

Originally, σ^N, also called NtrA, the gene product of the *rpoN* or *ntrA* gene (see Table 7.7), was identified as a factor required for the expression of the glutamine synthetase gene (*glnA*). It is now known to be responsible for a set of pleiotropic genes, including those for formate hydrogenase (*fdhF*) or the phage shock protein (*psp*). However, most of the σ^N-controlled genes encode proteins important for the nitrogen metabolism. σ^N belongs to a single class of σ factors which distinctly deviates in structure and function from the σ^{70} class. σ^N-dependent promoters do not have conserved –10 and –35 recognition sequences, instead, a GG dinucleotide around position –24 and a GC dinucleotide located around –12 are highly conserved (see Table 2.4). There is no detectable sequence similarity between primary σ factors (σ^{70}) and the σ^N family. In fact, several amino acid sequence motifs within the σ^N family are reminiscent of eukaryotic transcription factors. The N-terminal domain of σ^N is rich in glutamine. The protein contains two hydrophobic heptad repeats forming a potential **leucine zipper** motif (see Box 2.2). The middle domain is acidic and, in analogy to eukaryotic transcription factors, could provide an 'activation' surface. In addition, there are several putative HTH elements which are probably involved in binding the conserved promoter sequences. Unlike σ^{70}, σ^N is able to bind to promoter DNA in the absence of RNA polymerase. Although $E\sigma^N$ can form stable closed complexes with σ^N-dependent promoters no transcription initiation can be observed in the absence of additional activator proteins such as NtrC. In the absence of an activator, the $E\sigma^N$ enzyme is thus completely inactive in melting the promoter from a closed to a strand-opened complex. Isomerization to the open complex does not only require the presence of the activator, it also requires the hydrolysis of ATP by the activator proteins. The activator proteins, in turn, are regulated by phosphorylation. Thus, transcription from σ^N-dependent promoters has much in common with eukaryotic transcription initiation, where the activity of transcription factors is controlled by phosphorylation, and open complex formation depends on the hydrolysis of ATP. There is a further parallel between σ^N-dependent and eukaryotic transcription initiation. Binding sites for the activators at σ^N-dependent promoters are usually too far upstream to permit direct contact between the activator and RNA polymerase unless the intervening DNA is looped out. However, activators still retain their function even if the binding sites are positioned more than 1000 base pairs away from the site of polymerase interaction. This is a typical feature of eukaryotic **enhancer elements**. (More detail of the transcription of nitrogen-regulated genes is described in Section 7.6.2.)

2.3.5 A minor σ factor, FecI (σ^{19})

The iron citrate transport system in *E. coli* is transcriptionally activated by FecI, the *fecI* gene product. FecI shows remarkable sequence similarity to members of the σ^{70} family and, in combination with the core RNA polymerase, exhibits σ factor function. Holoenzyme containing FecI transcribes the *fecA* operon, consisting of five genes which are necessary for the citrate-dependent iron transport. FecI belongs to a subgroup of σ factors that respond to extracytoplasmic stimuli (see section on σ^E). The activity of FecI as a σ factor is very likely to be controlled by complex formation with the anti-sigma factor FecR.

2.3.6 Phage-specific σ factors, gene products 55, 28 and 33/34

Deprogramming the specificity of host RNA polymerase is a general feature of phage infection. In some cases the natural properties of RNA polymerase are corrupted by phage-specific σ factors which inhibit host-specific transcription and modulate expression in favour of the phage genes. Examples can be found in the *E. coli* phage T4 or in the *B. subtilis* phages SP01 and SP82. Apparently, phage T4 uses a whole set of tools to suppress host transcription and to program RNA polymerase for the transcription of the T4-specific genes. The ADP-ribosylation of the CTD of the α subunits was discussed in Section 2.2.1. As a second means for reprogramming the host RNA polymerase, the T4 gene product 55 (gp55) acts as an alternative σ factor, providing specificity for the transcription of the late phage genes. Promoters recognized by gp55 are characterized by a simple −10 element and do not have any −35 consensus sequence. Accordingly, there is only a conserved domain 2 and no sequence similarity with domain 4 of the σ^{70} family, which is known to be implicated in the recognition of the −35 promoter element (see Section 2.2.4 above). Hence, gp55 has been classified as a 'minimal' σ factor.

Activation of late T4 transcription provides a nice example for a mechanistic coupling between replication and transcription. Activation of T4 late promoters requires mobile components of the replication machinery that track along the DNA. In other words, late T4 transcription is dependent on the concurrent replication of the T4 DNA. The tracking system for T4 replication consists of the trimeric T4 phage protein gp45 that forms a **sliding clamp** of the T4 DNA polymerase. This protein, which is topologically linked to the DNA but not fixed by binding, is known to interact with two proteins that constitute part of the transcription complex with host RNA polymerase. Binding occurs to the T4 phage-specific coactivator protein gp33 and the late T4 σ factor gp55. Thus, transcription of T4 late genes is directly linked to replication through binding of the sliding clamp protein gp45 to the RNA polymerase attached phage proteins gp55 (σ factor) and gp33 (coactivator).

The related *B. subtilis* phages SP01 and SP82 use phage-encoded σ factors for the expression of their middle and late genes. The SP01 gp28 has been shown to act as a σ factor for the transcription of middle genes. Interestingly, SP01 late gene expression requires a combination of two phage proteins which act in concert as a single σ factor. The factor is formed by the two gene products 33 and 34 (gp33 and gp34).

2.3.7 Anti-sigma factors

Changing the pattern of expression by alternative σ factors can only be efficient if the cellular concentration or activity of the respective σ factors is controlled properly. Without such a control there would be continuous competition with the principal σ factor (σ^{70}) for the core enzyme of RNA polymerase. No coordinated response of the subset of genes under the control of the alternative σ factors would thus be possible. Hence, alternative σ factors must be precisely regulated so that they are only present (or active) when they are needed. As has been mentioned above, the concentration of different σ factors in the cell is often regulated transcriptionally, at the translational level or at the level of protein stability through the action of proteases. However, some alternative σ factors are present in the cell in rather high and constant concentrations, even under conditions when the genes that they control are not needed. In those cases the activity of the σ factors is often controlled through the action of anti-sigma factors. Anti-sigma factors bind to σ factors and inhibit transcriptional activity by interfering with initial binding and open complex formation at bacterial promoters.

Regulation of σ factor activity by anti-sigma factors seems to be a general mechanism. Anti-sigma factors have been characterized in *E. coli*, *B. subtilis*, *Salmonella typhimurium* and in the *E. coli* T4 phage. The first anti-sigma factor reported was a 10-kDa (90 amino acids) bacteriophage T4 protein known to be a part of the phage regulatory system, which shuts off transcription of host genes after infection. This protein, now called **AsiA, was shown** *in vitro* to inhibit $E\sigma^{70}$ transcriptional activity by binding to region 4.2 of σ^{70} (see above). Region 4.2 is responsible for binding the –35 promoter element. Promoters that lack a –35 region are therefore less sensitive to AsiA inhibition. Transcription of the T4 DNA can be divided into three classes which are characterized by their different promoters. There are promoters for the early, middle and late genes which are used sequentially by host RNA polymerase. The T4 AsiA protein acts as anti-sigma factor by binding to σ^{70}. This interaction triggers the transition from early to middle transcription because AsiA binding to σ^{70} inhibits both T4 early and host promoters but stimulates T4 middle transcription. Transcription of the middle genes is further promoted by the action of **MotA**, another T4 phage protein. MotA binds to middle promoters at a site centred at –30, termed the MotA box. Together with the Asia protein this interaction causes an efficient reprogramming of the host RNA polymerase in favour of T4 middle gene transcription. The AsiA anti-sigma factor furthermore supports late bacteriophage development since it allows transcription of the late phage-specific genes by the T4 σ factor gp55 (see above), while $E\sigma^{70}$-dependent transcription is inhibited. Normally, $E\sigma^{gp55}$ is not successful in competition with $E\sigma^{70}$. It was considered therefore that the AsiA protein helps to redirect *E. coli* RNA polymerase to transcribe the late T4 phage genes. Recent studies have shown however that the effect is only modest. (Kolesky *et al.*, 1999)

Another example of regulation by anti-sigma factors is known to occur during the expression of σ^E-dependent genes. The alternative *E. coli* σ factor σ^E, which functions as a heat shock factor and controls the extracytoplasmic stress response, is encoded as the first gene of a four-gene operon. The second gene product of the operon, the 27-kDa protein RseA (*r*egulator for *s*igma *E*), was shown both *in vitro* and *in vivo* to bind to σ^E and to inhibit σ^E-directed transcription. Based on these genetic and biochemical studies RseA has also to be considered as an anti-sigma factor for σ^E.

A large number of σ factors are involved in the developmental process of sporulation in *B. subtilis*. Some of these σ factors are regulated in their activity by anti-sigma factors. An example is the 16.5-kDa *spoIIAB* gene product, which inhibits the transcription from *B. subtilis* σ^F- and σ^G-dependent promoters (σ^F is required for the early sporulation genes, whereas σ^G controls the late sporulation genes, see Table 2.4). Inhibition almost certainly involves interaction between SpoIIAB and *B. subtilis* σ^F or σ^G. The activity of SpoIIAB itself is regulated by SpoIIAA, which is able to bind to SpoIIAB in the presence of ADP, but not ATP, which phosphorylates SpoIIAA, rendering it inactive for SpoIIAB interaction. Thus the ratio of ADP:ATP in the cell determines whether SpoIIAB will bind to SpoIIAA or to the σ factors σ^F or σ^G. Hence, SpoIIAA functions as an **anti**

anti-sigma factor and is modulated by the energy status of the cell, which determines the ADP:ATP ratio. Note that the cell-type specific gene expression in the sporulating *B. subtilis* cell is highly complex. It is regulated by a cascade of σ factors, which are activated or inactivated by anti-sigma and anti anti-sigma factors.

Similarly, a general stress response σ factor σ^B is involved in regulation of stress genes in *B. subtilis*. The anti-sigma factor RsbW inhibits σ^B-dependent transcription by binding to σ^B but not to RNA polymerase. The activity of RsbW in turn is inhibited in a similar manner as SpoIIAB by interaction with the anti anti-sigma factor RsbV. This complex prevents interaction of RsbW with *B. subtilis* σ^B.

A further example of the action of anti-sigma factors is the complex biosynthesis of the flagellar organelle in *S. typhimurium*. Transcription of the flagellar genes is directed by the holoenzyme containing the sigma factor FliA (σ^F). The 10.6-kDa anti-sigma factor FlgM was characterized; by association with σ^F it inactivates transcription from the σ^F-dependent promoters. FlgM shows homology to the **DnaK chaperone** protein from several bacterial species, which acts as a negative regulator of the heat shock regulon by binding to the heat shock σ factor σ^H, thereby initiating its inactivation through proteolysis (see Section 8.2). The two proteins apparently contain similar domains for the recognition and binding of σ factors. Once the basic elements of the flagellar organelle are synthesized FlgM is exported through the flagellar hook-basal body structure and, with decreasing concentrations of FlgM, transcription from σ^F-dependent promoters is restored.

2.4 The three-dimensional structure of RNA polymerase

What is known about the overall architecture of RNA polymerase? Although no high resolution structure is available yet, considerable progress has been made in the past few years by three-dimensional reconstructions from electron micrographs of flattened helical or two-dimensional crystals. Low resolution structural information (≈ 23 Å) is thus available for the *E. coli* core and holoenzyme, and also for *yeast* RNA polymerase I and II at a slightly higher resolution. The overall dimensions of the *E. coli* core enzyme, which has an asymmetric structure, are 85 Å × 105 Å × 140 Å. It is remarkable that the structure of the eukaryotic yeast enzyme, which has a much higher subunit complexity, bears a significant resemblance to that of *E. coli*. Both have in common a surface cleft, which forms a groove or channel with dimension about 25 Å × 45 Å. This channel has the right dimensions to accommodate double helical DNA. It is surrounded by a thumb-like projection which could either form an open structure or a closed conformation like a ring that completely encloses the channel. While the open structure has been found for the *E. coli* holoenzyme, the closed conformation is characteristic for the yeast and the *E. coli* core enzymes. It is assumed that the two conformations define promoter binding and elongating species of RNA polymerase. Based on the different structures the following process for promoter binding and elongation can be inferred: the deep and open groove of the holoenzyme provides a suitable surface for binding of the double-stranded promoter DNA. When the enzyme turns into elongation mode, after the release of σ, the thumb-like structure closes, surrounding the DNA. This stabilizes the elongation complex, which translocates along the template and which is characterized by a high processivity (see Section 4.1). A similar mechanism is known to occur during DNA replication. The stability of DNA replication complexes is maintained by **sliding clamps** which act as processivity factors by forming ring-shaped structures. Interestingly, a similar thumb-like structure is also found in the single subunit phage-specific RNA polymerases as well as in DNA polymerase I or HIV-I reverse transcriptase. This thumb-like structure may alternatively function in DNA melting or as a flexible element in chain elongation or translocation. The length of the channel that is visible on the polymerase surface could accommodate more than two turns of double-stranded DNA. Its path on the protein surface is bent. This is in line with the known bending of DNA in both the initiation and elongation complexes (see Sections 4.1 and 6.2). Furthermore, from the length of the DNA protected in RNA polymerase complexes it is clear that DNA and RNA polymerase must be wrapped around each other.

Presumably the major structural features visible in the structural model of RNA polymerases are determined by the two largest subunits, β and β′. In this respect it is noteworthy that in a recent study a fusion protein of β and β′ could be assembled into a functional RNA polymerase. Note also that in eubacteria and archaea the two subunits are generally encoded next to each other in a single transcription unit. It therefore appears that within the active centre of RNA polymerase the C-terminus of β must be close to the N-terminus of β′. The structural information derived from the low resolution X-ray studies is certainly helpful in accommodating current biochemical data on the architecture and function of RNA polymerase. Clearly, however, more high resolution information is needed to understand the individual steps of the transcription reaction at a molecular level.

2.5 Is RNA polymerase modified during different stages of transcription?

Although the constitutive components of RNA polymerase have been defined above there are a large number of proteins that interact transiently with RNA polymerase, modulating its activity or specificity. Many of these proteins can be classified as activators or repressors which act transiently during the stage of transcription initiation. However, some transcription factors, like the **Nus factors**, that associate during elongation are also known; they affect pausing, termination or antitermination (see Chapters 4 and 5).

The small 10-kDa ω protein has already been mentioned as being permanently associated with RNA polymerase. Its function is not completely clear, however. The ω factor was first claimed to be involved in the stringent control (see Section 8.5). This assumption has been challenged since cells devoid of the functional gene for ω were not affected in this type of regulation. More recent studies suggest a more structural than functional role for ω. It was shown that the presence of ω facilitates the renaturation of denatured RNA polymerase subunits. Therefore a function of ω in the overall stability of the enzyme has been suggested.

Two additional proteins, **GreA** and **GreB**, have been identified which are tightly associated with RNA polymerase. They stimulate a phosphohydrolase activity within the core enzyme. The two proteins have been shown to play a role in the escape of arrested transcription complexes ('dead end' complexes), which are impaired on elongation. The two proteins can stimulate a phosphohydrolase activity whereby the growing RNA chain is cleaved near the 3′ end. After transcript cleavage the arrested transcription complexes can resume elongation. The proteins GreA and GreB are therefore also called **transcript cleavage factors**. They are also thought to be implicated in a putative proofreading mechanism during RNA transcription (see Section 4.4).

Recently, a 110-kDa RNA polymerase-associated protein has been purified and characterized by different groups. The protein has been independently designated as **HepA** or **RapA** (RNA polymerase-*a*ssociated *p*rotein). This protein forms a 1:1 complex with RNA polymerase. RapA has an ATPase activity which appears to be stimulated when bound to RNA polymerase. No significant change in the *in vitro* activity of RNA polymerase has been measured in the presence of RapA, and cells unable to express RapA are viable. The function of RapA is therefore not yet known. It is a homologue to a family of eukaryotic proteins, however, which are involved in transcription activation, nucleosome remodelling and DNA repair.

Finally, specific proteins involved in the termination of transcription are known to associate with the transcription complex. They are able to change the processivity of the enzyme and to convert it from elongation into termination mode, thereby directing the end of a transcription cycle. These proteins will be discussed in more detail in the chapters dealing with transcription factors and termination and antitermination, respectively (see Chapters 5 and 7).

It should be noted that, in comparison to *E. coli*, the *B. subtilis* RNA polymerase has some disparate features. For instance, the targets for the antibiotics rifampicin and streptolydigin are not the same as in *E. coli*, but change between the largest and the second largest subunits β and β′. Moreover, an additional 21-kDa protein factor, termed δ, is known to exist specifically in *B. subtilis*. It is assumed that δ functions in RNA polymerase–promoter complex formation. However, δ does not contribute directly to promoter recognition but enhances promoter selectivity by restricting unspecific interaction of *B. subtilis* Eσ with DNA. Its precise *in vivo* function is not known, however, and the disruption of the gene for δ does not confer any notable phenotype in *B. subtilis*. Instead, it has been shown that RNA polymerases, as well as the major σ factors from *B. subtilis* and *E. coli*, recognize identical promoter sequences and can be successfully substituted for each other.

From the above it is clear that the basic features of transcription by RNA polymerase can be modulated by the interaction of proteins during the transcription cycle. Reprogramming the initiation, elongation and termination properties of the basic transcription machinery by transcription factors is one of the main characteristics of transcription regulation. Many of the subsequent chapters will be devoted to discussion of where, to what extend, and by what mechanisms this modulation of the activity of RNA polymerase occurs. It is an astonishing fact that current knowledge on the regulation of cellular RNA polymerase concentration during different growth phases is limited. It appears, however, that the overall concentration of RNA polymerase is kept relatively constant during cell growth (about 2500–3000 molecules per cell). About 50% of the polymerase molecules are assumed to be engaged in active transcription. About the same amount is believed to be bound non-specifically to DNA, buffering the concentration of the free enzyme, and only 1% of the RNA polymerase molecules are considered to be free for initiation of a new round of transcription.

Apart from changes in the concentration of RNA polymerase, regulation of transcription can also be related to changes in the activity of the enzyme. Modification of RNA polymerase activity does not always require the action of protein factors. Regulation in biological systems frequently occurs by chemical modification, mostly phosphorylation. Although regulation via phosphorylation cascades is the hallmark of eukaryotic signal transduction and transcription regulation, it also plays an important role in the regulation of prokaryotic transcription. Many examples are firmly established in which phosphorylation determines the activity of transcription factors (see Section 7.6). In addition, direct modification of RNA polymerase subunits is likely to occur. A notable example, namely the ADP-ribosylation of the α subunits following T4 phage infection, has already been mentioned. In a similar way, it has been reported that the activities of RNA polymerase can be shifted between the logarithmic and the stationary growth phase. A change in the specificity of the enzyme is believed to be brought about by specific modification. This was concluded from the finding that, upon transition from exponential to stationary growth, several different isoforms of RNA polymerase could be isolated by phosphocellulose chromatography. This type of separation is dependent on the net charge of the separation products. *In vitro* transcription analysis with the different polymerase preparations has revealed a selective preference for different promoters. The different core polymerases have a higher negative charge when isolated at stationary phase, compared to the log phase enzyme. Surprisingly, the different enzyme forms can be reverted *in vitro* by incubation with ATP or ADP, but not with phosphatase. Hence, it was concluded that polyphosphates or oligophosphates are involved in the interconversion of the different RNA polymerase isoforms.

During one of the major regulatory pathways in bacteria, namely the stringent response, it is assumed that RNA polymerase is modified by the action of a small effector nucleotide in response to nutritional stress, e.g. amino acid deprivation. Several lines of evidence suggest that the effector molecule ppGpp mediates RNA polymerase activity at stringently controlled promoters. Modulation is brought about by amino acid starvation and also probably during changes in the growth rate of the cells. Several studies suggest that the effector ppGpp binds to the RNA polymerase (most likely to the β subunit), thereby changing the activity of the enzyme. More details on the mechanism of stringent control and growth rate regulation can be found in Section 8.5.

Summary

RNA polymerase initiates transcription at DNA structures termed promoters. Promoters are characterized by two conserved hexameric DNA sequences, which centre 35 and 10 base pairs upstream of the transcription start site. These sites are called the *–35* and *–10 regions*. The –35 and –10 promoter elements controlled by the RNA polymerase specificity factor σ^{70} are separated by a non-conserved stretch of 17±1 base pairs, which is called the *spacer*. Together these three elements define the *core promoter*. The degree of correspondence of the –35 and –10 elements with their consensus sequences TTGACA and TATAAT, largely correlates with the promoter efficiency. The spacer poses the –10 and –35 regions at the correct distance and at an optimal angular position for binding to the RNA polymerase surface, and thus contributes to promoter specificity and efficiency. The DNA conformation and contour of the complete promoter is essential for overall RNA polymerase affinity. Many strong promoters are characterized by intrinsic DNA curvature, which extends RNA polymerase-DNA contacts and which may facilitate DNA melting at the transcription start site. A high number of AT base pairs within the promoter sequence supports the transition from a double-stranded to the strand-opened transcription complex. Strong promoters are often flanked by specific sequences upstream and downstream from the core region. These flanking sequences can contribute significantly to promoter efficiency. Most transcripts are initiated with ATP as the first nucleotide. The exact start site, however, is not defined by a unique sequence or a specific nucleotide but by the complete promoter context. For a given promoter, the start site is normally precisely defined, and only in rare cases can it vary by one or two nucleotides. At different promoters, however, initiation can generally occur at any position between five or nine nucleotides downstream from the –10 region. Transcription units are often controlled by more than one promoter. According to the direction of transcription such *composite promoters* can be arranged in *tandem*, in *divergent* or in *convergent* orientation.

Bacterial RNA polymerase is a multisubunit enzyme. Four subunits constitute the *core enzyme*, which has the composition $\alpha_2\beta\beta'$. The core enzyme is sufficient to carry out the elongation and termination reactions of transcription. A specificity factor σ is required for the correct initiation. RNA polymerase capable of specific initiation is termed the *holoenzyme* and has a subunit composition $\alpha_2\beta\beta'\sigma$. Individual subunit structures are highly conserved between different bacteria. They are also homologous to the larger subunits of eukaryotic RNA polymerases, although the structures of eukaryotic polymerases have generally a much higher complexity.

Based on sequence comparisons, analysis of mutants, and affinity label or cross-linking studies numerous functional sites have been assigned within the primary sequences of the RNA polymerase subunits. The α subunits can be divided into two functional domains, the N-terminal (αNTD) and the C-terminal domain (αCTD). The two domains are connected by a stretch of unstructured amino acids which functions as a flexible linker. The αNTD is involved in the assembly of the RNA polymerase core enzyme. The αCTD contains binding sites for a number of transcription factors and specific upstream promoter sequences termed UP sites. The β and β' subunits together constitute the active centre of RNA polymerase. They contain domains involved in DNA binding and are responsible for the resistance against the antibiotics rifampicin and streptolydigin, which inhibit bacterial transcription initiation or elongation, respectively. Both subunits contain amino acid structures important for transcription termination and RNA polymerase pausing. There is evidence indicating that the stringent control regulator guanosine tetraphosphate (ppGpp) binds to sites within the β subunit and thus affects the activity of RNA polymerase. The specificity factor for initiation of the housekeeping genes at exponential growth is σ^{70}. σ factors contain amino acid domains which directly bind the –10 and –35 core promoter elements. A cluster of aromatic amino acids within the primary structure of σ^{70} participates in the process of melting the promoter DNA. The C-terminal domain of σ^{70} is involved in binding specific transcription factors.

In addition to σ^{70}, sets of *alternative σ factors* have been characterized in different bacteria. Alternative σ factors often recognize specific promoter sequences and direct the transcription of different regulons. Specific σ factors are required, for instance, for the transcription of the heat shock genes (σ^{H}, σ^{E}), for genes important at stationary phase growth (σ^{S}), for the expression of the nitrogen fixation genes (σ^{N}), flagellar protein expression (σ^{F}), or the formation of spores of Gram-positive bacteria (e.g. *B. subtilis* σ^{F}). In addition, several phages encode their own specific σ factors which support phage gene transcription while suppressing transcription of host genes. The function of alternative σ factors is normally controlled through their concentration at the level of synthesis (transcription and translation) and degradation. Often, alternative σ factors are modulated in their activity either by proteolytic removal of prosequences or through the interaction with *anti-sigma factors*.

No high resolution information on the three-dimensional structure of bacterial RNA polymerase is yet available. However, low resolution structures based on electron microscopy of two-dimensional crystals indicate significant structural similarity between the RNA polymerases of prokaryotes and eukaryotes, but clear differences between prokaryotic RNA polymerase core and holoenzymes. Elongating core RNA polymerase probably contains a ring-like structure similar to the *sliding clamp* structures of DNA polymerase complexes during replication. Differences in the physical properties of RNA polymerase suggest that the activity of the enzyme may be modulated during different growth phases. Such a modulation may occur through the association of different protein factors or it may involve the chemical modification of RNA polymerase subunits, possibly by phosphoryl group-carrying ligands. The stringent response and growth rate effector ppGpp may be a candidate for such a modification.

Reference

Kolesky, S., Ouhammouch, M., Brody, E.N., Geiduschek, E.P. (1999) *Journal Molecular Biology* **291**: 267–81.

Further reading

Atkinson, M. R. and Ninfa, A. J. (1994) Mechanism and regulation of transcription from bacterial σ^{54}-dependent promoters. In: Conaway, R. C. and Conaway, J. W., (eds) *Transcription: Mechanisms and Regulation*. New York: Raven Press, 323–42.

Beck, C. F. and Warren, R. A. J. (1988) Divergent promoters, a common form of gene organization. *Microbiological Reviews* **52**: 318–26.

Blatter, E. E., Ross, W., Tang, H., Gourse, R. L. and Ebright, R. H. (1994) Domain organization of RNA polymerase α subunit: C-terminal 85 amino acids constitute a domain capable of dimerization and DNA binding. *Cell* **78**: 889–96.

Borukhov, S., Lee, J. and Goldfarb, A. (1991) Mapping of a contact for the RNA 3′ terminus in the largest subunit of RNA polymerase. *Journal of Biological Chemistry* **266**: 23932–5.

Bown, J. A., Barne, K. A., Minchin, S. D. and Busby, S. J. W. (1997) Extended −10 promoters. In: Eckstein, F. and Liley, D. M. J. (eds) *Nucleic Acids and Molecular Biology*. Vol. 11. Berlin-Heidelberg: Springer Verlag, pp. 41–52.

Brown, K. L. and Hughes, K. T. (1995) The role of anti-sigma factors in gene regulation. *Molecular Microbiology* **16**: 397–404.

Brunner, M. and Bujard, H. (1987) Promoter recognition and promoter strength in the *Escherichia coli* system. *EMBO Journal* **6**: 3139–44.

Chamberlin, M. J. (1982) Bacterial DNA-dependent RNA polymerases. *The Enzymes* **15**: 61–86.

Colland, F., Orsini, G., Brody, E. N., Buc, H. and Kolb, A. (1998) The bacteriophage T4 AsiA protein: a molecular switch for sigma 70-dependent promoters. *Molecular Microbiology* **27**: 819–29.

Darst, S. A., Roberts, J. W., Malhotra, A., Marr, M., Severinov, K. and Severinova, E. (1997) Pribnow box recognition and melting by *Escherichia coli* RNA polymerase. In: Eckstein, F and Liley, D. M. J. (eds) *Nucleic Acids and Molecular Biology*. Vol. 11. Berlin-Heidelberg: Springer Verlag, pp. 27–40.

Dombroski, A. J., Walter, W. A., Record, M. T., Siegele, D. A. and Gross, G. A. (1992) Polypeptides containing highly conserved regions of transcription factor sigma 70 exhibit specificity of binding to promoter DNA. *Cell* **70**: 501–12.

Fassler, J. S. and Gussin, G. N. (1996) Promoters and basal transcription machinery in eubacteria and eukaryotes: concepts, definitions, and analogies. *Methods in Enzymology* **273**: 3–29.

Geiduschek, P. E., Fu, T.-J., Kassavetis, G. A., Sanders, G. M. and Tinker-Kulberg, R. L. (1997) Transcriptional activation by a topologically linkable protein: forging a connection between replication and gene activity. In: Eckstein, F. and Lilley, D. M. J. (eds) *Nucleic Acids and Molecular Biology*, Vol. 11. Berlin-Heidelberg: Springer-Verlag, pp. 135–50

Gross, C. A., Chan, C. L. and Lonetto, M. A. (1996) A structure/function analysis of *Escherichia coli* RNA polymerase. *Philosophical Transactions of the Royal Society London* **351**: 475–82.

Haldenwang, W. G. (1995) The sigma factors of *Bacillus subtilis*. *Microbiological Reviews* **59**: 1–30.

Harley, C. and Reynolds, R. (1987) Analysis of *E. coli* promoter sequences. *Nucleic Acids Research* **15**: 2343–61.

Hawley, D. K. and McClure, W. R. (1983) Compilation and analysis of *Escherichia coli* promoter DNA sequences. *Nucleic Acids Research* **11**: 2237–55.

Helmann, J. D. (1994) Bacterial sigma factors. In: Conaway, R. C. and Conaway, J. W. (eds) *Transcription: Mechanism and Regulation*. New York: Raven Press, pp. 1–17.

Helmann, J. D. and Chamberlin, M. J. (1988) Structure and function of bacterial sigma factors. *Annual Reviews of Biochemistry* **57**: 839–72.

Hernandez, V. J. and Cashel, M. (1995) Changes in conserved region 3 of *Escherichia coli* σ^{70} mediate ppGpp-dependent functions *in vivo*. *Journal of Molecular Biology* **252**: 536–49.

Hertz, G. Z. and Stormo, G. D. (1996) *Escherichia coli* promoter sequences: analysis and prediction. *Methods in Enzymology* **273**: 30–42.

Heyduk, T., Heyduk, E., Severinov, K., Tang, H. and Ebright, R. H. (1996) Determinants of RNA polymerase α subunit for interaction with β, β' and σ subunits: hydroxyl-radical protein footprinting. *Proceedings of the National Academy of Sciences USA* **93**: 10162–6.

Igarashi, K. and Ishihama. A. (1991). Bipartite functional map of the *E. coli* RNA polymerase α subunit: involvement of the C-terminal region in transcription activation by cAMP-CRP. *Cell* **65**: 1015–22.

Igarashi, K., Fujita, N. and Ishihama, A. (1991) Identification of a subunit assembly domain in the alpha subunit of *Escherichia coli* RNA polymerase. *Journal of Molecular Biology* **218**: 1–6.

Ishihama, A. (1993) Protein–protein communication within the transcription apparatus. *Journal of Bacteriology* **175**: 2483–9.

Ishihama, A. (1997) Promoter selectivity control of RNA polymerase. In: Eckstein. F. and Lilley, D. M. J. (eds) *Nucleic Acids and Molecular Biology*, Vol. 11. Berlin–Heidelberg: Springer Verlag, pp. 53–70.

Jeon, Y. H., Negishi, T., Shirakawa, M., *et al.* (1995) Solution structure of the activator contact domain of the RNA polymerase α subunit. *Science* **270**: 1495–7.

Jeon, Y. H., Yamazaki, T., Otomo, T., Ishihama, A. and Kyogoku, Y. (1997) Flexible linker in the RNA polymerase alpha subunit facilitates the independent motion of the C-terminal activator contact domain. *Journal of Molecular Biology* **267**: 953–62.

Jin, D. and Gross, C. A. (1988) Mapping and sequencing of mutations in the *Escherichia coli rpoB* gene that lead to rifampicin resistance. *Journal of Molecular Biology* **202**: 45–58.

Jin, D. J. and Zhou, Y. N. (1996). Mutational analysis of structure–function relationship of RNA polymerase in *Escherichia coli*. *Methods in Enzymology* **273**: 300–19.

Kumar, A., Malloch, R. A., Fujita, N., Smillie, D. A., Ishihama, A. and Hayward, R. S. (1993) The minus 35-recognition region of *Escherichia coli* sigma 70 is inessential for initiation of transcription at an "extended minus 10" promoter. *Journal of Molecular Biology* **232**: 406–18.

Lesley, S. A. and Burgess, R. R. (1989) Characterization of the *Escherichia coli* transcription factor sigma 70: localization of a region involved in the interaction with core RNA polymerase. *Biochemistry* **28**: 7728–34.

Lisser, S. and Margalit, H. (1993) Compilation of *E. coli* mRNA promoter sequences. *Nucleic Acids Research* **21**: 1507–16.

Lonetto, M., Gribskov, M. and Gross, C. A. (1992) The σ^{70} family: sequence conservation and evolutionary relationships. *Journal of Bacteriology* **174**: 3843–9.

Malhotra, A., Severinova, E. and Darst, S. A. (1996) Crystal structure of a σ^{70} subunit fragment from *E. coli* RNA polymerase. *Cell* **87**: 127–36.

Missiakas, D. and Raina, S. (1998) The extracytoplasmic function sigma factors: role and regulation. *Molecular Microbiology* **28**: 1059–66.

Mukherjee, K. and Chatterji, D. (1997) Studies on the ω subunit of *Escherichia coli* RNA polymerase. Its role in the recovery of denatured enzyme activity. *European Journal of Biochemistry* **247**: 884–9.

Mustaev, A., Zaychikov, E., Severinov, K., *et al.* (1994) Topology of the RNA polymerase active center probed by chimeric rifampicin–nucleotide compounds. *Proceedings of the National Academy of Sciences USA* **91**: 12036–40.

Severinov, K., Markov, D., Severinova, E., *et al.* (1995) Streptolydigin-resistant mutants in an evolutionarily conserved region of the β′ subunit of *Escherichia coli* RNA polymerase. *Journal of Biological Chemistry* **270**: 13926–9.

Severinov, K., Kashlev, M., Severinova, E., *et al.* (1994) A non-essential domain of *Escherichia coli* RNA polymerase required for the action of the termination factor Alc. *Journal of Biological Chemistry* **269**: 14254–9.

Siegele, D. A., Hu, C. J., Walter, W. A. and Gross, C. A. (1989) Altered promoter recognition by mutant forms of the σ^{70} subunit of *Escherichia coli* RNA polymerase. *Journal of Molecular Biology* **206**: 591–603.

Tanaka, K., Shiina, T. and Takahashi, H. (1988) Multiple principal sigma factor homologs in eubacteria: identification of the 'rpoD box'. *Science* **18**: 1040–2.

von Hippel, P. H. and Berg, O. G. (1989) Facilitated target location in biological systems. *Journal of Biological Chemistry* **264**: 675–8.

von Hippel, P. H., Bear, D. G., Morgan, W. D. and McSwiggen, J. A. (1984) Protein–nucleic acid interaction in transcription. *Annual Reviews of Biochemistry* **53**: 389–446.

Weilbaecher, R., Hebron, C., Feng, G. and Landick, R. (1994) Termination-altering amino acid substitutions in the β′ subunit of *Escherichia coli* RNA polymerase identify regions involved in RNA chain elongation. *Genes & Development* **8**: 2913–27.

Wösten, M. M. S. M. (1998) Eubacterial sigma-factors. *FEMS Microbiology Reviews* **22**: 127–50.

Zhang, G. and Darst, S. A. (1998) Structure of the *Escherichia coli* RNA polymerase α subunit amino-terminal domain. *Science* **281**: 262–6.

3

Initiation of transcription

As explained in the introduction, the roughly 4000 genes in *E. coli* are expressed in frequencies spanning four orders of magnitude. Much of this variability is caused by differences in the rate of transcription initiation. Most of the mechanisms that are involved in transcription regulation at the initiation steps. The following considers details of transcription initiation at σ^{70}-controlled promoters.

This chapter describes in single steps the pathway from the first encounter of RNA polymerase with a DNA template to the formation of a processive elongation complex. This pathway is presented schematically in simplified form in Fig. 3.1. From the scheme it becomes clear that transcription initiation itself is a multistep process. Although more than one intermediate may exist for each step, the following four main steps can be defined:

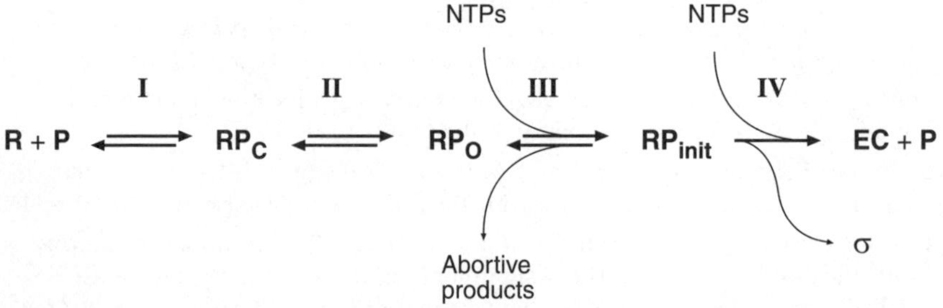

NTPs NTPs

$$R + P \underset{}{\overset{I}{\rightleftharpoons}} RP_C \underset{}{\overset{II}{\rightleftharpoons}} RP_O \underset{}{\overset{III}{\rightleftharpoons}} RP_{init} \overset{IV}{\longrightarrow} EC + P$$

Abortive products σ

Figure 3.1 Kinetic scheme of the major steps in initiation. The four main steps (I–IV), from the first binding of RNA polymerase (R) to a promoter (P), formation of a closed binary RNA polymerase–promoter complex (RP_C), isomerization to the binary open complex (RP_O), formation of a ternary initiating complex after binding of the first NTPs (RP_{init}), and finally formation of an elongating complex (EC) under release of the promoter, are shown schematically. The latter step involves the loss of the sigma (σ) subunit which may be reused for the next initiation cycle. Before the elongating complex is formed and productive transcription is started, the initiating complex may undergo multiple rounds of abortive cycling upon which short transcripts (abortive products) are repetitively released.

I. DNA binding and recognition of the promoter sequence by RNA polymerase
II. Isomerization of the initial complex into an open complex with the DNA in a strand-separated conformation
III. Binding of substrate NTPs and formation of the first phosphodiester bonds
IV. Transition from initiating to elongating complexes and the escape from the promoter sequence

A description of each substep, and a summary of the structural analysis of the complex intermediates during initiation will be given. Attention will also be drawn to reactions taking place at the end of the initiation cycle, namely the formation of abortive products and the steps leading to promoter clearance. At the end of the chapter a short presentation on how promoter strength can be defined will be given, and a procedure to determine kinetic parameters for the comparison of the initiation reactions of different promoters will be described.

Each of the substeps I to IV above can be rate-limiting and can thus determine the efficiency of the initiation cycle. It will be seen that specific controls act at each substep.

3.1 Promoter location

One puzzling problem in understanding the first step in transcription initiation emerges from the very fast rates that have been observed for the binding of RNA polymerase to strong promoters. The problem is related to the fact that the concentration of the target sites (promoters) relative to random non-promoter DNA sites is very low. The rate of promoter location by RNA polymerase among a vast excess of random DNA of a given genome by 'trial and error' through three-dimensional diffusion should thus be a rather slow process. However, quantitative analysis of the second-order rate constants (k_a) for the binding of *E. coli* RNA polymerase to several promoters has revealed that this step is much faster ($k_{measured} \approx 10^{10}$ $\text{M}^{-1}\text{s}^{-1}$) than one could expect from a one-step diffusion-controlled reaction ($k_{calc.} \approx 10^{8}$ $\text{M}^{-1}\text{s}^{-1}$) (Box 3.1).

A one-step binding model where RNA polymerase finds the promoters by simple three-dimensional diffusion can thus not account for the very fast rates observed for many of the strong *E. coli* promoters. Moreover, increasing the DNA length does not reduce the rate of promoter finding as one would expect if competition between specific and non-specific DNA sites occurs. The opposite is the case. Non-specific binding seems to facilitate the promoter search. It follows that non-specific binding appears to be an important intermediate in promoter location. The fast binding of RNA polymerase to promoters can therefore be described in terms of a two-step binding mechanism. The first step involves free

Box 3.1 Promoter location by linear diffusion

What is the fastest rate at which a molecule of RNA polymerase may encounter a promoter. Like any chemical reaction, the bimolecular association of RNA polymerase and a promoter within a DNA sequence has to follow thermodynamic and kinetic rules. Very fast bimolecular reactions are limited by the frequency of the correct contact of the reactants. Assuming a reaction each time the two molecular species (RNA polymerase and promoter DNA) come within an interactive distance, the rate will be limited by diffusion. Maximal rate constants k_a for such bimolecular diffusion-controlled reactions can be estimated according to the Debye–Smoluchowski equation:

$$k_{a,calc} = 4\pi\kappa\, f_{elec} b(D_P + D_R)N_O/1000$$

In this equation κ and f_{elec} are steric and electrostatic interaction factors, respectively. The interacting radius (in cm) is b, and D_P and D_R are three-dimensional diffusion constants for promoter and RNA polymerase, given in cm^2/sec. N_0 is the Avogadro constant. An estimation for the value $k_{a,calc}$ has been reported by von Hippel and coworkers (Berg *et al.*, 1981), assuming the interacting radius to be 80 Å, which roughly corresponds to half the diameter of RNA polymerase and half the diameter of DNA. The value for k_a apparently cannot exceed $10^8 M^{-1}s^{-1}$. Measured numbers for k_a, however, suggest values exceeding $10^{10} M^{-1}s^{-1}$, which clearly indicates a faster rate of location of the promoter by RNA polymerase than controlled by diffusion. It follows that the process of promoter location by RNA polymerase must proceed by a different, more complex mechanism than bimolecular diffusion-controlled reactions. In addition, the large excess of unspecific DNA binding sites within a given genome compared to specific promoter sites could be expected to slow down promoter binding through competition (there are about 4×10^6 non-specific binding sites in *E. coli* if one assumes that specific binding requires perfect sequence match and can be discriminated by a single base pair mismatch). However, there is no apparent competition by additional length of the DNA as suggested by the high forward rate of RNA polymerase promoter interaction. Obviously, non-specific binding to random DNA facilitates the transfer of RNA polymerase to the specific promoter site.

The observation that some DNA binding proteins find their target DNA faster than the upper limit estimated for a diffusion-controlled process has led to the proposal that such proteins initially interact in a non-specific way with an extended target. According to the suggestions made by von Hippel and coworkers (Berg *et al.*, 1981) the specific binding site is located by two mechanisms, namely sliding and intersegment transfer. While sliding can be viewed as a one-dimensional diffusion of the protein along the DNA in a non-sequence-specific binding mode, intersegment transfer speeds up target location in a different way. If it is assumed that the protein can bind non-specifically to two DNA sites, which are far apart in the sequence but close together in three-dimensional space, because of the coiled structure of the DNA molecule, the protein might be transferred from one site to the other without existing in a non-bound conformation. Depending on the lifetime of the non-specifically bound state a series of such intersegment transfer steps will contribute to a faster target location. Evidence for sliding and intersegment transfer has been demonstrated experimentally for a number of systems, such as the *lac* repressor–operator interaction, restriction enzyme target location and also for RNA polymerase-promoter binding. Fig. B3.1 indicates the different steps that a DNA-binding protein may undergo to locate a specific site within a large DNA molecule. The figure is adapted from von Hippel *et al.* (1989).

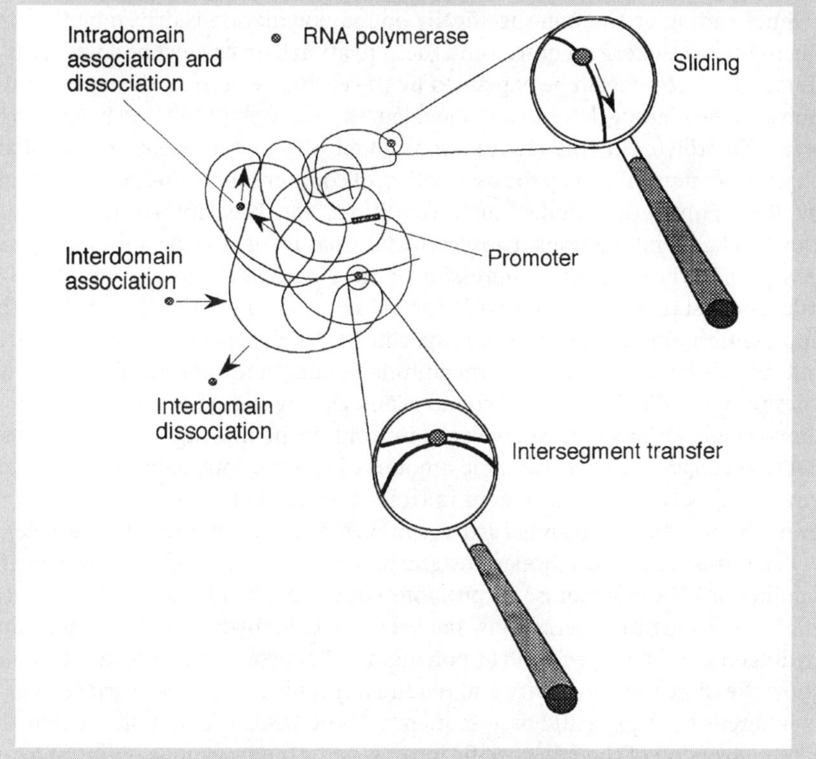

Fig. B3.1 Schematic illustration of different mechanisms for promoter location

diffusion to an extended target, namely to the complete DNA molecule where binding occurs non-specifically and is entirely of electrostatic nature. The second step comprises a series of transfer reactions on the DNA molecule, driven by thermal fluctuation. During this, RNA polymerase is translocated in a non-specifically bound form to the promoter. This second step largely represents a **linear diffusion** process of reduced dimensionality which speeds up the search time since the search volume is reduced. The RNA polymerases have to search along the contour length of the DNA instead of a random search through a three dimensional space. Reducing the search volume may be largely explained by RNA polymerase **sliding** along the DNA (other processes such as *hopping* or *intersegment transfer* may also contribute to the rate enhancement, see Box 3.1).

Facilitated promoter search by sliding can be described as an RNA polymerase transfer event along the contour length of the DNA molecule. Sliding occurs while the polymerase remains bound non-specifically and represents a one-dimensional random walk along the DNA. Such sliding in fact occurs very rapidly with diffusion rates in the range of approximately 10^{-9} cm^2/s, corresponding to a random walk rate along the DNA of 10^6 base pairs/s.

Since sliding by the non-specifically bound polymerase is driven by thermal fluctuation the energy required and the activation barrier must be small. This is in fact the case and can be explained by the entirely electrostatic nature of the non-specific binding. DNA in a monovalent salt solution (0.1–0.2 M) behaves as a weak electrolyte. In this environment about 88% of the negative phosphate charges are neutralized by the association of counter-ions. Non-specific binding by RNA polymerase under such conditions involves approximately eight charge–charge interactions. The electrostatic nature of non-specific binding can be shown by the complete suppression of this binding by high concentrations of salt. For instance, a tenfold variation in salt concentration can change the apparent binding constant for non-specific RNA polymerase holoenzyme–DNA interaction by seven orders of magnitude. Sliding does not require extensive energy for the displacement of counter-ions. During sliding a certain number of monovalent cations will be displaced in front of the moving RNA polymerase. Because replacement of the same amount of counter-ions behind the moving enzyme is as fast as displacement in front, there is no net ion difference, however. Sliding can thus be regarded as a movement on an isopotential surface.

In contrast to the totally electrostatic nature of non-specific binding, specific binding of RNA polymerase to promoters involves (in addition to the electrostatic component) interactions between specific hydrophobic groups and hydrogen bonding. Specific RNA polymerase–DNA complexes therefore do not show the same salt sensitivity and require a specific recognition surface, which is inherent to the specific base sequence of the DNA. It is important that the interconversion of the non-specific into a specific binding mode occurs at a rate comparable to the sliding rate when a promoter structure is encountered.

3.2 The closed RNA polymerase–promoter complex

From the observations in the preceding section it is obvious that binding of RNA polymerase to the promoter does not always occur in a one-step reaction. It is also clear that specific binding does not yield a strand-separated open DNA–

polymerase complex in one step. Instead, a specific intermediate, the closed complex is formed first. Recognition of the promoter DNA sequence by the polymerase in the closed complex does not occur through base hydrogen bonds, normally found in the formation of complementary Watson–Crick base pairs. Instead, non-Watson–Crick-type interactions involving specifically placed clusters of hydrogen bond donors and acceptors in the major and minor grooves of the DNA which match the acceptors and donors at the surface of the polymerase molecule must occur. The specific contacts between DNA and RNA polymerase may thus be different for various promoters in the closed complex. They probably include interactions between the –35 recognition element as well as promoter upstream DNA regions (USR or UP elements) and subregion 2.4 of the σ subunit or the α C-terminal domain (αCTD). Closed complexes, although strengthened by specific hydrogen bonds or hydrophobic interactions, still show a considerable amount of electrostatic interactions, e.g. between the negatively charged phosphate groups of the DNA and positively charged amino acid side chains of lysine or arginine residues (for instance, in the β′ subunit). The stability of closed complexes is therefore generally low and efficient competition occurs at high salt concentrations or at a molar excess of polyanionic competitor substances like heparin.

Detailed kinetic analyses of several promoters, such as the λP_R and the *lac* UV5 promoters, have revealed the existence of more than one closed complex. In both cases two distinctly different complexes have been characterized. The two complexes differ in their association and dissociation kinetics and in their extend of DNA protection during limited DNase I hydrolysis (see Box 2.1). In the first type of closed complex (RP_c1) RNA polymerase protects DNA from nucleotide positions –55 to –5, but the second type of complex (RP_c2) shows a strong downstream extended protection from sequence positions –55 to +20. The similar extent of DNA protected from DNase I digestion compared to open complexes (see below) indicates that the second class of closed complexes may represent intermediates where the nucleation for the formation of open complexes is already initiated. However, according to chemical reactivity, single-stranded nucleotides cannot be identified in this second class of complexes.

3.3 Formation of open complexes

Conversion of closed to open complexes involves the site-specific opening of base pairs comprising about one helical turn of the promoter DNA (positions around –9 to +2 are accessible to single strand-specific modifications). Base pair opening within this stretch of DNA occurs usually at temperatures below the melting point (T_m). Hence the necessary energy for strand separation must be provided by the binding free energy of the RNA polymerase to the promoter.

How can the fact that strand separation may occur at temperatures below T_m be explained? Certainly it must be assumed to be a multistep process. First, untwisting of the spacer DNA by binding to the surface of the polymerase may occur. Untwisting may be the result of extensive wrapping of DNA around RNA polymerase. Such an untwisting will distort the DNA double strand and destabilize AT base steps in the –10 region, from which strand separation nucleates. Strand opening is facilitated by RNA polymerase interactions with bases of the non-template strand. Such interactions have been shown to occur, for instance, with aromatic amino acid side chains of subregion 2.3 of the σ^{70} subunit (see Section 2.2.4).

It is not surprising that melting is nucleated at the AT-rich –10 promoter hexamer, which is known to be thermodynamically the least stable part of the whole promoter sequence. Since the isolated promoter DNA does not pre-exist in a strand-opened conformation under normal transcription conditions, interactions with functional groups of the RNA polymerase must facilitate this process.

How is such a process of facilitated DNA melting conceivable? It is known that within a continuous DNA molecule hydrogen bonds of individual base pairs break and reform on a millisecond time scale. The overall probability of base pair opening has been determined. This probability is in the range of 10^{-5}. Despite the thermodynamic stability DNA base pairs are kinetically labile. An individual base pair has an average lifetime in the range of 10^{-2} seconds. The lifetime of open bases within a DNA duplex is much shorter, about 10^{-7} seconds. The resulting phenomenon has been described as 'DNA breathing'. It may provide a mechanistic explanation of how RNA polymerase can access bases which on average are engaged in a double helical DNA structure but which are not permanently locked in base-paired form. It is clear from this consideration that changes in DNA topology or conformation (supercoiling or bending), which have an effect on these breathing motions, will further affect the probability of base pair opening (see Chapter 6).

The architecture of the promoter recognition sequence also contributes to make strand separation much easier. Consider that the centres of the –10 and –35 hexamer sequences, if separated by the optimal spacer length of 17 base pairs, are 23 base pairs apart. Assuming further that the promoter is in a B-DNA helical arrangement (10.5 base pairs per helical turn), would mean that the centres of the two promoter hexamers are separated by more than two helical turns. In a cylindrical projection they appear over-rotated by 68° with respect to each other (Fig. 3.2).

The –10 and –35 regions are bound by subregions 2.4 and 4.2 of the σ^{70} subunit (see Section 2.2.4). Binding of RNA polymerase in the open complex is known to occur on one side of the DNA helix. A conformational change of the RNA polymerase that rotates the angular orientations of the bound –10 and –35 elements such that they point to the same side of the DNA helix will cause untwisting of the spacer DNA. This torque distorts the DNA duplex and AT

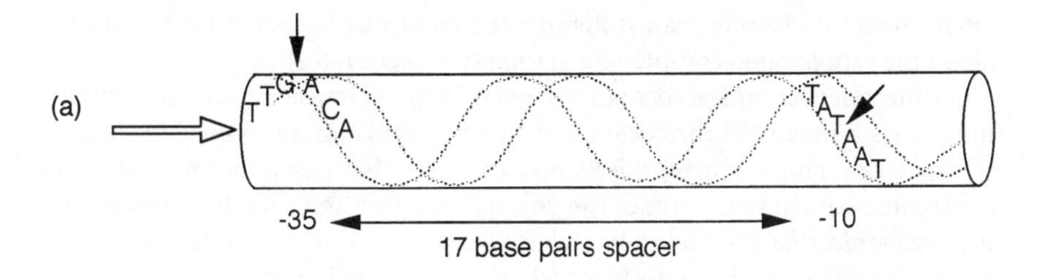

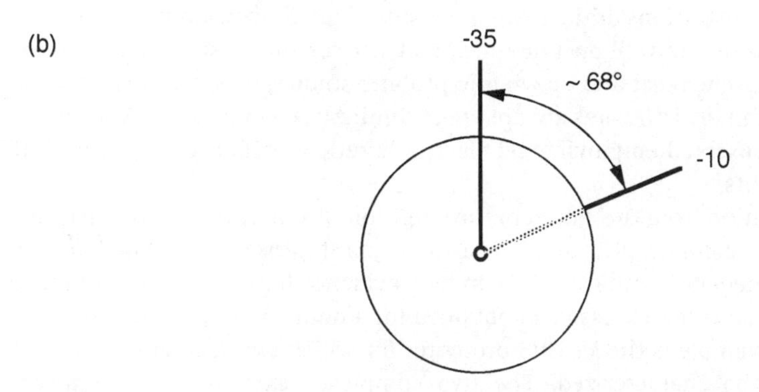

Figure 3.2 Angular orientation of promoter elements. (a) The promoter core elements are represented as a cylinder. The −10 and −35 consensus hexamer sequences are schematically shown on one strand of a helical B-DNA structure and their central positions are marked by arrows. They are separated by a 17 base pair spacer, which brings the centre of both promoter elements 23 base pairs apart. (b) A projection with a view along the helix axis (open arrow in (a)). The central positions of the −10 and −35 hexamer sequences (dark arrows in (a)) are indicated by dark lines. This distance results in a dihedral angle of approximately 68° given a 23 base pair distance and 10.5 base pairs per helical turn.

clusters within the −10 region will be significantly destabilized. The promoter consensus region thus gains a high probability of strand opening through breathing of the helix. Opening nucleates at the AT-rich region and is propagated until the initial transcription bubble is formed and 12 ± 2 nucleotides between positions −10 and +3 are finally separated. Favourable interactions between the RNA polymerase and bases of the non-template strand will support this propagation of the strand opening. (Remember the involvement of the aromatic residues of the subregion 2.3 and 2.1 of the σ^{70} subunit in promoter melting outlined in Section 2.2.4.)

Parameters affecting the twist of the spacer will affect the efficiency of the melting process. DNA supercoiling, for instance, which is modulated by environmental changes, can influence the twist of the spacer DNA and thus affect the rate of open complex formation (see Section 6.2).

During open complex formation the RNA polymerase–promoter complex undergoes a major conformational change. There is limited knowledge about the specific changes within the RNA polymerase, but based on low resolution three-dimensional structures of the Eσ^{70} holoenzyme and core RNA polymerase it is proposed that the clamp-like cleft closes around the DNA-like 'jaws' (see Section 2.4). For some promoters, such as the lambda P_R promoter, such a conformational change where the jaws of RNA polymerase close around the downstream promoter region occurs already during the conversion from the closed 1 to the closed 2 complex (Fig. 3.3).

Structural distortions of the DNA as a result of the conversion from closed to open complexes are well documented. Evidence for such distortions has been derived from chemical and enzymatic probing studies, where changes in the reactivity of many DNA sites are apparent. Similarly, changes in DNA structure have been observed employing gel electrophoretic mobility studies or optical measurements.

The transition from the closed to the open complex may result principally in a single open complex (RP_o) or, alternatively, in a stepwise reaction with one or more intermediate complexes (RP_i). In fact, evidence for the existence of more than one open complex has been obtained for a number of promoters studied. A notable example is the *lac* UV5 promoter for which two separate open complexes can be characterized. The two complexes exist in a temperature-

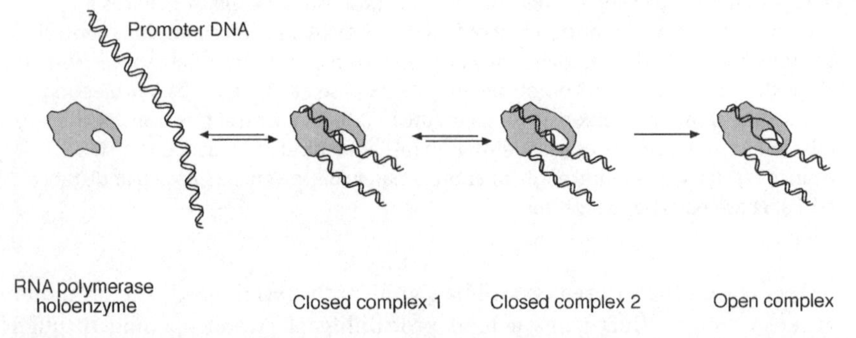

Figure 3.3 RNA polymerase pathway to the open complex. The pathway from free RNA polymerase holoenzyme to the open complex is presented as a cartoon. The shape of the enzyme is adapted according to electron crystallographic data. The downstream edge consists of jaws or thumb-like structural features which close tightly around the DNA, forming a sliding clamp during transition from the closed complex 1 to the closed complex 2. The DNA, presented as a helical structure, is wrapped around the holoenzyme consistent with the extended DNase I footprint data. Wrapping of the DNA is considered to facilitate strand separation of the promoter region during formation of the open complex. The figure is arranged according to Polyakov *et al.* (1995) and Craig *et al.* (1998).

dependent equilibrium which means they are interconvertible. They are stable enough to be isolated on polyacrylamide gels and can thus be analysed separately. Such analyses have shown that the two complexes differ in their rates of abortive *versus* productive product formation (see below). Interestingly, stepwise promoter melting is also known to occur during the initiation cycle of eukaryotic RNA polymerase III.

The existence of more than one open complex is also known for the T7 A1 and the phage λ P_R promoters. Several independent studies with the latter promoter have shown that the process leading from initial melting to the fully opened transcription complex requires Mg^{2+} ions. There is an uptake of three Mg^{2+} ions during the final extension of the transcription bubble from 12 to 14 base pairs. A more detailed study employing *E. coli* RNA polymerase has revealed that amino acids Glu464, Asp460 and 462 of the β′ subunit, and probably Glu1272 and 1274 of the β subunit are involved also in Mg^{2+} binding.

Clusters of A nucleotides (A tracts) located in the upstream region of many promoters are known to bend the DNA towards the minor groove (see Chapter 6). Such bending may serve to wrap DNA around the RNA polymerase. In line with this, footprinting studies have shown that positions –38 and –48 of the *lac* UV5 or λ P_R promoters are involved in DNA bends which accompany the conversion of the two open complexes. In addition, a bend of 45–54° at position –1 in the open complex has been localized as a result of wrapping. This indicates again that bending and wrapping of the DNA around the RNA polymerase seems to be a topological prerequisite for open complex formation. A similar conclusion has been reached by a comparison of the size of RNA polymerase and the length of DNA protected from footprinting studies. According to the crystal structural analysis the dimensions of the $E\sigma^{70}$ holoenzyme are $90 \times 95 \times 160$ Å. This is clearly too small to accommodate about 80 base pairs (greater than 270 Å) which are protected from hydroxyl radical cleavage in the open complex. It is compatible, however, with the view of DNA wrapped around the RNA polymerase.

3.4 Binding of substrate NTPs—the ternary complexes

With the binding of the first NTP substrates the open *binary complexes* are converted to so called *ternary complexes*. It is unclear whether there are different sites for different NTPs but there is more likely to be only a single NTP substrate site. The first NTP (which will be the 5′ terminal nucleotide of the growing transcript) is bound without catalytic conversion. Note that the first NTP which forms the 5′ end of the newly synthesized RNA chain still exists as a 5′ triphosphate. Only those NTPs are bound sufficiently stable which match the sequence of the template strand in the catalytic site through complementary base pairing. Correct binding of a subsequent NTP into the catalytic site of the enzyme initiates the formation of a phosphodiester linkage to the free 3′ hydroxyl group of the first (or the preceding) nucleotide. Pyrophosphate is released during this step and a dinucleoside monophosphate is formed as the shortest initial RNA fragment. Ternary complexes, after several rounds of phosphodiester linkage reactions, consist of RNA polymerase holoenzyme, the bound template DNA *and* a growing RNA chain that is gradually elongated in the 3′ direction. Hence, it is called a ternary complex. Within such ternary complexes the growing transcript can reach a length of up to 10 nucleotides. The RNA polymerase remains bound to the promoter DNA as long as the σ factor is not released from the complex. Probably the binding of the –10 and –35 promoter elements through the σ subunit will hold the RNA polymerase in position at the transcription start site. The active centre of RNA polymerase, however, is flexible, and apparently able to move up to around 10 nucleotides downstream within the ternary complex.

For some promoters it is known that stable open complexes are only formed in the presence of initiating substrate NTPs. Examples are the RNA promoters (e.g. *rrnB* P1 and P2). Normally, open complexes are stable compared to the respective closed complexes. In the latter case, however, NTP substrates for at least the first two nucleotides of the transcript sequence have to be present to yield stable open complexes. Hence, for these promoters the presence of initiating NTPs appears to be an obligatory requirement for stable open complex formation. The binary open complexes, if existing, are too unstable to be isolated. In this case it is not entirely clear whether the first phosphodiester bonds of the initiated transcript have to be formed or whether binding of the respective NTPs to the catalytic site of the RNA polymerase is sufficient to allow stable complex formation and strand opening.

Ternary complexes, in contrast to closed binary complexes, are resistant to salt and heparin. This property suggests the contribution of a large non-electrostatic component to the binding energy between RNA polymerase and the promoter DNA in the ternary complex. Possibly hydrogen bonding and hydrophobic stacking interactions between exposed bases and aromatic amino acid side chains of the σ^{70} subdomain 2.3 are involved in complex stabilization. To date no single amino acid residue within this domain has been identified as an essential component of the initiating process, however. This conclusion is based on a mutational study where aromatic residues in the σ^{70} subdomain 2.3 were altered but the resulting mutants retained some function in transcribing from strong phage promoters. Hence no single aromatic amino acid within subdomain 2.3 seems to be essential for the formation of functional initiation complexes. More likely, several such aromatic residues together contribute to the stabilization of the ternary complex.

The dimensions of the closed, open and typical ternary complexes, as determined by various footprinting techniques, are summarized in Fig. 3.4.

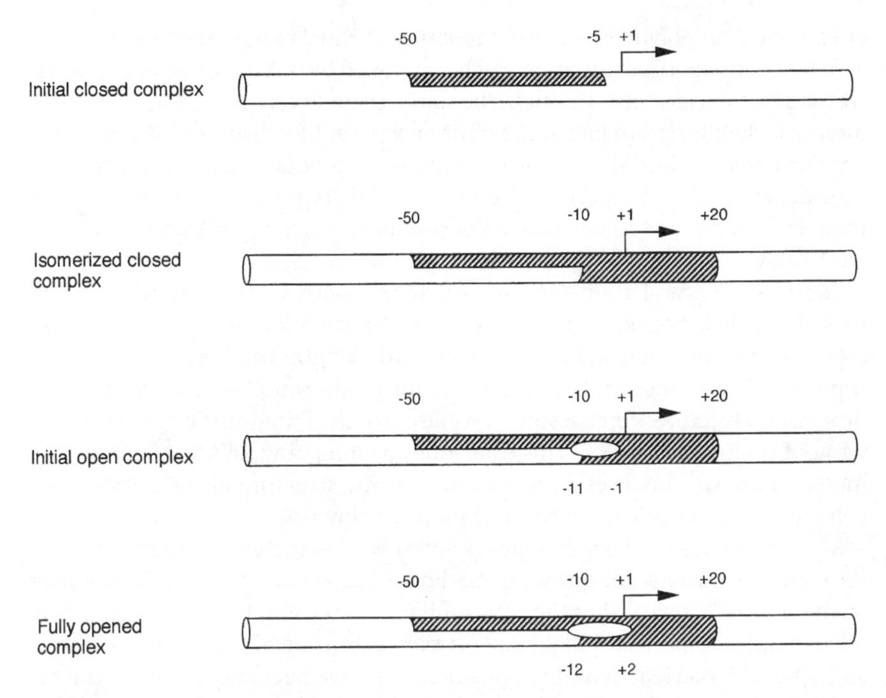

Figure 3.4 RNA polymerase–DNA contacts during initiation complex formation. DNA containing an arbitrary promoter region is schematically illustrated as a cylinder. The transcription start site (+1) is indicated by an arrow. The numbers represent sequence positions relative to the start position. Shaded areas correspond to regions protected from DNase I digestion or reduced accessibility towards chemical modification of the DNA. Note that much of the protected area is localized on one face of the helical DNA molecule. The white oval denotes a region of single-stranded DNA characterized by its sensitivity to single strand-specific reagents.

3.4.1 Abortive initiation

The initial polymerization reactions of the ternary complexes, which are usually termed **initial transcribing complexes** (ITC), do not immediately yield long productive transcripts. Instead, short oligomeric products are released in an abortive way and chain initiation is restarted by the same complex in several repetitive rounds of reactions. RNA polymerase in the ternary complexes during this **abortive cycling** remains stably bound to the promoter site. During abortive initiation short RNAs from two to about 10 nucleotides in length are synthesized and released as abortive products. In fact, at many promoters the majority of the nascent transcripts are released from the ternary complexes as abortive products. The extent of formation of abortive rather than productive transcripts is different for different promoters, however. Abortive initiation is a generally occurring natural property of all RNA polymerases capable of *de novo* RNA synthesis. It has been observed in prokaryotic, phage-specific or eukaryotic RNA polymerases where it occurs at almost all promoters. During *in vitro* reactions abortive initiation can be provoked by omitting one of the necessary NTPs to transcribe through the early sequence region. The RNA polymerase is then forced to halt at the position where the substrate NTP is missing, and the nascent transcript is released as abortive product. Measurements of the time-dependent accumulation of abortive products provoked in this way has frequently been used to determine the promoter efficiency *in vitro* (see Section 3.6.1 below).

There is a natural limit beyond which no abortive products will be synthesized. This is normally the case after 8–10 nucleotides have been incorporated into the growing RNA chain. Once this length has been achieved the transcript is no longer released and the initiating complex is converted to a processive structure which is now capable of transcribing very long RNA chains without premature release. This transition to elongating RNA polymerase promoter complexes involves a gross change in the structure of the enzyme and coincides with the release of the σ factor (see below).

RNA polymerase ternary complexes have been characterized during abortive initiation by various footprinting methods. The results from such footprint analyses of ITCs reveal that they are similar in extension to the corresponding open binary complexes. A slight downstream extension of about four base pairs and a novel hyperreactive site at position –22 have been described for the tac promoter. No difference in the extension of the footprint along the DNA is recorded, however, for complexes where abortive products differ in size between five and eight nucleotides. Note again that the σ subunit is always present in ternary complexes undergoing abortive initiation.

3.4.2 Transcript slippage

A phenomenon termed **transcript slippage** or **reiterative transcription** is frequently observed when at least three consecutive identical nucleotides are located within the initiation region of the template. In these cases slippage between the nascent transcript and the template can occur. During transcript slippage, the active centre of RNA polymerase does not move in the downstream template direction, although a continuous polymerization reaction occurs and RNA transcripts with long homopolynucleotide sequences are generated. Obviously, the catalytic centre of the enzyme undergoes repetitive cycles of nucleotide addition (each time the same template base is correctly decoded). This is similar to abortive initiation. However, the catalytic site of the enzyme is not translocated in register with the polymerization steps. It seems to slip back after each phosphodiester bond formation, presenting the same template base again and again. During the reaction a fraction of the synthesized chains is released, giving rise to a regularly spaced ladder of products when analysed on polyacrylamide gels. In contrast to the abortive initiation reaction (where the size of the products is limited to about 10 nucleotides), the synthesized transcripts during the slippage reaction can be very long, and homopolymer sequences up to several hundred nucleotides can be formed. It should be emphasized again that transcript slippage does not occur through non-specific nucleotide incorporation. Although the reiterative sequence of the transcript no longer matches the sequence of the template strand, the individual reaction cycles of nucleotide addition occur in a template-directed way. Only those nucleotides are incorporated for which a complementary template base is positioned in the active centre of the polymerase. In contrast to normal transcription, the template base in the active centre is not changed between two nucleotide addition cycles. At some point during the slippage reaction the reiterative cycle is stopped and normal transcription, with the corresponding downstream movement of the active centre is resumed. It is unclear which signal exactly triggers and stops the slippage reaction, but probably the concentration of the reiteratively incorporated nucleotide plays a critical role.

Reiterative transcription or slippage reactions are not just an artificial or exotic phenomenon of transcription. In recent years the slippage phenomenon has been observed in a number of cases, and today it is clear that transcript slippage plays an important regulatory role during the biosynthesis of several operons. It is known that the intensity of the slippage reaction depends on the cellular concentration of exactly that NTP which is reiteratively incorporated. High NTP concentrations cause a large amount of slippage products while low concentrations of the respective NTP diminish the slippage reaction. The efficiency of transcription thus correlates with the cellular NTP concentration. Notable examples where transcript slippage, in conjunction with the cellular concentration of pyrimidines, provide the basis for a control system are found for operons or genes encoding enzymes which are involved in the biosynthesis of pyrimidines (e.g. *pyrBI*, encoding aspartate transcarbamylase subunits) or pyrimidine salvage (e.g. *codBA*, encoding cytosine permease and cytosine deaminase or *upp*, encoding uracyl phosphoribosyltransferase).

Transcript slippage is not only restricted to the initiation phase of transcription but can also be observed during elongation when a stretch of more than 10 identical nucleotide residues (A or T) is present on the template. In these cases, the reiterative incorporation of more or less nucleotides (10–20 times) into the growing RNA chain may occur.

A phenomenon related to transcript slippage has been reported as **primer shifting**. Primer shifting is observed for promoters where a similar sequence can be found adjacent to the start sequence. Examples are the λ P_L promoter or the ribosomal *rrnB* P1 promoter. The latter has the start site sequence of the non-coding strand, 5′-CACCACTG-3′, with the emboldened A indicating position +1. Normally a transcript, 5′pppACUG . . . 3′ is synthesized. When the NTPs are limited, however, during initiation the situation depicted in Fig. 3.5 occurs.

After the first dinucleotide sequence AC has been correctly synthesized starting at position +1 the primed transcript is shifted back to fit the template positions −3 and −2. From there the synthesis is continued, yielding the overall

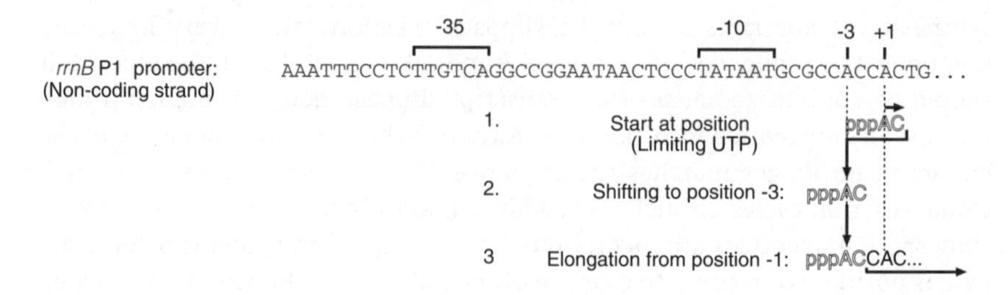

Figure 3.5 Primer shifting at the *rrnB* P1 promoter. The non-coding strand of the *rrnB* promoter is shown. At limiting UTP concentration the primary transcript pppAC initiated at +1 is shifted backwards to position −3. From this position elongation is continued in a different register.

sequence of the transcript, 5′pppACCAC .... 3′. Primer shifting, as depicted in Fig. 3.5 for the *rrnB* P1 promoter, occurs when transcription initiation is performed at substrate NTP limitations with only ATP and CTP present during the reaction. *In vivo*, primer shifting at the *rrnB* P1 promoter has not been observed.

3.5 Promoter clearance and the formation of elongation complexes

Promoter strength, taken as a measure of the efficiency of transcript formation, is not simply determined by the affinity of RNA polymerase to bind to the promoter. The rate of open complex formation can be rate-limiting in many cases. In addition, the rate of formation of ternary complexes may contribute to the overall efficiency. There is, however, at least one additional step which may contribute to the rate at which an initial transcribing complex is converted to a transcribing (elongating) complex. This step is characterized by the rate at which the ternary initiation complex leaves the promoter to allow binding of the next RNA polymerase for a second round of transcription. This process has been described as **promoter clearance**. At some point during the conversion of open complexes into initial transcribing ternary complexes the contacts between RNA polymerase and the promoter DNA must be released, otherwise no productive transcription can occur. In Section 2.2.4 it was shown that the −35 and −10 promoter sequences are both bound by the σ subunit of RNA polymerase. Movement of the transcribing RNA polymerase away from the promoter requires that these contacts be broken, either simultaneously or sequentially. If these interactions are very strong it might be disadvantageous for promoter efficiency. It is immediately clear that a high affinity of the RNA polymerase for the promoter and a resulting strong interaction may be counterproductive for efficient promoter clearance. Footprinting analyses during the early steps of transcription (open, abortive and productive complexes) with different promoters have revealed that the DNA regions, protected by RNA polymerase, are not continuously shifted downstream when open complexes are converted into initial transcribing complexes. Instead, the upstream borders of the RNA polymerase–DNA complex stay at first unchanged, while the downstream contacts are extended. When the length of the abortive transcript exceeds around 10 nucleotides, RNA polymerase escapes abortive transcription and shifts to a productive transcription complex. During this step the upstream contacts on the DNA are suddenly lost. The downstream border of the transcribing complex is not shifted in the same register, however, so that the total range of DNA protection in the productive transcribing complex is reduced from about 65 base pairs to about 45 base pairs. This indicates that the RNA polymerase-DNA complex undergoes a major conformational change. Obviously, contacts to the upstream promoter elements (–35 region and/or UP elements; see Sections 2.1 and 2.2.1) are kept during the synthesis of abortive products while the enzyme stretches out along the DNA in the downstream direction. This situation has been described as a **stressed intermediate**. When the abortive complex changes into a productive complex the upstream contacts are released and the enzyme advances along the DNA at a more compact conformation (Fig. 3.6). The transition is further characterized by the release of the σ subunit. The resulting complex has the properties of a typical elongating complex, with an average range of DNA protection that is reduced to about 35 base pairs. There is a simultaneous enlargement of the length of the transcription bubble from 12 to 14 nucleotides in the open complex to about 18 nucleotides in the elongating complex. As soon as the promoter contacts are broken the RNA polymerase is free to advance downstream along the template. At this point a new initiation cycle can be started. This phase of transcription is usually described as **promoter clearance**. The elongating RNA polymerase forms a highly stable complex. Probably a ring-shaped structure holds the DNA like a sliding clamp. The enzyme is therefore very processive, and elongation ceases only when a transcriptional terminator or a pausing signal is encountered.

An alternative view to flexible RNA polymerase and the stressed intermediate would be that it is the DNA in the transcription bubble that 'scrunches' (compaction of DNA single strands within the active site). DNA scrunching is supported by structural studies of the T7 phage RNA polymerase (Cheetham and Steitz, 1999).

Escape of the RNA polymerase from the promoter hexamers (–35 and –10 regions) and release of the σ subunit does not always mean that the promoter is automatically free for the next round of initiation. The transcribing complex often pauses within the initial transcribed sequences and thus blocks the formation of new open complexes which require free DNA up to position +20 (see Fig. 3.3). Situations like this are more appropriately described as **promoter escape** rather than by promoter clearance. It should be clear from the above

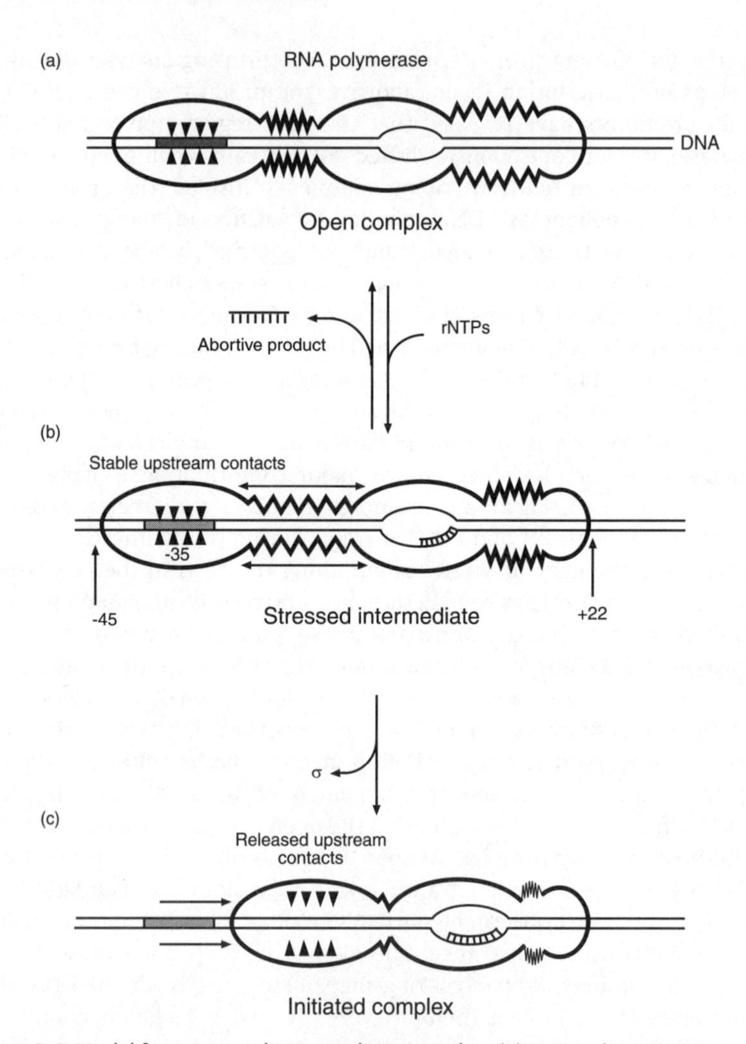

Figure 3.6 Model for a stressed intermediate complex. (a) RNA polymerase is schematically presented with two flexible regions indicated by zigzag lines. Arrowheads in the upstream segment represent tight binding to the −35 promoter region shown as a dark segment on the DNA. The DNA forms a transcription bubble which characterizes the open complex. (b) The addition of NTPs causes the central part of RNA polymerase with the active centre to move upstream. Numbers indicate the upstream and downstream borders of DNase I protection. The stressed conformation of the enzyme is indicated by the extended zigzag lines and the horizontal arrows. A short transcript is formed which is presented as a heteroduplex with the DNA template strand. The structure can return to the open complex with release of abortive transcripts. Alternatively, stress can be released when the enzyme moves forward, losing the upstream contacts to the −35 region. This movement, accompanied by the loss of σ, results in a productive initiation complex (c). The figure is arranged according to Straney and Crothers (1987).

considerations that promoter clearance or promoter escape can be the *rate-limiting* steps of transcription initiation. This is actually the case for many promoters. A notable example is the *lac* promoter, although (or better, because) this promoter has a high affinity for RNA polymerase it is limited at the step of promoter clearance.

There is a natural limit to the rate at which promoter clearance or escape can occur. This is given by the maximal rate of elongation. To allow for the next round of transcription initiation RNA polymerase has to translocate about 60 base pairs. This is the equivalent range that RNA polymerase open complexes protect on the DNA (see Fig. 3.3). Promoter clearance is thus limited to the time required for the RNA polymerase to move a minimum of 60 base pairs. As will be discussed in Section 4.2 the rate of RNA polymerase during transcription elongation does normally not exceed 60 nucleotides per second. The fastest time at which a promoter can be cleared for the next round of transcription initiation is therefore about 1 second. In other words, the highest frequency of transcription initiation at bacterial promoters is 1 per second. This rate is actually achieved by the strong promoters of bacterial rRNA genes at exponential growth (see Section 8.7).

If clearance limits the initiation cycle it is feasible that this step is also a target for regulation. This has been shown for some promoters where clearance is dependent on accessory factors, like CRP at the *malT* promoter or CRP at the *lac* and *gal* promoters (see Chapter 7). Promoter clearance can also be affected by the salt concentration or the temperature. Furthermore, the intracellular NTP concentration is an important parameter for the clearance step at several promoters. UTP concentration, for instance, regulates transcription from the *pyrB* promoter by a slippage mechanism (see above and Section 5.3).

3.6 The kinetics and thermodynamics of transcription initiation

The above sections have shown that RNA polymerase undergoes a series of different steps that finally lead to the elongating complex. To understand the kinetics of these conversions better, a formal description of the complete process is desirable. In a strongly simplified model a scheme describing binding and isomerization has been developed for the determination of the kinetic parameters that lead to the reactive species, the open complex (Fig. 3.7). The following analysis is restricted to this minimal description. It has proven to be very useful, however, for a comparison of the promoter strength of many different promoters.

It is clear from the sections above that later steps, such as promoter escape,

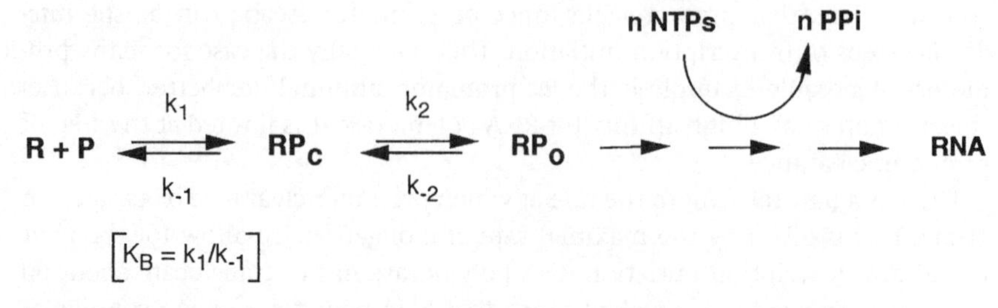

Figure 3.7 Simplified kinetic scheme leading to open complexes. The reaction is dissected into two essential steps: first, formation of a closed complex (RP_c) and, second, isomerization to the open complex (RP_o). R and P denote the concentrations of free RNA polymerase and free promoter, respectively. Formation of the closed binary complex (RP_c) can be characterised by the association and dissociation rate constants k_1 and k_{-1}, respectively. K_B represents the binding equilibrium constant of this reaction. The isomerization rate constant (k_2) leads to the open complex (RP_o) while the reverse of the reaction is described by k_{-2}. Under saturating concentrations of NTPs the rate of RNA product formation is assumed to be proportional to the concentration of RP_o. During this reaction an equivalent of pyrophosphate (PPi) is produced for every NTP incorporation.

for instance, may also limit the rate of transcription initiation. Moreover, the process of transcription initiation can be of much higher complexity, involving intermediates not considered to be rate-limiting in this description. For a more sophisticated analysis the reader is referred to the Further Reading at the end of the chapter.

3.6.1 Analysis of the rate-limiting steps of productive transcription complexes

For a quantitative analysis of the kinetic steps leading to the open complex the following simplifying assumptions are made:

1. RNA polymerase (R) should be in excess of the total promoter concentration (P_t): $[R] >> [P_t]$; although binding of RNA polymerase to a promoter is a bimolecular reaction these conditions reflect a pseudo first-order approximation.
2. The concentration of the open complex $[RP_o]$ is assumed to be in steady state equilibrium.
3. Limitations by promoter clearance are disregarded.
4. An additional assumption is made (valid for most promoters), namely that the forward reaction to the open complex is much faster than the reverse reaction, making this step almost irreversible: $k_{-2} << k_2$.

Under these conditions, the rate of product formation, hence the promoter efficiency, is proportional to the concentration of the open complex $[RP_o]$. According to the above kinetic description a rate equation can be written as:

$$\frac{dRPo}{dt} = k_2[RP_c] - k_{-2}[RP_o] \qquad \text{(eqn 3.1)}$$

The solution to this equation is:

$$[RP_o] = [Pt]\cdot(1 - e^{-k_{obs}}) \qquad \text{(eqn 3.2)}$$

$$k_{obs} = \frac{k_1\cdot[R]\cdot(k_2 + k_{-2}) + k_{-1}\cdot k_{-2}}{k_1\cdot[R] + k_{-1} + k_2} \qquad \text{(eqn 3.3)}$$

With $k_{-2} << k_2$ eqn 3.4 can be derived:

$$\frac{1}{k_{obs}} = \tau_{obs} = \frac{1}{k_2} + \frac{k_{-1} + k_2}{k_1\cdot k_2\cdot[R]} \qquad \text{(eqn 3.4)}$$

or with $K_B = k_1/k_{-1} + k_2$, eqn 3.4 can be written as:

$$\tau_{obs} = \frac{1}{k_2} + \frac{1}{K_B\cdot k_2\cdot[R]} \qquad \text{(eqn 3.5)}$$

τ represents the average time required for open complex formation at the given RNA polymerase concentration. It is thus a measure for the isomerization time. K_B characterizes the *rapidly reversible* RNA polymerase concentration-dependent binding step to the promoter DNA. k_2 gives the rate constant for the *irreversible* step which is independent of the RNA polymerase concentration. A plot of τ_{obs} *versus* 1/[R] according to eqn 3.5 is called a **τ-plot**. It yields a straight line with $1/k_2$ as the intercept on the ordinate and $1/K_Bk_2$ as the slope. K_Bk_2, the reciprocal slope, gives the apparent bimolecular association rate constant k_a ($M^{-1}s^{-1}$) of the reaction (Fig. 3.8).

How can these parameters be determined? Two independent sets of reactions must be performed *in vitro*. First, RNA polymerase is preincubated with promoter DNA to allow formation of the open complex. The reaction is then started by the addition of the necessary NTPs. The steady-state rate of product synthesis (this can be abortive products) is measured over a range of several minutes. The results from such measurements can be plotted as shown in Fig. 3.8a, reaction 1. In a second reaction NTPs are preincubated with DNA, and the reaction is started by the addition of RNA polymerase as the last component. A lag in the product synthesis can be observed which corresponds to the time needed for isomerization. This lag time can be obtained, for instance, by extrapolation to the time axis of the curve representing the product formation of the second reaction (reaction 2), as shown in Fig. 3.8a. The intercept of the time axis represents τ_{obs}. It is a measure for the isomerization time at the RNA polymerase concentration used. The assay has now to be repeated with different RNA polymerase concentrations. With increasing concentrations of RNA polymerase τ_{obs} gets smaller. The τ_{obs} values obtained are plotted against the reciprocal concentration of the RNA polymerase employed. An example of such a graph (a τ-plot) is given in Fig. 3.8b. K_B, **the equilibrium binding**

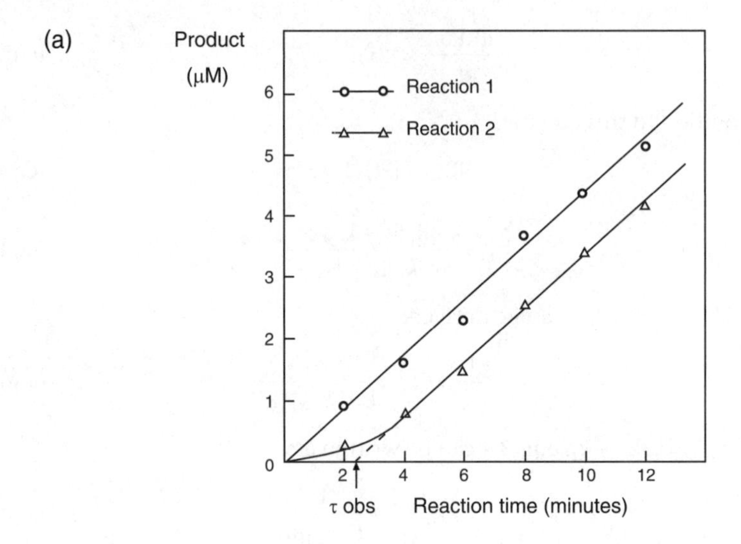

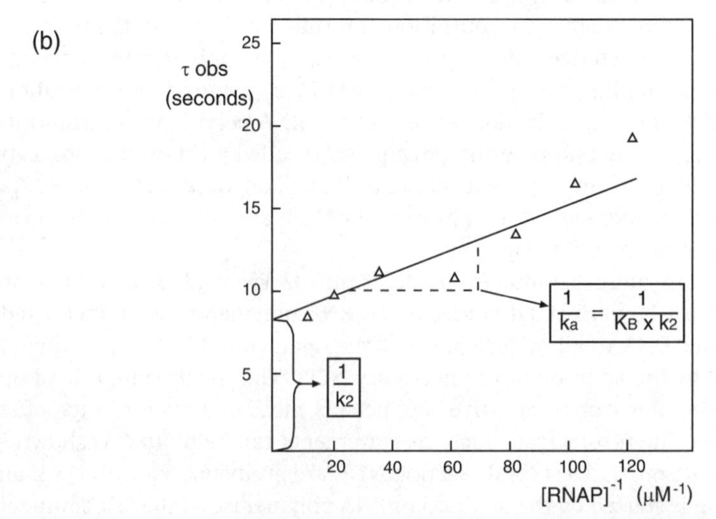

Figure 3.8 Kinetic analysis of promoter strength. (a) The time-dependence of RNA product formation of an *in vitro* transcription reaction is plotted against the reaction time. Reaction 1 is started by the addition of NTP substrates to preformed RNA polymerase–promoter complexes. Reaction 2 is started by the addition of RNA polymerase to a mixture of template and substrates. Extrapolation of the latter reaction on the time axis yields τ_{obs}, the time needed for the isomerization from a closed to an open complex. (b) τ_{obs} values derived from a series of reactions at different RNA polymerase concentrations are plotted against the reciprocal concentration of RNA polymerase ($[RNAP]^{-1}$). This should result in a straight line. From this graph, the kinetic constant $1/k_2$ can be obtained as the intercept of the time axis and the expression $1/k_a$ can be derived from the slope of the line.

constant for the RNA polymerase promoter interaction, k_2, **the isomerization rate constant**, as well as k_a, **the overall rate constant**, can be obtained from such a graph.

What do the parameters tell us? There is no doubt that the analysis provides useful information for the comparison of promoter strength for those promoters that are not limited by their clearance rates (and match the requirements specified above). Typical values obtained are in the range of 10^7–10^9 [M^{-1}] for K_B, and 10^{-1}–10^{-3} [s^{-1}] for k_2. Lower limits are $K_B \leq 5 \times 10^6\ M^{-1}$ and $k_2 \leq 10^{-3}\ s^{-1}$. No measurable initiation would be observed at physiological RNA polymerase concentrations at the average bacterial generation times with smaller kinetic parameters. The product of K_B and k_2 is k_a, which can generally be taken as a measure of the **promoter strength**.

The following inferences can be made directly from a τ-plot analysis. First, strong and efficient promoters have flat slopes and small Y-axis intercepts; second, weak promoters are characterized by steep slopes and high Y-axis intercepts; and, third, according to their constants, K_B and k_2, promoters may be classified as limited in their affinity for RNA polymerase binding (small values of K_B) or limited in their rate of isomerization (small values of k_2).

Both parameters, K_B and k_2, can contribute to promoter strength. Either one of the two constants can be affected individually or in combination by base change mutations in the promoter sequence. The binding constants for RNA polymerase (K_B) or the isomerization rates (k_2) may also be affected by the template topology or by transcription factors, however. Examples are given in Table 3.1.

The thermodynamic parameters of the transcription initiation steps have also been determined for a few promoters. Measurements of the overall forward rate constant k_a have been performed for the λ P_R promoter at different temperatures, for instance. From such measurements an Arrhenius activation energy of the association reaction of about 20 ± 5 kcal M^{-1} was obtained. It was furthermore shown that k_a decreases systematically with increasing potassium chloride concentration. A van't Hoff enthalpy (ΔH^0_{obs}) of 29 ± 9 kcal M^{-1} was estimated. Standard free energies were determined at different temperatures (−14.0 kcal M^{-1} at 25°C or −15.7 kcal M^{-1} at 37°C). From the values ΔG^0_{obs} and ΔH^0_{obs} the change in entropy ΔS^0_{obs} of 114 ± 30 e. u. was derived.

From the temperature-dependence of equilibrium constants $K_{eq} = k_a/k_d$ of several promoters (e.g. *lac* UV5, λ PR or *tetR*) van't Hoff enthalpies (ΔH^0) have been determined. They were endothermic (positive) in all cases. It must be concluded therefore that in all the examples above the reactions are entropically driven.

Table 3.1 Changes in K_B and K_2 for different promoters

Promoter	K_B (M^{-1})	k_2 (s^{-1})	$k_a = K_B \times k_2$ (M^{-1} s^{-1})	Effector
λP_{RM}	9.9×10^6	7.0×10^{-4}	6.9×10^3	$-\lambda$ cI protein
λP_{RM}	9.2×10^6	7.8×10^{-3}	7.2×10^4	$+\lambda$ cI protein
λP_{RM}up–1*	1.2×10^8	3.7×10^{-3}	4.4×10^6	$-\lambda$ cI protein
λP_{RM}up–1*	4.0×10^7	2.3×10^{-2}	9.2×10^5	$+\lambda$ cI protein
λP_R	6.7×10^8	1.0×10^{-2}	6.7×10^6	
$\lambda P_R \times 3$†	3.1×10^7	2.0×10^{-3}	6.2×10^4	
λP_L	2.85×10^8	4.4×10^{-3}	1.25×10^6	linear, +IHF
λP_L	8.9×10^7	4.4×10^{-3}	3.9×10^5	linear, –IHF
λP_L	2.0×10^8	1.0×10^{-1}	2×10^7	supercoiled, +IHF
λP_L	6.1×10^7	6.6×10^{-2}	4.0×10^6	supercoiled, –IHF
rrnB P1	2.1×10^9	1.8×10^{-3}	3.8×10^6	linear, +FIS
rrnB P1	1.2×10^8	3.8×10^{-3}	4.6×10^5	linear, –FIS
rrnB P1	1.2×10^8	2.3×10^{-2}	2.7×10^6	supercoiled, –FIS

IHF, DNA-binding protein; FIS, factor for inversion stimulation.
* λP_{RM} promoter with base change at position –31.
† λP_R promoter with base change in the –35 region.
The data are taken from Hwang *et al.* (1988), Giladi *et al.* (1992) and Zacharias *et al.* (1991)

Summary

The process of transcription initiation can be dissected into four main substeps, each of which may be rate-limiting. The first step consists of *promoter location* and binding of RNA polymerase to the promoter. For very efficient promoters this step may involve *linear diffusion* of non-specifically (electrostatically) bound RNA polymerase along the DNA. *Sliding* along the isopotential surface of DNA reduces the search volume and facilitates promoter location, which can be faster than driven by three-dimensional diffusion, therefore.

The first complex formed between RNA polymerase and the promoter is characterized as the *closed complex*, which indicates that the DNA within the promoter core is still double-stranded. The second step of the initiation cycle is characterized by *isomerization* from the closed to the *open complex*. During this transition about 12 base pairs of DNA are disrupted around the transcription start site. This process nucleates at the AT-rich sequence within the promoter core and extends up to a template position of around +3. Isomerization may involve several substeps, leading to more than one open complex. Strand opening is catalysed by the action of the σ subunit and is strongly influenced by the DNA twist, the spacer length and the template topology. Complete melting requires the presence of Mg^{2+} ions.

In the next step of the initiation cycle substrate NTPs are bound and the binary RNA polymerase–promoter complex is converted to a *ternary complex*. The correct NTPs are discriminated by base pair formation with the free bases of the template strand exposed in the catalytic site. Correctly bound nucleotides are linked to the 3′ hydroxyl group of the preceding nucleotide by a phosphodiester bond. Pyrophosphate is released during this reaction. The initiating ternary complex starts to synthesize short RNA chains up to a length of about 10 nucleotides. During this reaction the active centre of RNA polymerase moves downstream in register with every incorporated nucleotide. The rest of the enzyme, however, remains bound to the promoter and the upstream and downstream edges do not move. The short RNA chains are released as *abortive products*, and the active centre snaps back to template position +1. The process is normally repeated many times. At some point during the process of abortive cycling the initiating complex is converted from an abortive to a *processive complex*. Transcripts are synthesized exceeding a length of 10 nucleotides. Upstream DNA contacts of RNA polymerase are released and the enzyme moves in the downstream direction. The step is termed *promoter clearance* or *promoter escape*. When the transcript length exceeds around 10 nucleotides, the σ factor is released from the complex, which changes from an initiating to an elongating conformation.

In rare cases, initiation complexes may undergo a *slippage reaction*. This can be the case when the template contains a cluster of the same nucleotide within the early transcribed region. In such a case reiterative incorporation of the same nucleotide may occur many times without downstream movement of the active site. During the slippage reaction fairly long chains with a homopolymeric sequence can be synthesized, which are either released or productively elongated.

For a quantitative comparison of different promoters a convenient method, termed *τ-plot analysis*, has been developed. This enables the determination of the RNA polymerase concentration-dependent *binding step* (K_B) and the rate constant for the *isomerization reaction* (k_2). If the initiation reaction is not limited by promoter clearance the product of the two constants, the *bimolecular association rate constant* ($k_a = K_B \times k_2$), gives a measure of the relative *promoter strength*.

References

Berg, O. G., Winter, R. B. and von Hippel, P. H. (1981) Diffusion-driven mechanisms of protein translocation on nucleic acids. 1. Models and theory. *Biochemistry* **20**: 6929–48.

Cheetham, G. M. T. and Steitz, T. A. (1999) Structure of a transcribing T7 RNA polymerase initiation complex. *Science* **286**: 2305–9.

Craig, M. L., Tsodikov, O. V., McQuade, K. L. *et al.* (1998) DNA footprints of the two kinetically significant intermediates in formation of an RNA polymerase-promoter open complex: evidence that interactions with start site and downstream DNA induce sequential conformational changes in polymerase and DNA. *Journal of Molecular Biology* **283**: 741–56.

Giladi, H., Igarashi, K., Ishihama, A. and Oppenheim, A. B. (1992) Stimulation of the phage λ pL promoter by integration host factor requires the carboxy terminus of the α-subunit of RNA polymerase. *Journal of Molecular Biology* **227**: 985–90.

Hwang, J. J., Brown, S. and Gussin, G. N. (1988) Characterization of a doubly mutant derivative of the λ P_{RM} promoter. *Journal of Molecular Biology* **200**: 695–708.

Polyakov, A., Severinova, E. and Darst, S. A. (1995) Three-dimensional structure of *E. coli* RNA polymerase: promoter binding and elongation conformations of the enzyme. *Cell* **83**: 365–73.

Straney, D. C. and Crothers, D. M. (1987) A stressed intermediate in the formation of stably initiated RNA chains at the *Escherichia coli lac* UV 5 promoter. *Journal of Molecular Biology* **193**: 267–78.

von Hippel, P. H. and Berg, O. G. (1989) Facilitated target location in biological systems. *Journal of Biological Chemistry* **264**: 675–8.

Zacharias, M., Theißen, G., Bradaczek, C. and Wagner, R. (1991) Analysis of sequence elements important for the synthesis and control of ribosomal RNA in *E. coli*. *Biochimie* **73**: 699–712.

Further reading

Berg, O. G., Winter, R. B. and von Hippel, P. H. (1981) Diffusion-driven mechanisms of protein translocation on nucleic acids. 1. Models and theory. *Biochemistry* **20**: 6929–48.

Borukhov, S., Sagitov, V. and Goldfarb, A. (1993) Transcript cleavage factors from *E. coli*. *Cell* **72**: 459–66.

Brewer, B. J. (1988) When polymerases collide: replication and the transcriptional organization of the *E. coli* chromosome. *Cell* **53**: 679–86.

Buckle, M. and Buc, H. (1994) On the mechanism of promoter recognition by *E. coli* RNA polymerase. In: Conaway, R. C. and Conaway, J. W. (eds) *Transcription: Mechanisms and Regulation.* New York: Raven Press, pp. 207–25.

Bujard, H. (1980) The interaction of *E. coli* RNA polymerase with promoters. *Trends in Biological Sciences* **10**: 274–8.

Chamberlin, M. J. and Hsu, L. M. (1996) RNA chain initiation and promoter escape by RNA polymerase. In: Lin, E. C. C. and Lynch, A. S. (eds) *Regulation of gene expression in* Escherichia coli. Austin: Landes Company, pp. 7–25.

Chan, C. L. and Landick, R. (1994) New perspectives on RNA chain elongation and termination by *E. coli* RNA polymerase. In: Conaway, R. C. and Conaway, J. W. (eds) *Transcription: Mechanisms and Regulation.* New York: Raven Press, pp. 297–321.

deHaseth, P. L. and Helmann, J. D. (1995) Open complex formation by *Escherichia coli* RNA polymerase: the mechanism of polymerase-induced strand separation of double helical DNA. *Molecular Microbiology* **16**: 817–24.

Gussin, G. N. (1996) Kinetic analysis of RNA polymerase–promoter interactions. *Methods in Enzymology* **273**: 45–59.

Hawley, D. and McClure, W. R. (1980) *In vitro* comparison of initiation properties of bacteriophage λ wild-type P_R and x3 mutant promoters. *Proceedings of the National Academy of Sciences USA* **77**: 6381–5.

Horwitz, M. S. Z. and Loeb, L. A. (1990) Structure–function relationships in *Escherichia coli* promoter DNA. *Progress in Nucleic Acid Research and Molecular Biology* **38**: 137–64.

Hsu, L. M. (1996) Quantitative parameters for promoter clearance. *Methods in Enzymology* **273**: 59–71.

Leirmo, S. and Record, M. T. (1990) Structural, thermodynamic and kinetic studies of the interaction of σ^{70} RNA polymerase with promoter DNA. In: Eckstein, F. and Lilley, D. M. J. (eds) *Nucleic Acids and Molecular Biology*, Vol. 4. Berlin–Heidelberg: Springer Verlag, pp. 123–51.

McClure, W. (1980) Rate-limiting steps in RNA chain initiation. *Proceedings of the National Academy of Sciences USA* **77**: 5634–8.

McClure, W. R. (1985) Mechanism and control of transcription initiation in prokaryotes. *Annual Reviews of Biochemistry* **54**: 171–204.

Schickor, P., Metzger, W., Wladyslaw, W., Lederer, H. and Heumann, H. (1990) Topography of intermediates in transcription initiation of *E. coli*. *EMBO Journal* **9**: 2215–2220.

Straney, D. C. and Crothers, D. M. (1987) A stressed intermediate in the formation of stably initiated RNA chains at the *Escherichia coli lac* UV5 promoter. *Journal of Molecular Biology* **193**: 267–78.

4

Transcription elongation

This chapter will start with a description of the architecture of the transcription elongation complex or, better, what it is thought to look like owing to lack of high resolution data. The kinetics of the RNA chain elongation reaction, or what is called the rate of elongation of bacterial RNA polymerase are discussed. From the discontinuous progression of RNA polymerase along the DNA template models for the translocation of RNA polymerase have been developed. One of these, the 'inchworm' model, is presented below. Special emphasis is given in this chapter to the phenomenon of RNA polymerase pausing during transcription elongation, and to factors that have an influence on the RNA chain elongation rate. These factors include the structure of the nascent transcript but also the conformation and topology of the DNA template. In addition, the effects of transcription factors and small ligands on pausing and elongation of RNA polymerase are summarized. The involvement of transcription factors which are able to cleave transcripts in the elongating complex, and their implications in a putative proofreading mechanism will be discussed. At the end of the chapter a problem resulting from the coupling of the transcription and replication processes is explained. This section deals with the inevitable encounter of transcription and translation complexes during the life cycle of a bacterium.

The conversion of RNA polymerase complexes that leave the initiation phase and enter the elongation phase is characterized by three properties:

1. The specificity factor σ is released from the complex and becomes ready to recycle in a new round of initiation.
2. The translocating RNA polymerase is displaced from the promoter and the consensus recognition elements are free for the next RNA polymerase entering the initiation phase.
3. A conformational transition of the RNA polymerase occurs which causes a highly stable binding of the template DNA and the growing transcript.

The high stability of elongating complexes over time and distances ensures that very long transcripts can be produced in the cell (and also *in vitro*) without premature termination. Note that some bacterial transcription units exceed

20 000 base pairs and some human genes, for instance, require uninterrupted transcription of more than 2000 kilo base pairs. The stability of elongating complexes is crucial to transcription since any premature termination would be irreversible. Remember that RNA polymerase is an obligatory processive enzyme, which means that once the elongating complex has been dissociated the released RNA cannot be elongated further by the same or a different RNA polymerase. Instead, transcription has to be restarted from the beginning.

Although transcription elongation complexes are very stable there are at least three exceptions where the processivity of the complexes is modified and alternative reactions are enabled.

1. Situations where the transcribing complexes pause for a variable period of time before they resume transcription (**pausing sites**).
2. Sites where the transcription reaction is terminated (**termination sites**).
3. Sites where the complexes are arrested and are unable to resume elongation without additional factors (**arrest sites**).

This chapter aims to provide the structural information necessary to understand the different elongation reactions.

4.1 The architecture of elongating complexes—what is known about the different functional sites?

The following characteristic structural elements can be defined for elongating complexes (Fig. 4.1):

1. *The size of DNA covered within the elongation complex*: studies with different elongation complexes have revealed that they vary considerably in the extent of the DNA region protected from different nucleases (see Box 2.1). According to such nuclease protection studies transcription elongation complexes cover between 25 and 40 base pairs along the DNA.

2. *DNA binding sites*: binding of the RNA polymerase core to the DNA double strand occurs at *two* separate sites. One binding site is at the front and one at the rear edge of the RNA polymerase. Interaction at the front of the RNA polymerase appears to be of non-ionic character and involves between seven and nine intact base pairs of the DNA duplex. Binding at the rear site occurs at a stretch of six to seven nucleotides of the separated template strand. This interaction is thought to be ionic. Cross-linking studies have shown that the front interaction site is close to the conserved region A at the NH_2-terminus of the β′ subunit. This part of the β′ subunit contains the Zn-finger motif (see Section 2.2.3) known as a DNA-binding element. In addition, region B of the β subunit contributes to the front binding site. The rear binding site has been mapped to

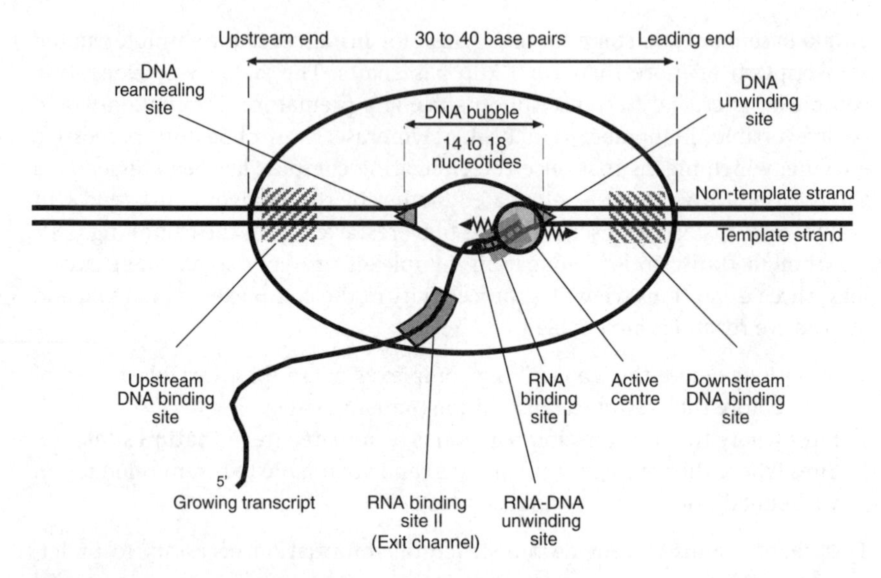

Figure 4.1 Model of a transcription elongating complex. RNA polymerase is represented by an oval covering 30–40 base pairs between the leading and the upstream end on the double-stranded DNA, which is schematically presented as two parallel lines with the template and the non-template strands indicated. The upstream and downstream DNA-binding sites are indicated by shaded areas at the upstream and leading edges of the polymerase, respectively. A DNA bubble of 14–18 nucleotides is shown which is flanked by unwinding and reannealing sites indicated by grey triangles. The catalytic centre of RNA polymerase is shown as a grey circle, with zigzag arrows in the upstream and downstream direction indicating structural flexibility at this site. The 3′ end of the growing transcript, represented by a dark line, is hybridized to the single-stranded DNA of the template strand within a short region of the transcription bubble (RNA-binding site I). After passing an RNA-DNA unwinding site, marked by a dark triangle, the growing transcript runs through the exit channel (RNA-binding site II). The figure is based on a model presented by Uptain *et al.*, (1997).

the COOH-terminal domain I of the β subunit. Comparison of this structural information with the three-dimensional model of core RNA polymerase, deduced from electron crystallography, suggests that the front binding site is identical to the ring-structured channel visible in the electron micrographs. This ring structure seems to enclose the DNA duplex and acts as a **sliding clamp** explaining the high processivity of the elongation complex.

3. *The transcription bubble*: the size of the transcription bubble with the single-stranded template and non-template strands varies between 14 and 18 nucleotides. It is located between the front and rear DNA-binding sites.

4. *DNA unwinding and reannealing sites*: movement of the elongation complex along the DNA requires a DNA unwinding centre in front of the transcription bubble and a site where the two strands are reannealed. It is assumed that

unwinding is a passive process and is not driven by separate energy consumption.

5. *the catalytic site*: The catalytic site, where NTP substrates are bound and the phosphodiester bonds are formed, must be localized within the transcription bubble, in the direct vicinity of the single stranded template. For the active recognition process the 3′ OH group of the growing transcript must be precisely positioned with the DNA template base and the incoming NTP substrate. For some elongation complexes the catalytic centre has been located at the downstream site of the transcription bubble, less than six base pairs from the front edge of the polymerase. It appears, however, that it is not always close to the front edge of the melted region (see Section 3.4.2 and below). Recent studies show that the location of the catalytic site within the elongating complex is highly dynamic and does not show a fixed position with respect to the ends of the transcription complex. Moreover, the catalytic centre is of modular composition, involving different elements of the RNA polymerase β and β′ subunits. This finding is based on hydroxyl radical mapping (see Box 2.1). Replacement of the Mg^{2+} ions in the catalytic centre by Fe^{2+}, which generates reactive hydroxyl radicals, has revealed that different domains, distributed over the primary structure of the RNA polymerase β and β′ subunits, contribute to the catalytic centre. This site therefore shows a modular organization probably involving the subregions D, E, F and H of the β subunit, and D, F and G of the β′ subunit (see Section 2.2).

6. *The product (RNA) binding site*: the growing transcript is considered to form an RNA:DNA hybrid with the complementary template strand when it leaves the catalytic site. The length of this RNA:DNA hybrid is not precisely known and is a matter of controversial discussion. Two alternative views have been proposed. According to one proposal, the length of the heteroduplex is considered to be between eight and 12 nucleotides. The opposing view suggests that there are only three or even less nucleotides involved in RNA:DNA heteroduplex formation. The residual nucleotides of the growing transcript, which are not yet free in solution, are believed to interact directly with the protein surface of the RNA polymerase. No matter what the exact length of the RNA:DNA hybrid actually might be, there must be a site where the growing transcript is separated from the template strand. Before the transcript leaves the elongation complex it probably passes an RNA exit channel. Such a putative RNA exit channel has tentatively been assigned on the surface of the RNA polymerase according to the structural models derived from electron crystallography (see Section 2.4). Binding sites for the 3′ end of the growing RNA chain have been mapped to the conserved regions D and H of the RNA polymerase β subunit as well as regions D and G of the β′ subunit.

7. *Non-template strand interaction*: the role of the non-template strand in stabilizing the elongation complex is not clear. However, there is now convincing evidence that not only the template strand, but also the non-template strand, interacts directly with RNA polymerase. It can be shown that the downstream

segment of the non-template strand binds to RNA polymerase. Accordingly, base change mutations in the non-template strand affect the transcription properties of elongation complexes (e.g. pausing, see below). The importance of the non-template strand for transcription elongation, pausing and termination has in particular also been shown for the T7 phage-specific RNA polymerase. This underlines that the non-template strand is generally an important component of the transcription elongation complexes.

These structural elements and their putative location within the elongation complex are depicted in the schematic model presented in Fig. 4.1.

4.1.1 Models for translocation

During translocation the elongation complex moves along the DNA double helix carrying an RNA chain of increasing length. During this movement the melted region of the transcription bubble has to be shifted in coordination with the downstream translocation of the elongation complex, allowing continuous recognition of the bases in the template strand. The number of DNA base pairs that are broken in front of the moving enzyme on average equals the number of base pairs reformed in the rear of the transcription bubble. The length of the DNA:RNA hybrid, although not yet precisely defined (see above), seems also to be more or less constant. It certainly does not exceed 12 base pairs. The forward movement of the elongation complex can thus be regarded as isoenergetic with respect to these interactions. Hence there is no requirement for extra energy supply during translocation. In fact, inspection of the energetics of RNA synthesis (the difference of the free energy of NTP phosphoanhydrid hydrolysis and the formation of the internucleotide phosphodiester bond) yield a standard free energy difference $\Delta G^{0'}$ of about –3 kcal/mol. About that much energy should be available for every nucleotide addition step of the translocation reaction.

Early models of translocation assumed a homogeneous monotonic single nucleotide step mechanism. However, this view has been revised in light of recent findings. Studies of the accessibility of DNA in elongation complexes have revealed a periodic expansion and contraction of the contact domains for the DNA and the growing transcript. According to these findings it now seems clear that the movement of RNA polymerase along the DNA is *discontinuous* and does not occur in register of single nucleotide steps. The prerequisite for the discontinuous movement is the existence of two separate DNA-binding sites within the RNA polymerase, one in front and one at the back of the enzyme (see above). The front and rear edges of the elongation complex undergo independent motions. Translocation has, therefore, been compared with the *inchworm movement.* According to the **inchworm model** of translocation the RNA polymerase moves in the following way. Several steps of nucleotide addition occur during which the front edge of RNA polymerase remains tightly bound to a specific DNA site. When the nucleotide addition site (catalytic centre) comes within a distance of approximately 6–8 nucleotides of the downstream border, tight binding at the front edge is released and the enzyme slides forward by about 10 nucleotides. During this step the rear edge and the catalytic centre maintain a fixed contact to the DNA. The next step involves new binding of the front edge of RNA polymerase to upstream DNA, and release of the back grip. The catalytic site now approaches the front side by a stepwise series of nucleotide addition cycles. At the critical distance of 6–8 nucleotides the front edge binding is released, the enzyme slides forward, and the translocation cycle is repeated (Fig. 4.2).

It should be noted, however, that the inchworm movement does not have a precise periodic pattern and asynchronous jumps of different length as well as phases of uniform single nucleotide step movements may randomly alternate. The inchworm model elegantly explains the observed pattern of compressions and extensions in the range of DNA protection found with limited nuclease digestion studies of elongation complexes.

An alternative view to the inchworming model has also been put forward. It is based mainly on cross-linking results which suggest that the elongation complex in fact contains an 8–9 base pair RNA:DNA heteroduplex. The model proposes that translocation is monotonic and discontinuous advancements are considered to be side pathways. The characteristic contractions and expansions of the inchworming model are explained by reverse and forward sliding of the elongation complex (**backtracking**). This involves cycles of unwinding and rewinding of the RNA:DNA hybrid. The RNA:DNA hybrid is believed to keep the active centre in register with the template. Weak interactions at the 3′ end of the hybrid cause backtracking of the elongation complex and may explain the reversible loss of the catalytic activity of the complex. The stability and processivity of the complex is explained in accordance with the inchworm model by a strong front end DNA-binding site and a tight RNA-binding site at the rear end.

Two extreme models of translocation movement have been proposed. According to one model the forward movement is driven randomly by thermal energy (Brownian motion). The correct binding of NTPs to the catalytic site prevents backward movement (*Brownian ratchet* mechanism). The forward movement is explained by diffusional sliding. The affinity of the active site of RNA polymerase for binding NTP at the upstream position on the template determines the direction. The other extreme explanation of the translocation mechanism is that of RNA polymerase as a mechanoenzyme, like the motor proteins kinesin or myosin. The necessary force derives from the incorporation of the free energy of hydrolysis of the nucleotides into the growing RNA chain. In fact, electron microscopic methods to visualize a single tethered RNA polymerase molecule bound to DNA have been developed and used to determine translocation with high precision. Similar measurements, where the system is fixed by a laser trap technology, have been employed to determine the resulting force of RNA polymerase movement. Such experiments have shown that

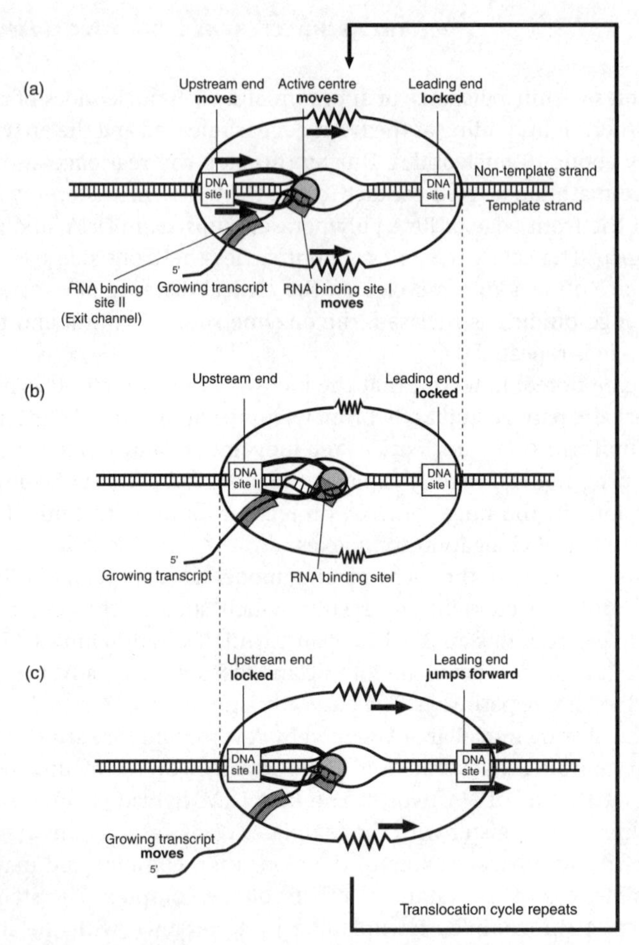

Figure 4.2 Inchworm model of RNA polymerase translocation. An inchworm-like movement of RNA polymerase along the template DNA is schematically presented. The movement results as a consequence of alternate binding and translocation steps of the upstream and downstream edges of the enzyme with one side locked and the other side sliding along the DNA. (a) The cycle starts with the leading end locked to the DNA via DNA-binding site I. The catalytic centre, located at the upstream end of the transcription bubble, moves in small increments with every nucleotide added to the growing RNA chain which extends the RNA:DNA hybrid (RNA-binding site I) in the downstream direction. The upstream end of RNA polymerase, including DNA-binding site II, follows this movement, which causes the RNA polymerase footprint to shrink. Black arrows indicate parts of the moving enzyme. (b) After the addition of between four and 10 nucleotides the leading end (DNA site I) gets locked to the downstream DNA. The catalytic centre is now located close to the downstream end of the transcription bubble. (c) The upstream end of the polymerase (DNA site II) gets locked to the DNA. The grip at the leading end (DNA site I) is released. The enzyme stretches in the downstream direction, as indicated by the extended zigzag line and the black arrows. The RNA polymerase front edge moves forward by increments of four to 10 nucleotides. The catalytic site is now again located close to the upstream end of the transcription bubble.

translocating RNA polymerase can act as a molecular motor, generating forces of 25–30 pN. These forces are thus several-fold larger than the forces achieved by the ATP-consuming motor proteins kinesin or myosin. Interestingly, the measured forces exceed the calculated forces that arise upon transcription-induced supercoiling (see Section 6.3).

4.2 The rate of elongation

The rate of transcription elongation (average number of nucleotides incorporated into a growing RNA chain per second) varies considerably among different polymerases and inversely correlates with the complexity of the enzymes. The single polypeptide phage T7- or SP6-specific RNA polymerases, for instance, transcribe at high rates of 200–400 nucleotides/s *in vitro*. Bacterial RNA polymerases have elongation rates of about 50 nucleotides/s *in vivo*. The corresponding rates *in vitro* are between 10 and 35 nucleotides/s. However, for some genes which are not translated, such as the ribosomal RNA genes, higher rates have been determined. In these cases transcription elongation probably approaches up to 100 nucleotides/s *in vivo* (see below). In contrast, the elongation rates of the complex eukaryotic RNA polymerases are estimated to be between 20 and 30 nucleotides/s *in vivo*.

The rate of transcription elongation in bacteria is thus about 20 times slower than the rate at which DNA is elongated during replication (see Section 4.5). The transcription elongation rate does, however, match the rate of mRNA translation. Translation rates in bacteria amount to roughly 16 codons/s. This corresponds to a forward rate along the mRNA of 48 nucleotides/s. As pointed out in the introduction, this is a very important fact. It explains that the two processes, transcription and translation, which have the same directionality, can occur simultaneously and in the same compartment. Changes in the synchronization of the two processes have strong implications on the efficiency of overall gene expression, as discussed below.

Transcription elongation rates are not constant. They vary greatly for different sites of a transcription unit or a gene. Furthermore, RNA chain elongation rates are subject to regulation. At reduced growth rates or under conditions of nutritional deprivation (e.g. amino acid starvation) transcription elongation rates are downregulated. This downregulation is linked to the accumulation of the small effector molecule guanosine tetraphosphate (see Section 8.5.5). Increasing concentrations of this effector are known to reduce the transcription elongation rate. Different effects have been found, however, for the transcription elongation of mRNA or rRNA genes (see Section 4.3.4).

It is important to know that the rate of transcription is often crucial for the later function of the transcripts. This is especially valid for stable RNA transcripts which are not translated. These molecules have to be folded into precise structures to execute their functions. Apparently, the kinetics of RNA folding and structure formation seem to be linked to the transcription rate. This became clear when several stable RNA genes from *E. coli* were cloned under the control of the promoter specific for phage T7 RNA polymerase. These genes were then transcribed in bacteria in which the gene for T7 polymerase had been induced. Transcription by the T7 RNA polymerase occurs at a much faster rate than that of *E. coli* RNA polymerase (see above). The resulting RNAs had the correct length and sequence but were not functional because of defects in their secondary or tertiary structures. This shows that the transcription rate and the higher order structure of stable RNAs are closely coupled (see also Section 8.7.7).

4.3 RNA polymerase pauses during transcription

The inchworm model of transcription elongation described above already suggests that elongation rates are not monotonous. In fact, the step time for each nucleotide addition is not a constant. Instead, the nucleotide addition rates observed are discontinuous and may vary considerably from one position to the next. It has been shown that this variation can be over 1000 fold from one template position to the next. During its path along the template, RNA polymerase may encounter positions where it hesitates and pauses for a variable period of time. These sites are commonly designated **pause sites**. Transcriptional pausing is a very widespread phenomenon which is observed in prokaryotic, eukaryotic and viral transcription and occurs *in vitro* as well as *in vivo*. Transcriptional pausing has been correlated to a number of different functions. It is, for instance, an important prerequisite for all termination events. It should be noted that every termination is preceded by a pause (see Chapter 5). The reverse, however, is not true. Not every pause leads to termination.

Paused complexes are able to resume transcription without the action of external factors. This distinguishes them from **arrested complexes** or **dead end complexes**, which are functionally impaired (see Section 4.4). During pausing the RNA polymerase in the elongating complex remains functional and remains bound to the RNA product. Pausing is thus reversible, and paused complexes differ from **termination complexes**, as will be seen later (Chapter 5).

Pausing can be characterized by two parameters: the *pausing half-life* and the *efficiency of pausing*. The first parameter defines the time that an RNA polymerase molecule which has entered a pausing site and undergoes pausing will remain there before transcription elongation is resumed. Pausing half-lives can vary from seconds to several minutes. The second parameter indicates that not every

RNA polymerase that encounters a potential pausing site will necessarily pause there. The fraction of pausing-sensitive RNA polymerase molecules can vary from a few molecules to 100%. Several different methods have been developed to determine and compare pauses. A useful parameter termed **pausing strength** can be defined, which takes into consideration pausing half-lives and pausing efficiency. The pausing strength is given as the relative occupancy of RNA polymerase at a defined pausing site integrated over the time the pausing phenomenon lasts (Theissen *et al.* 1990).

4.3.1 Factors influencing RNA polymerase pausing

Although RNA and DNA elements have been recognized as being responsible for pausing it is not yet possible to predict where in a DNA sequence a pause will occur. In other words, there are presently no consensus elements known which reliably indicate where a pause will occur. In addition, understanding is incomplete as to how different parameters will affect the pausing strength. Such factors are the DNA template or the transcript structure, but also external factors (see below). It seems clear, however, that more than one mechanism accounts for the occurrence and strength of RNA polymerase pausing. In the following sections three distinct features are described which are known to function in some way as pausing determinants.

4.3.2 Effects of the transcript structure

Many pauses can be associated with the capacity of the nascent transcript to form a stem-loop structure (hairpin) (Box 4.1).

In these cases the stability of the hairpin correlates with the pause strength. Generally, mutations which disrupt or weaken the stem structure reduce the pausing strength, whereas compensatory mutations restore the pausing strength. Similar effects are observed if modified NTPs, which have a characteristically different tendency to form stable base pairs, are used as substrates for transcription. For instance, if inosine triphosphate (ITP) is substituted for GTP during the transcription reaction, stable GC base pairs in the transcript are replaced by the much weaker IC base pairs. This is because GC base pairs have three hydrogen bonds, whereas IC base pairs have only two. Likewise, substitution for CTP by the 5-bromo or 5-iodo derivative stabilizes base pair formation and thereby causes more stable hairpin structures (Fig. 5.2). Substitutions with these nucleotides within a potential RNA stem structure lead to enhanced pausing. From such studies it can be concluded that the secondary structure, not the primary sequence of the transcript, determines the pausing strength.

For some pauses that occur at hairpin structures it has been shown that base substitutions close to the 3′ end of the transcript, but outside the hairpin structure, also have an effect on pausing.

Box 4.1 RNA secondary structural elements

In contrast to the few conservative helical structures that dominate DNA, an amazing variety of secondary or tertiary structural elements determine the overall structure of RNA molecules. Single-stranded regions, double-stranded helical structures and many different secondary structural elements or higher

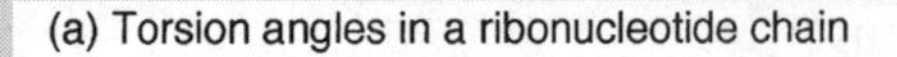

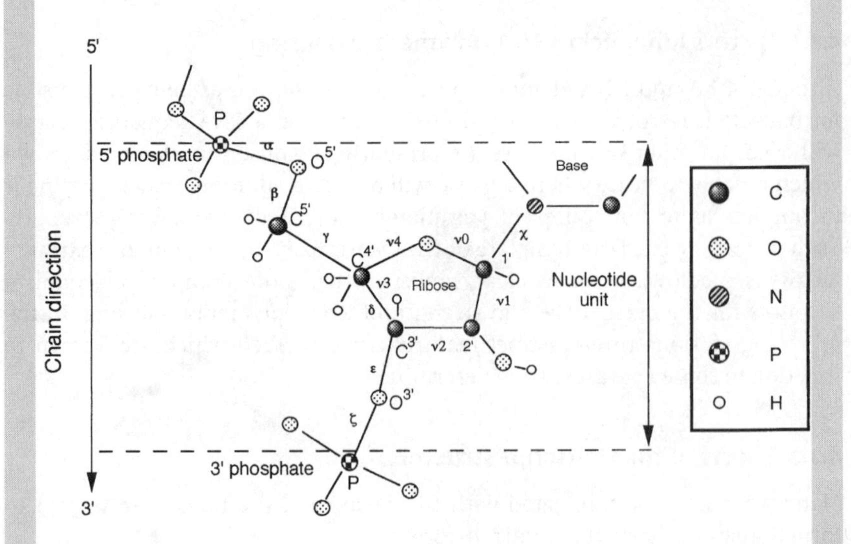

(b) Definition of torsion angles:

α	$_{(n-1)}O_{3'}$-P-$O_{5'}$-$C_{5'}$
β	P-$O_{5'}$-$C_{5'}$-$C_{4'}$
γ	$O_{5'}$-$C_{5'}$-$C_{4'}$-$C_{3'}$
δ	$C_{5'}$-$C_{4'}$-$C_{3'}$-$O_{3'}$
ε	$C_{4'}$-$C_{3'}$- $O_{3'}$-P
ζ	$C_{3'}$-$O_{3'}$-P-$O_{5'(n+1)}$
χ	O_4'-$C_{1'}$-N_1-C_2 (Pyrimidines); $O_{4'}$-$C_{1'}$-N_9-C_4 (Purines)
ν0	$C_{4'}$-$O_{4'}$-$C_{1'}$-$C_{2'}$
ν1	$O_{4'}$-$C_{1'}$-$C_{2'}$-$C_{3'}$
ν2	$C_{1'}$-$C_{2'}$-$C_{3'}$-$C_{4'}$
ν3	$C_{2'}$-$C_{3'}$-$C_{4'}$-$O_{4'}$
ν4	$C_{3'}$-$C_{4'}$-$O_{4'}$-$C_{1'}$

Fig B4.1a Torsion angles in a ribonucleotide chain

order structural arrangements are found in various combinations. The combination of single-strand and double-strand elements, for instance, causes various types of loops, such as hairpin loops, internal loops, bulge loops or multibranched loops (see Fig. B4.1c). The high structural diversity is a prerequisite for the fact that RNA molecules are able to accomplish catalytic functions and act as enzymes (ribozymes). It is remarkable that an apparently small molecular change, namely the substitution of a hydrogen atom at the 2′ position of deoxyribose with an OH group in ribose, causes such a fundamental structural difference. Rules to predict the higher order RNA structure are very complicated

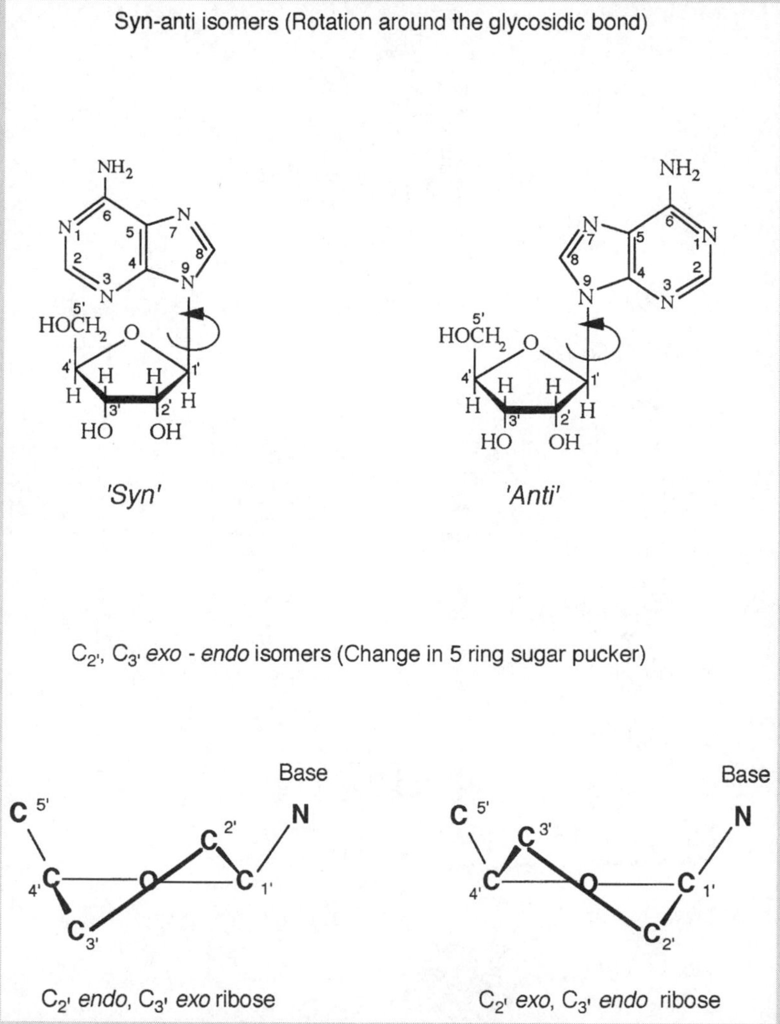

Syn-anti isomers (Rotation around the glycosidic bond)

'Syn' *'Anti'*

$C_{2'}$, $C_{3'}$ *exo - endo* isomers (Change in 5 ring sugar pucker)

$C_{2'}$ *endo*, $C_{3'}$ *exo* ribose

$C_{2'}$ *exo*, $C_{3'}$ *endo* ribose

Fig B4.1b Conformational isomers of nucleic acid building blocks and require more information to define single- or double-stranded regions on the basis of base pair complementarity. The higher-order or tertiary structure is, to a large extent, the result of an enormous conformational variability. This can best be explained by the rotational degree of freedom that specifies the backbone conformation of a single nucleotide within an RNA chain. As can be seen in Fig.

Schematic representation of secondary structural motifs within an arbitrary RNA molecule

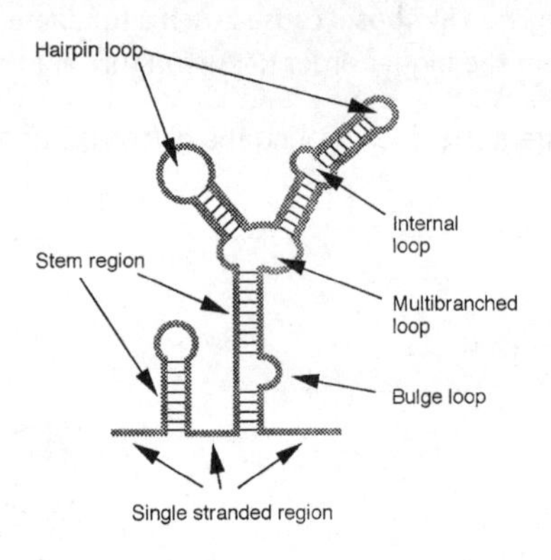

Special RNA structures

1) Tetra loops:

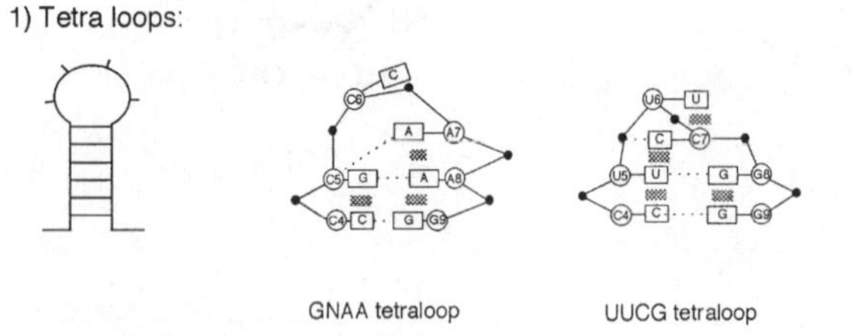

2) Pseudoknots:

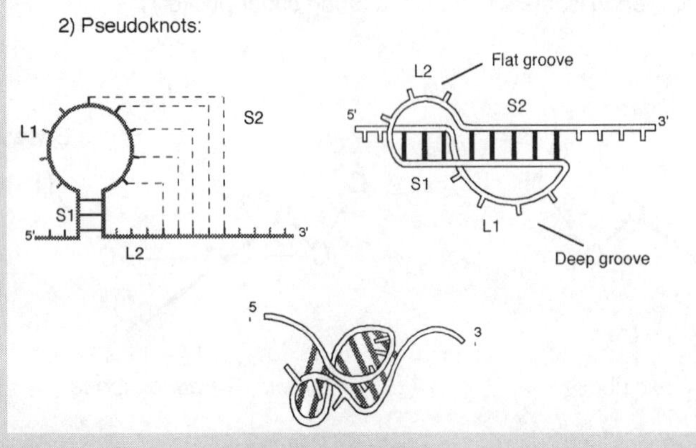

Fig B4.1c Structural elements of RNA

B4.1a, seven different torsional angles of the sugar–phosphate backbone and five torsional angles within the ribose ring can be defined. These give rise to a large conformational freedom (note, however, that the angular orientations are normally restricted to small changes only).

Typical isomers resulting from torsional changes are the ***syn*** and ***anti*** isomers, which result from rotations around the glycosidic bond (χ) These two nucleotide conformers are characterized by two preferred positions of the aromatic base pointing either away from the $C_{5'}$ position (*anti*) or, after rotation of 180° pointing in the $C_{5'}$ direction (*syn*). Torsional deflections within the five-membered ribose ring give rise to ***exo*** or ***endo*** conformers of the $C_{2'}$ and $C_{3'}$ atoms. If the C atom points above the plane of the ribose (in the direction of the glycosidic bond) it is defined as *endo*. In contrast, C atoms of the ribose ring located below the plane of the sugar are designated as *exo* (Fig. B4.1b).

Some recurring RNA secondary structural elements are defined below. Certain conserved sequences within four-membered hairpin loops are often found in functional RNA molecules. These loops are termed **tetraloops**, and they are characterized by a number of special structural features, such as unconventional hydrogen bonds and base-stacking interactions, which lend great stability to the structures. Depending on the sequence of the bases in the loop two different types of tetraloops are distinguished, the UUCG structures and the family of GNAA tetraloops. Another frequently occurring structure, actually a tertiary structural element, is the **pseudoknot**. A pseudoknot is formed when bases of a hairpin loop are paired with bases outside the hairpin stem. The resulting structure is stabilized by coaxial stacking of the resulting two helical structures (Fig. B4.1c).

A quasi-continuous A-form helix is formed by stem 1 (S1) and stem 2 (S2). The two loops (L1 and L2) cross the major (L1) or minor grooves (L2) of this contiguous helical structure.

4.3.3 Effects of DNA structure and topology on elongation

It has been shown in several cases that DNA sequences up to 17 base pairs downstream of the pause site are responsible for the pause. This may in part be explained by changes in the stability of the DNA structure. A higher GC content, for instance, increases the melting temperature of the DNA and can thus inhibit strand opening. As a consequence, the translocation rate of the elongation complex may be slowed down, resulting in a pause. However, increasing the GC content of the downstream sequence, which clearly affects DNA stability, does not completely correlate with the observed changes in the pausing strength. The exact cause is probably more complex and therefore very likely resides in the overall conformation of the RNA polymerase–DNA complex.

Specific sequences of the non-template strand can also affect pausing. This has been documented in at least one case. The early transcribed region of the phage λ late gene promoter ($P_{R'}$) is known to contain a strong pausing site at position +16. (This site is crucial for the Q protein-mediated **transcription antitermination**, see Section 5.4.3). Analysis employing a heteroduplex DNA with base changes at position +2 and +6, in the non-template strand only, has shown that the pause is eliminated. These results are in accordance with the earlier conclusion that the RNA polymerase interaction with the non-template strand within the transcription bubble has a direct effect on elongation.

In Chapter 6 it will be seen that **superhelical DNA** also has a strong influence on transcription. The superhelical status or topology also affects the pausing behaviour. *In vitro* studies have shown that the pause strength on linear or superhelical DNA varies greatly. It is known that the energy stored in superhelical turns of DNA can be converted into base pair opening (melting). A high degree of supercoiling should thus facilitate strand opening of the DNA. It might be expected, therefore, that supercoiling might reduce the pausing strength. Surprisingly, the reverse is the case. For a number of strong pauses in a ribosomal RNA operon a direct correlation between the pause strength and the negative **superhelical density** (see Box 6.4) can be demonstrated. It must be assumed, therefore, that base pair opening is not the limiting parameter for pausing in these cases. It seems plausible, however, that the superhelical DNA affects the conformation of the elongation complex, e.g. by different wrapping of the DNA around the polymerase or other conformational distortions of the RNA polymerase structure.

4.3.4 Effects caused by substrate NTPs, transcription factors or small ligands

Pauses can generally occur at high or low levels of substrate NTPs. Reducing the concentration of all four substrate NTPs certainly reduces the average step time of transcription but does not necessarily affect pausing in a specific way. Pauses are specifically affected, however, when the concentration of the next nucleotide to be incorporated at a pause site is selectively decreased. Moreover, the 3′ terminal nucleotide of a paused transcript seems to be of special importance. In many (but not all) pauses analysed, this nucleotide appears to be a pyrimidine (U or C) which is followed by a G. Evidently, phosphodiester formation is somewhat hampered when GTP has to be arranged for binding to a 3′ terminal U or C in the active centre of RNA polymerase.

As seen in Chapter 5, low concentrations of UTP in the cell function as a regulatory signal for the transcription of the pyrimidine biosynthesis operon *pyrBI*. Transcription regulation of *pyrBI* involves a pausing/termination site in the early transcribed region of this operon.

Some pauses are specifically affected by proteins (transcription elongation factors). One of the most important of these factors is **NusA**. The NusA protein, which has a molecular weight of 55 kDa, is an essential protein in *E. coli*. NusA is known to be involved in *slowing down* transcription elongation at several operons. It plays a role, therefore, in synchronizing the rates of transcription and translation. Clearly, NusA affects pausing (and termination, see Chapter 5). Many studies suggest that the pauses affected by NusA all belong to one class in which the RNA hairpin plays an essential role. This is in line with findings that NusA is able to form complexes with specific RNA sequences. Furthermore, cross-linking studies with paused complexes have shown that NusA interacts directly with the nascent transcript in the exit domain of the elongation complex. In addition, NusA is able to bind the core enzyme of RNA polymerase, but not to the holoenzyme. It is assumed, therefore, that NusA can replace the σ factor during elongation, and very probably occupies the same binding site at the RNA polymerase as the σ subunit.

Apart from its role in regulating the efficiency or duration of pauses, NusA is a pleiotropic transcription factor with different functions. Its participation in termination and antitermination mechanisms is described in more detail in Chapter 5.

A second transcription elongation factor which modulates pauses is **NusG**. The NusG protein of *E. coli* has a molecular weight of 21 kDa. In contrast to NusA, NusG *enhances* the rate of transcription and is able to reduce pausing *in vitro* and *in vivo*. NusG has also been shown to be able to form complexes with RNA polymerase. It is assumed that NusG helps to restructure the elongation complex from a pausing-proficient to an elongation-proficient conformation. Like NusA, NusG also functions in termination and antitermination (see Chapter 5).

As described above (Section 4.2), the small effector nucleotide guanosine tetraphosphate (ppGpp) affects the transcription elongation rate. This is partly due to a small reduction in the average step time of the translocation cycle (transcription rate). In addition, ppGpp has a profound effect at specific pauses. *In vitro* studies have shown that ppGpp does not affect all pauses in the same way. Some pauses are considerably enhanced, others are reduced, and some pauses are unchanged by the presence of ppGpp. Furthermore, the effect of ppGpp depends on the particular promoter and the early transcribed region, which RNA polymerase has passed before a pausing site is encountered. The following conclusions can be drawn from these findings. Obviously, the transcription start sites, including the early transcribed sequences of some promoters, are able to fix a conformation of the elongation complex that is refractory to the action of ppGpp. The fact that different pauses react differently in the presence of ppGpp supports the view that different mechanisms are responsible for these pauses. The general importance of pausing in transcription regulation has been recognized previously. Since such studies are still in their infancy much more information can be expected to be discovered in the near future.

4.4 Transcript cleavage factors

Transcription elongation complexes can be stalled either by deprivation of NTPs or permanently paused, for instance, by the stable association of a DNA-binding protein immediately downstream of the elongation complex. Such artificially halted complexes can gradually undergo transitions to arrested complexes which are unable to resume elongation. Two proteins are known which can facilitate the reactivation of arrested complexes at a specific location. The two proteins are **GreA** and **GreB**. Both proteins are 158 amino acids long (19 kDA) and have a similar primary structure (35% amino acid identity). A more detailed study with the GreA protein shows that it consists of two domains, of which the N-terminal domain (NTD) possesses the activity necessary for transcript cleavage, whereas the C-terminal domain (CTD) increases this activity, but does not contain any nucleolytic activity by itself. The CTD is instead responsible for binding GreA to RNA polymerase. Both GreA and GreB are able to induce a cleavage reaction close to the 3′ end of the transcript in an arrested ternary complex. The reaction is not the enzymatic reversal of the synthesis (pyrophosphorolysis) but cleavage occurs endonucleolytically, creating two fragments. The 5′ fragment containing a free 3′ OH-group remains bound to the ternary complex while the 3′ fragment with a 5′ terminal phosphate is released. It is assumed that the reaction is catalysed at the RNA polymerase active centre that also catalyses synthesis and pyrophosphorolysis. The reaction does not depend on particular promoters or genes and is independent of the position within the template DNA. According to their amino acid structure GreA and GreB do not have any similarity with known nucleases, and experiments to detect an independent nuclease activity have failed. This has led to the conclusion that the cleavage activity resides in the RNA polymerase and is only activated or stimulated by the two proteins. In fact, cleavage reactions within arrested complexes have been demonstrated to occur with RNA polymerases isolated from *greA*/*greB* mutants.

The cleavage reactions induced by GreA and GreB have two important implications for transcription elongation. First, transcriptional arrest, if it occurs in the cell, can be overcome and the formation of **dead end complexes** can be prevented or rescued. This can be achieved in the following way: the Gre factor-dependent cleavage is preceded by a backward translocation of RNA polymerase on the DNA template. With the release of the cleaved 3′ fragment a new 3′ terminus is created, presenting a new 3′-OH terminus of the transcript, which is still bound to the catalytic centre of RNA polymerase. The arrested complex can thus resume elongation from a new RNA 3′ end. This reaction can in principle be repeated several times until a productive forward transcription is restored.

There is a second implication ascribed to the Gre-factor-mediated cleavage reaction. The cleavage reaction is considered to be part of a mechanism to decrease the error frequency during elongation, and thus to play a role in proofreading of transcription. Misincorporation of nucleotides during transcription elongation can occur at frequencies of about 1 per 10^3 to 10^5 residues (see Section 4.4.1). GreA or GreB can prevent or remove such errors, respectively. Error correction may occur by removal of the 3′ transcript ends of a complex that is arrested because of nucleotide misincorporation. Alternatively, the wrong nucleotide may be removed before the complex changes to an arrested state.

GreA and GreB do not function in exactly the same way. There is a clear difference in their mode of cleavage. GreA usually creates only small cleavage fragments (dinucleotides or trinucleotides), whereas GreB produces longer oligonucleotides, up to a length of nine nucleotides. It has been concluded, therefore, that GreA cleavage occurs close to the catalytic site, while cleavage induced by GreB probably occurs at a site located more 5′ to the end of the growing RNA chain. The two proteins deviate in their activity in a second respect. Arrested complexes can only be reactivated for elongation by GreA when the protein is present before the site of arrest is reached. This is not the case for GreB. GreB can reactivate arrested elongation complexes even when transition of the elongating complex to an arrested state occurred before GreB was present.

Transcript cleavage factors are not specific for prokaryotic transcription elongation. Proteins with similar activity are also known to act during the transcription reactions of higher organisms. For example, the eukaryotic transcription elongation factor TFIIS (or SII) has been shown to induce very similar transcript cleavages as GreA or GreB at RNA polymerase II elongation complexes.

4.4.1 Are there proofreading steps during transcription elongation?

It is clear that, during the flow of genetic information in a cell, only a very small number of errors can be tolerated. The synthesis steps leading to the macromolecular compounds in the cell must therefore be of high **fidelity**. The fidelity or precision of a reaction is commonly defined as the inverse of its **error frequency**, which indicates the number of mistakes per correct synthesis step. It is known, for instance, that the error frequency of DNA replication is very low, with less than one erroneous nucleotide incorporation per 10^{10} synthesis steps. This low error frequency is essential for genetic stability, since DNA represents the permanent form of genetic information which should be faithfully transferred from one generation to the next. The observed high fidelity of DNA replication is mainly due to an efficient **proofreading** mechanism. This proofreading involves a 3′ to 5′ exonuclease activity of the DNA polymerases, which enables the efficient removal of misincorporated nucleotides from the synthesized DNA strands.

RNA polymerases generally lack a 3′ to 5′ exonuclease activity. A similar proofreading mechanism, as observed in DNA replication, is therefore questionable.

In addition, it was believed that transcription, in contrast to replication, does not necessarily need to proceed at the same high fidelity. Errors in transcription should not cause a *permanent* genetic change. It is clear, on the other hand, that a single error in a primary transcript can completely destroy the function of the resulting RNA. If one considers that a single mRNA molecule is translated about 40 times, it follows that an erroneous mRNA could rapidly lead to the accumulation of detrimental products in the cell. In fact, the observed error frequency of transcription elongation is significantly lower (less errors are made) than might be expected from the thermodynamic energy difference of the correct *versus* the incorrect nucleotide addition steps according to the stability of Watson–Crick base pairs. In principle, the error frequency is determined by the thermodynamic difference in stability of a correct or incorrect bound NTP to the catalytic centre of the RNA polymerase. Incorrect NTPs are bound less stably to the active site than correct NTPs. Thus, in the absence of proofreading, the maximum free energy difference (ΔG_{max}) of binding a correct *versus* an incorrect NTP at the elongation site determines the upper limit of accuracy. The free energy differences between the equilibrium binding of correct or incorrect NTPs can be determined based on the known hydrogen bonding energies of base pair formation. These measure about 1–3 kcal, corresponding to an expected maximal error frequency of 10^{-2} to 10^{-3}. The actual error frequencies observed for transcription elongation vary substantially from this number and are estimated to be in the range of 10^{-3} to 10^{-5}. Moreover, the overall error frequency in protein synthesis, measured as amino acid substitutions, is estimated to be between 10 and 40×10^{-5}. This error combines mistakes made during transcription and translation. The error frequency of transcription alone should therefore be smaller. Thus, error frequencies of 10^{-5}, as estimated above, seem to be plausible. It is likely, therefore, that the accuracy of RNA transcription is increased by one or more types of proofreading steps which act during elongation.

Accuracy can in principle be improved by mechanisms which involve **error prevention** or mechanisms of **error correction**. Observations supporting both types of mechanisms during transcription elongation have been presented.

First, it was found that RNA polymerase contains an NTPase activity. This activity converts NTPs to the corresponding nucleoside diphosphates (NDPs) which can no longer be used as substrates for RNA polymerase. This NTPase reaction was shown to be template-dependent. This means that significant NDP synthesis could only be observed for NTPs for which no complementary bases were present in the template. RNA polymerase mutants have been isolated which were defecient in this NTPase activity. The corresponding mutations mapped between amino acid positions 565 and 576 of the conserved region D of the β subunit (see Section 2.2.2). The mutant RNA polymerases had a clearly reduced accuracy during elongation, indicative of the involvement of the NTPase reaction in proofreading. The accuracy of transcription can be improved by conversion of the incorrectly bound NTPs to the non-substrate

NDPs, before the incorrect nucleotide is enzymatically added to the growing RNA chain. The resulting NDP is rapidly released from the active centre of RNA polymerase, enabling binding of the correct NTP in a subsequent reaction cycle. The generated NDP is probably quickly recycled to the corresponding NTP. In this way mistakes can be avoided before the nucleotide addition step becomes irreversible. Increase of the transcriptional fidelity through such a mechanism is termed *error prevention*. A branched kinetic model with the irreversible hydrolysis of incorrect (non-cognate) NTPs to NDPs as an error-preventing proofreading step is depicted in Fig. 4.3

If an incorrect NTP is occasionally incorporated into the RNA chain the error can be corrected by hydrolytic cleavage or the reversal of the polymerization reaction. Error correction by reversal of the nucleotide addition step is termed **pyrophosphorolysis**. In fact, it has been described that, at high concentrations of pyrophosphate (PPi), a template-dependent reversal of nucleotides can be observed. An effect on accuracy, similar to the 3′ to 5′ exonucleolytic proofreading of DNA polymerases, is thus feasible. The rate of removal of the 3′ terminal nucleotide from the nascent RNA by pyrophosphorolysis may be equally efficient for correct and incorrect nucleotides. Nevertheless, accuracy may be improved by this step. It is known that a mismatch between the 3′ terminal nucleotide and the DNA template will disturb the next copying step

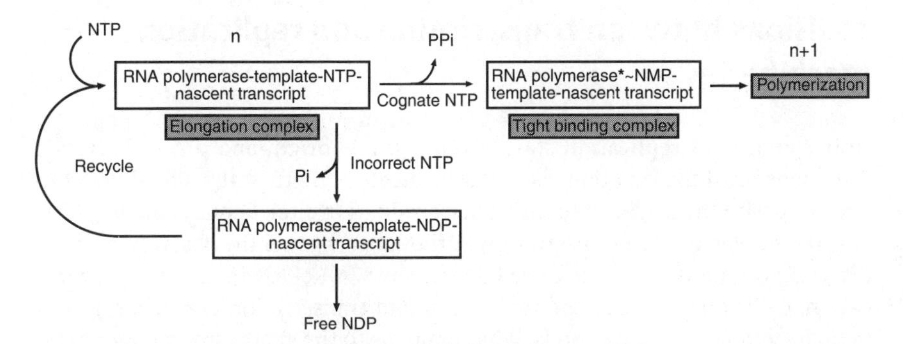

Figure 4.3 Error prevention proofreading steps in RNA transcription. Scheme explaining increased transcriptional fidelity through error prevention mediated by the NTPase activity of RNA polymerase. RNA polymerase ternary elongating complexes at template position n (elongation complex) bind NTP. If the incoming NTP is correct (cognate) the elongation complex undergoes a transition to a tight binding conformation indicated by an asterisk. The bound nucleotide is incorporated into the growing RNA chain and pyrophosphate (PPi) is released. The elongation complex moves to template position n+1 (polymerization). If an incorrect (non-cognate) NTP is bound to the catalytic site of the elongation complex it is hydrolysed by the inherent NTPase activity to the corresponding NDP and orthophosphate (Pi). The NDP is not a substrate for polymerization and is released from the complex and probably rephosphorylated during subsequent steps in the cell. The ternary complex, still at position n, is ready for a new cycle of NTP binding. The figure is adapted according to Libby and Gallant (1991).

and cause the elongation complex to pause. Additional time is thus provided for the *error correction step*. Removal of any incorrect nucleotide will thus exceed a similar reaction at a site of correct incorporation.

The transcript cleavage reactions induced by GreA and GreB have led to a new proposal for a proofreading mechanism consistent with the observed fidelity of transcription. According to this mechanism the transcription complex switches between an active and an inactive conformation. The inactive conformation is induced when the elongation rate is slowed down. This is the case, for instance, in the absence of correct NTPs or when an incorrect nucleotide has been incorporated at the 3′ end of the growing transcript. The inactive conformation supports GreA- and/or GreB-induced cleavage of the transcript containing the misincorporated nucleotide. The catalytic centre of the elongation complex is shifted backward and the active state of the complex is restored. Transcription elongation can now be resumed from the new 3′ end. The process explains the observed fidelity of transcription and provides a mechanism as to how the formation of inactive or dead end complexes can be avoided in the cell.

4.5 'Traffic problems' at the bacterial genome: collisions between transcription and replication machineries

In bacterial DNA replication starts from a single origin and proceeds in two directions until the two replication forks meet each other at a site almost directly opposite the origin of replication (termination site). Replication is at least 15–20 times faster than transcription. Although it is generally observed that the direction of intensively transcribed operons is parallel to the direction of replication, collisions between the replication and transcription apparatus appear to be inevitable. The question is, what happens to the synthesizing complexes? Are they decomposed by a collision? Are the premature products released? It should be clear that incomplete or prematurely terminated DNA strands or transcripts might be harmful to the cell. At the least, such products would represent an extreme energetic waste.

From early electron microscopic studies it was concluded that, during a transcription–replication encounter, replication may either cause displacement or stall in front of an elongating transcription complex. In a recent series of elegant *in vitro* studies employing the bacteriophage T4 replication system and elongating *E. coli* RNA polymerase, surprising results were obtained. Active T4 replication complexes were assembled on the same DNA where transcription elongation complexes were either steadily transcribing or stalled by the omission of one of the four NTPs. The fate of the transcription complexes and the replication forks was analysed during collision. During codirectional movements, where both polymerases use the same DNA strand as template, the replication fork passes the transcription complex after a brief pause. Surprisingly, the ternary transcription complex remains bound to the DNA. Moreover, the transcription bubble is still in its original site after the collision. The transcript is not released and the fully active complex resumes elongation after the replication fork has passed. The same results were obtained when the transcription complex was actively transcribing and not artificially stalled. The results strongly support models of elongation complexes with at least two independent DNA-binding sites and a separate RNA-binding site. In this way it is feasible that, during invasion of the replication fork into the transcription complex, one domain of RNA polymerase can always be kept attached to the DNA while the replication complex passes the transcription complex.

Similar experiments were conducted to determine what happens when transcription and replication complexes move towards each other, provoking a head-on collision. Similar to that observed in the previous codirectional case, the collision did not cause destruction of either of the two complexes. A most surprising finding from the experiment was that the RNA polymerase switched from the original template strand to the newly synthesized daughter strand after the collision. Elongation of the transcript was resumed from there at the same sequence position at which the RNA polymerase had left the original template. It is assumed that the correct position of the transcript end with respect to the template is guided by complementary base pairing between the 3′ end of the bound RNA and the new DNA daughter strand. The transcript itself must be fixed at the catalytic site by RNA polymerase interactions which exist during all collisions.

These findings enable a new insight into the dynamics of macromolecular interactions in the cell, and indicate that elaborate mechanisms have evolved to minimize damage and destruction, and also to prevent wasteful reactions during the intimately coupled processes of transcription and replication.

Summary

An essential feature of transcription elongation complexes is their *high processivity* and stability, which enables transcription of long sequences without interruption. The processive character of the elongation complex is probably caused by a ring-like closure of the RNA polymerase structure around the DNA. Elongating complexes are further characterized by multiple sites of interaction between the RNA polymerase, the DNA, and the growing transcript. Two *independent DNA-binding sites* are present at the front and the rear edge of RNA polymerase. Moreover, binding of the non-template strand to the RNA polymerase stabilizes the transcription bubble. The growing transcript is fixed to the active centre by base pair formation with DNA bases of the template strand. The length of this heteroduplex may range from three to eight base pairs. The growing RNA chain is additionally bound to the surface of the RNA polymerase at an *RNA exit site*. The catalytic centre, where incoming nucleotides are bound and linked to the 3′ end of the growing RNA chain, does not have a fixed position relative to the borders of the elongation complex. Instead, it can move within the transcription complex several base pairs upstream or downstream along the template strand. The transcription elongation complex does not move monotonously along the template but translocates in an *inchworm-like* fashion. This is supported by periodic binding of the upstream and downstream RNA polymerase edges and the accompanying conformational changes of the enzyme. The size of DNA protection by RNA polymerase during elongation thus varies between 25 and 40 base pairs.

The energy for the translocation of the elongating complex is provided by the phosphohydrolysis of incorporated NTPs. The applied force during a translocation cycle has been shown to be several-fold higher than for standard motor proteins, such as myosin or kinesin. The average transcription elongation rate of bacterial RNA polymerases *in vivo* is in the range of 50 nucleotides per second. This rate closely matches the rate at which ribosomes translate along an mRNA sequence. Transcription rates are discontinuous, however, and RNA polymerases often pause at certain template positions for several seconds or even minutes. Such *RNA polymerase pausing* can be affected by the transcript secondary structure, the DNA conformation and topology, the nucleotide concentration, the effector molecule guanosine tetraphosphate, or by transcription factors. RNA polymerase pausing is an important mechanism for transcription regulation. It is not restricted to prokaryotic transcription but can be observed for all RNA polymerases known. Frequently paused complexes can undergo transition to *arrested or dead end complexes*, unable to resume transcription. Such complexes can be rescued by the action of the *transcript cleavage factors* GreA and GreB, which remove a short RNA fragment from the 3′ end of the arrested transcript, such that the catalytic site of RNA polymerase is transferred to a position from which it can resume transcription. The two proteins are furthermore considered to be *proofreading* factors, enhancing the fidelity of transcription by the cleavage of erroneous transcript 3′ ends. In addition, an intrinsic NTPase activity of bacterial RNA polymerase has been shown to act as a proofreading tool.

During the life cycle of bacteria, encounters between transcription and replication complexes appear to be inevitable. Surprisingly, the two reactions do not cause mutual destruction of the respective complexes. During a codirectional encounter the faster replication complex passes the transcription complex, which resumes transcription after a brief pause without release of the transcript. The two complexes remain intact even after a head-on collision. In this case the RNA polymerase complex switches transcription from the parent template strand to the newly synthesized daughter strand. The transcript is not released during the collision. This mechanism can be explained by the existence of multiple independent DNA- and RNA-binding sites within the transcription complex.

References

Libby, R. T. and Gilbert, J. A. (1991) The role of RNA polymerase in transcriptional fidelity. *Molecular Microbiology* **5**: 999–1004.

Theissen, G., Pardon, B., and Wagner, R. (1990). A quantitative assessment for transcriptional pausing of DNA-dependent RNA polymerase *in vitro*. *Analytical Biochemistry*, **189**: 254–61.

Uptain, S. M., Kane, C. M. and Chamberlin, M. J. (1997) Basic mechanisms of transcription elongation and its regulation. *Annual reviews of Biochemistry* **66**: 117–72.

Further reading

Erie, D. A., Yager, T. D. and von Hippel, P. H. (1992) The single-nucleotide addition cycle in transcription: a biophysical and biochemical perspective. *Annual Reviews in Biophysics and Biomolecular Structure* **21**: 379–415.

Gelles, J. and Landick, R. (1998) RNA polymerase as a molecular motor. *Cell* **93**: 13–16.

Guajardo, R. and Sousa, R. (1997) A model for the mechanism of polymerase translocation. *Journal of Molecular Biology* **265**: 8–19.

Heumann, H., Zaychikov, E., Denissova, L. and Hermann, T. (1997) Translocation of DNA-dependent *E. coli* RNA polymerase during RNA synthesis. *Nucleic Acids and Molecular Biology*, In: Eckstein, F. and Lilley, D. M. J. (eds.) Vol. 11 Berlin–Heidelberg, Springer Verlag, pp. 151–177.

Kabata, H., Kurosawa, O., Arai, I., Washizu, M. Margarson, S. A., Glass, R. E. and Shimamoto, N. (1993) Visualization of single molecules of RNA polymerase sliding along DNA. *Science* **262**: 1561–3.

Kassavetis, G. A. and Geiduschek, P. E. (1993) RNA polymerase marching backward. *Science* **259**: 944–5.

Libby, R. T. and Gallant, J. A. (1991) The role of RNA polymerase in transcriptional fidelity. *Molecular Microbiology* **5**: 999–1004.

Libby, R. T. and Gallant, J. A. (1994) Phosphorolytic error correction during transcription. *Molecular Microbiology* **12**: 121–9.

Metzger, W., Schickor, P. and Heumann, H. (1989) A cinematographic view of *Escherichia coli* RNA polymerase translocation. *EMBO Journal* **8**: 2745–54.

Mooney, R. A., Artsimovitch, I. and Landick, R. (1998) Information processing by RNA polymerase: recognition of regulatory signals during RNA chain elongation. *Journal of Bacteriology* **180**: 3265–75.

Nudler, E., Goldfarb, A. and Kashlev, M. (1994) Discontinuous mechanism of transcription elongation. *Science* **265**: 793–6.

Nudler, E., Avetissova, E., Markovtsov, V. and Goldfarb, A. (1996) Transcription processivity: protein–DNA interactions holding together the elongation complex. *Science* **273**: 211–17.

Nudler, E., Mustaev, A., Lukhtanov, E. and Goldfarb, A. (1997) The RNA–DNA hybrid maintains the register of transcription by preventing backtracking of RNA polymerase. *Cell* **89**: 33–41.

Nudler, E., Gusarov, I., Avetissova, E., Kozlov, M. and Goldfarb, A. (1998) Spatial organization of transcription elongation complex in *Escherichia coli*. *Science* **281**: 424–8.

Record, M. T., Reznikoff, W. S., Craig, M. L., McQuade, K. L. and Schlax, P. J. (1996). *Escherichia coli* RNA polymerase (Eσ^{70}), promoters, and the *kinetics of the steps of transcription initiation*. In: Neidhard, F. C., Curtiss III, R., Ingraham, J. L., *et al.* (eds.) *Escherichia coli and Salmonella Cellular and Molecular Biology*. Vol. 1. Washington DC: ASM Press, pp. 792–821.

Reines, D. (1994) Nascent RNA cleavage by transcription elongation complexes. In: Conaway, R. C. and Conaway, J. W. (eds.) *Transcription: Mechanisms and Regulation.* New York: Raven Press; pp. 263–78.

Richardson, J. P. and Greenblatt, J. (1996) Control of RNA chain elongation and termination. *Escherichia coli and Salmonella Cellular and Molecular Biology*, Vol. 1. In: Neidhard, F. C., Curtiss III, R., Ingraham, J. L. *et al.*, (eds.) Washington DC: ASM Press, pp. 822–48.

Uptain, S. M., Kane, C. M. and Chamberlin, M. J. (1997) Basic mechanisms of transcript elongation and its regulation. *Annual Reviews of Biochemistry* **66**: 117–72.

von Hippel, P. H. (1998) An integrated model of the transcription complex in elongation, termination, and editing. *Science* **281**: 660–5.

Wang, H.-Y., Elston, T., Mogilner, A. and Oster, G. (1998) Force generation in RNA polymerase. *Biophysical Journal* **74**: 1186–202.

Yager, T. D. and von Hippel, P. H. (1991) A thermodynamic analysis of RNA transcript elongation and termination in *E. coli*. *Biochemistry* **30**: 1097–118.

Yin, H., Landick, R. and Gelles, J. (1994) Tethered particle motion method for studying transcript elongation by a single RNA polymerase molecule. *Biophysical Journal* **67**: 2468–78.

Yin, H., Wang, M. D., Svoboda, K., Landick, R., Block, S. M. and Gelles, J. (1995) Transcription against an applied force. *Science* **270**: 1653–7.

5

Termination of transcription

This chapter attempts to shed light on the processes that terminate the transcription cycle. It starts with a description of the mechanism of intrinsic termination, triggered by a specific transcript secondary structure. The general effect of the intrinsic terminator structures on stability is subsequently explained. There follows a section on the structure and function of the major bacterial transcription termination factor Rho, and its role in the phenomenon of transcription polarity. A section on the importance of the auxiliary termination factor NusG, and alternative termination factors is provided. Alternative functions of Rho in stabilizing mRNA structures are also reported. In the subsequent sections the two important mechanisms of attenuation and antitermination are discussed in some detail. Examples are provided for antitermination in the regulation of phage λ gene expression, but also for the transcription of rRNA operons. At the end of the chapter some representative examples of transcription regulation by alternative antitermination mechanisms independent from *nut* site sequences are reported.

Because of the high stability and processivity of transcription elongation complexes special mechanisms to stop transcription at the end of an operon or gene are vital for the cell. Consequently, specific termination systems act to release the transcript and RNA polymerase from the template. Proper termination is essential for bacteria as the regions between transcription units are generally rather small. However, termination does not only determine the end of a transcription cycle. Changing the site or efficiency of termination often means an important regulatory step. As seen below, the expression of many genes is precisely regulated by the control of transcription termination.

5.1 Factor-independent termination

Termination can occur either spontaneously or involve the action of cellular factors (proteins). Factor-independent termination is first discussed.

Comparison of many different sites where spontaneous termination occurs has revealed a set of specific sequence elements which characterize factor-independent or *intrinsic termination* sites (Fig. 5.1).

Factor-independent terminators can be characterized by the following sequence elements:

1. They contain a region of dyad symmetry.
2. This region is located about 30 base pairs upstream of the termination point.
3. The transcribed sequence of this region has the propensity to fold in a GC-rich stem-loop structure. The average stability of such a stem-loop structure is in the range of –20 to –30 kcal/mol. The signals for the formation of a stable hairpin structure are similar to the sequences responsible for the secondary structures of the transcripts found at many pausing sites.
4. A run of four to eight A residues in the template strand is found approximately eight to 10 nucleotides downstream from the centre of the dyad symmetry. This A cluster in the template strand is transcribed into a stretch of uracils in the growing RNA chain.

It was not immediately clear when the common characteristics of the terminator sequence were first recognized whether the sequence elements noted for many terminator sites were determinants that act at the level of DNA or, after transcription, at the level of RNA. This question was solved by base substitution experiments in both the DNA and the RNA. It was shown that base change mutations stabilizing the potential hairpin structure of the transcript enhanced the termination efficiency while destabilizing mutations had the reverse effect. Base changes were only effective when positioned in the template strand. This underlines the fact that the structure of the transcript, and not the structure of the DNA template, is important for termination. Finally, replacement of nucleotide analogues, which either destabilize RNA secondary structures (inosine instead of guanosine) or stabilize it (5-fluorocytidine or 5-iodocytidine instead of cytidine), clearly demonstrated that the stem-loop structure formed by the transcript is responsible for termination activity (Fig. 5.2).

The hairpin structure preceding the termination site causes transcribing RNA polymerase to pause. The stability of the hairpin correlates with the observed pausing strength (see Section 4.3) and simultaneously with the *termination efficiency*. However, the spatial distance between the stem-loop structure and the 3′ boundary of the termination structure determines the *site of termination* and transcript release.

Although important for the determination of the site and efficiency of termination, the RNA secondary structure is not the only signal-affecting intrinsic termination. Sequences up to 20 base pairs downstream of the termination site, which have not yet been transcribed, can considerably affect termination. This can be explained by different energy requirements for unwinding and/or differences in the stability of binding of the front edge of RNA polymerase to the

(a)

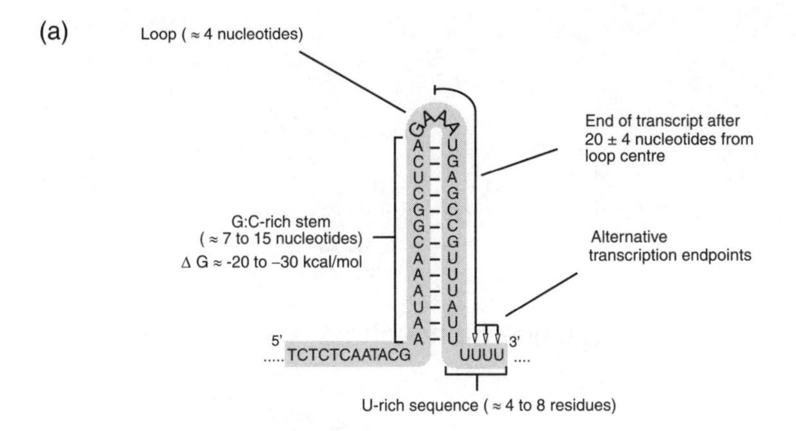

(b)

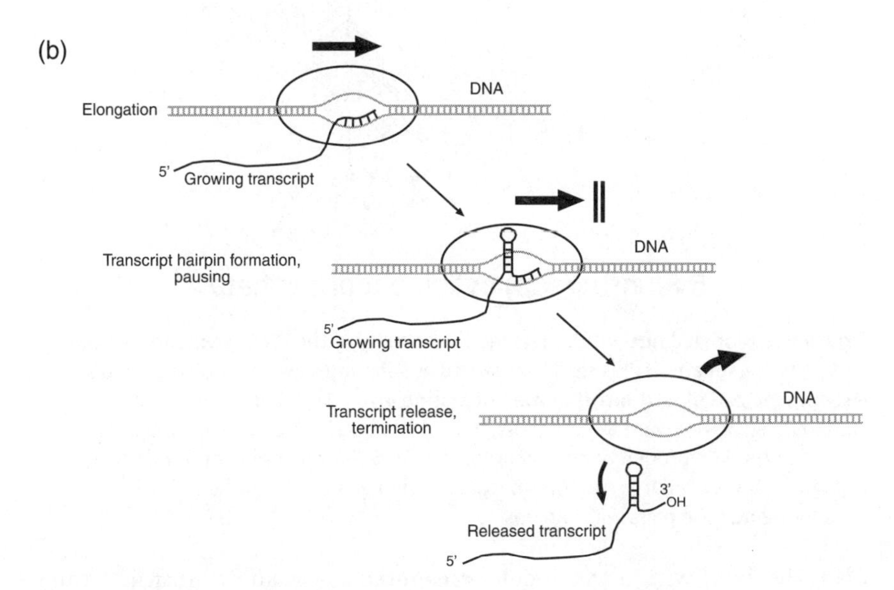

Figure 5.1 Structure of factor-independent terminators. (a) The Rho-independent transcription termination structure of the *E. coli* S10 operon is presented as an example of factor-independent terminators. A typical hairpin structure is formed by the transcript. ΔG of approximately –20 to –3 kcal/mol gives the average thermodynamic stability for typical terminator hairpin structures. Termination occurs at the positions indicated by open arrows. (b) The mechanism of factor-independent termination is schematically depicted. Elongating RNA polymerase moves along the template DNA. Within the transcription bubble the growing transcript and the single-stranded DNA of the template strand form a heteroduplex. After transcription through a terminator site a typical hairpin structure is formed and RNA polymerase pauses (indicated by the double vertical lines). Weak interactions between the uridine residues at the 3′ end of the growing RNA and the template DNA cause release of the transcript. RNA polymerase falls off the template.

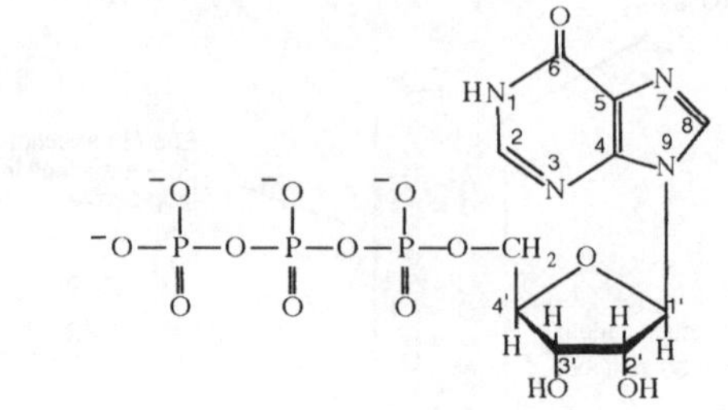

Inosine triphosphate

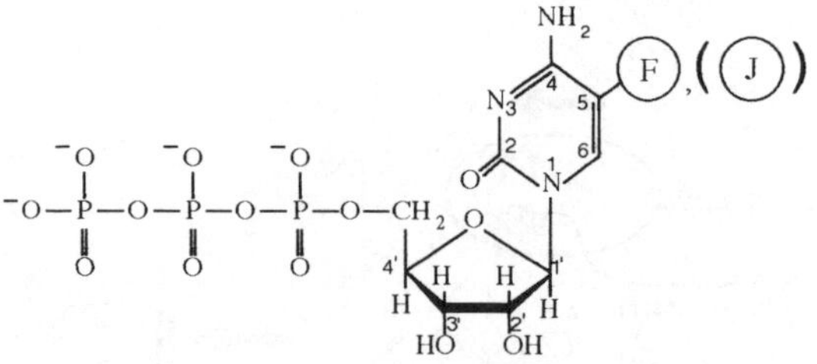

5 Fluoro (iodo) cytidine triphosphate

Figure 5.2 Modified nucleotides change the stability of the RNA secondary structures. Inosine triphosphate (ITP) is readily accepted as substrate by many polymerizing enzymes instead of GTP. Incorporation of inosine instead of guanosine destabilizes secondary structure since IC base pairs, in contrast to GC base pairs, have only two instead of three hydrogen bonds. Incorporation of 5-fluorocytidine or 5-iodocytidine stabilizes RNA secondary structures because of higher stability of the corresponding base pairs with guanosine.

DNA. The exact way in which downstream sequences affect intrinsic termination is unclear, however, and no consensus elements can be drawn from the collection of known intrinsic termination sites. These findings have led to the conclusion that the stability and/or the conformation of the DNA flanking the transcript hairpin region can also affect termination.

Termination efficiency can furthermore be influenced by sequences close to the promoter region. This striking effect has been described for combinations of several factor-independent termination sites fused to different promoters. In each case the efficiency of termination was dependent on the passage of the RNA polymerase through a specific sequence of the early transcribed region, comprising about 25–30 nucleotides distal to the promoter site. The efficiency of termination was affected several hundred nucleotides downstream. This finding cannot easily be reconciled with the assumption that a terminator cassette at the end of a transcription unit suffices to define the properties of termination. The results rather point to the conclusion that transcription complexes may undergo a conformational transition, leading to different activities with respect to the decision between elongation, pausing or termination. Such transitions may be triggered by the readthrough of specific sequences and may persist over a large distance and a significant period of time. Because the above observations were made with purified components *in vitro* the change in termination efficiency must reside in the structure of the elongating complex alone. No additional cellular factor(s) seem to be required for the transition between termination-deficient or -proficient transcription elongation complexes.

A similar phenomenon of a long range influence ('chemical memory') of the transcription complex is known to occur during eukaryotic RNA polymerase II transcription. This enzyme transcribes mRNAs and small nuclear RNAs (snRNAs). The structures at the 3′ ends of the two types of transcript are identical. Surprisingly, the transcripts initiated at the snRNA promoters are processed at a specific signal at the 3′ ends, while transcripts from mRNA promoters are not processed, although the same downstream signal is present. Obviously transcription initiation at the snRNA promoters causes a change in the RNA polymerase complex, which enables the processing enzymes to become effective or not.

While the way in which terminator sequences trigger RNA polymerase conformations or activities to respond to termination signals is not completely understood, the responsible hairpin structures of potential intrinsic terminator sites can be readily localized and identified from large sequence files. The sequences are screened for the specific RNA secondary structures. Several computer-assisted programs are available for the prediction of factor-independent termination sites. Thus intrinsic bacterial terminators can be searched for, or predicted, from the sequence data of transcriptionally uncharacterized DNA in a fairly reliable way.

Since the signals for many pausing sites have similar features to intrinsic termination sites, what distinguishes a pause site from a termination site? It has already been emphasized that all termination events are preceded by a pause or a reduction in the elongation rate. On the other hand, a reduction of this rate or pausing does not always cause termination. Two structural features of the intrinsic termination sites are different compared to the hairpin-type of pauses. First, the distance of the centre of the hairpin to the active centre of RNA polymerase, where the transcript is either released or elongation is resumed, differs for pauses or termination sites. This distance appears to be about eight nucleotides for terminators and about 11 nucleotides for pausing sites. Second, the stem-loop structures of terminators end with a run of U residues. This cluster of uracils seems to be of prime importance for the release of the transcript. After several of the U nucleotides have been incorporated into the growing transcript they form a heteroduplex with the adenosines of the template strand. This RNA:DNA hybrid, as does any RNA:DNA hybrid, adopts a helical structure similar to the DNA A-form (see Box 6. 1 on DNA structure). It has been shown by physicochemical measurements, however, that clusters of deoxyadenosines do not fit the DNA A-form very well. Runs of adenosines within a helical DNA A-form are thermodynamically unfavourable. The resulting helical oligo(riboU:deoxyriboA)heteroduplex is therefore exceptionally unstable. This destabilization of the RNA:DNA hybrid greatly facilitates the release of the transcript.

A model for the decision between elongation and termination at intrinsic terminators has been suggested. It is based on the thermodynamic and kinetic competition between the elongation and termination pathways. In addition, it relies on an extended RNA:DNA hybrid within the transcription complex, which is not universally accepted (see Section 4.1). According to this model there is a fixed probability at any template position of the elongation complex for termination or elongation. This probability is determined by the rate constants for either process: $k_{forward}$ for the elongation pathway or $k_{release}$ for the termination pathway. At non-terminator sites the activation energy barriers for the termination pathway are significantly higher than the corresponding activation energy barriers for elongation. Consequently, when $k_{forward}$ exceeds $k_{release}$, the decision is clearly in favour of elongation. When the elongation complex is destabilized at a termination site the activation energy barriers for the termination or the elongation pathway will be similar, however. Minor changes in the heights of the activation barriers can easily favour the termination pathway ($k_{release}$ exceeds $k_{forward}$) (Fig. 5.3).

Such changes may be the result of the structural alterations of the elongation complex which encounters a termination site. The riboU:deoxyriboA hybrid at the termination site, for instance, has only a very small contribution to the stability of the transcription complex. The complex will instead be destabilized by the formation of an RNA hairpin. Formation of the RNA hairpin is energetically favoured over the RNA:DNA hybrid. It will therefore displace part of the existing RNA:DNA heteroduplex. The free energy of the transcription complex is thus reduced, which in turn increases the termination efficiency.

5.1.1 The role of the RNA 3′ end structure in transcript stability

The formation of a stable stem-loop structure at the 3′ end of mRNA transcripts is not only important for the correct factor-independent termination at the end of the transcription unit. Apparently, it has a profound function for the RNA lifetime. Most nucleases are active on single stranded RNAs and strong base pairing, as in the helical stem regions of intrinsic terminators, will prevent degradation from the 3′ end. Stable RNA secondary structures at the ends of RNA molecules can therefore serve as barriers to exonucleolytic digestion. For this reason they are of special importance for the lifetime of

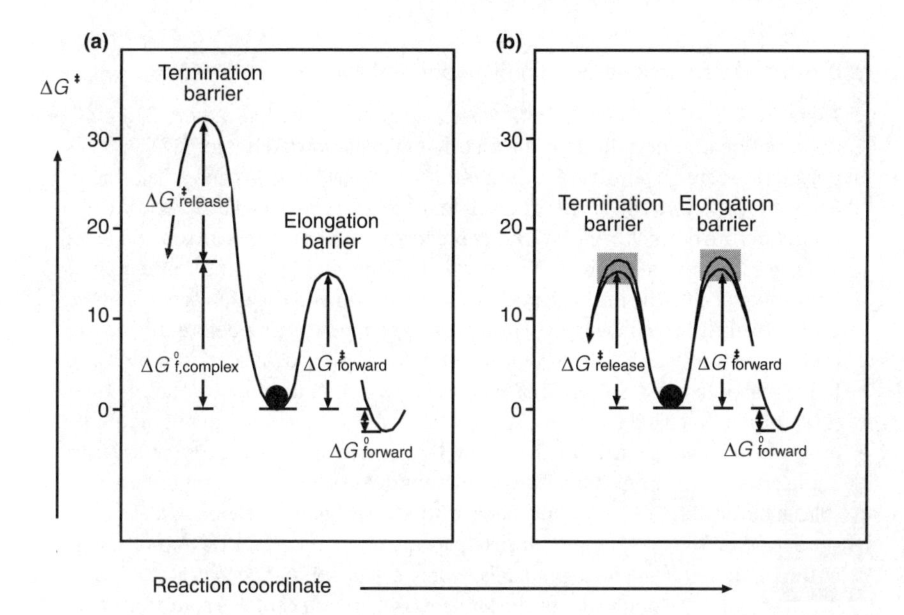

Figure 5.3 Thermodynamic model for the elongation/termination decision. Schematic diagrams for the relative activation energy barrier heights of an elongation complex at a typical elongation site (a) or a typical termination site (b) are shown. The alternative peaks with the lower activation energy in (b) correspond to a tenfold change in pause time. The shaded area corresponds to a tenfold change in termination efficiency. $\Delta G^{\ddagger}_{forward}$ and $\Delta G^{\ddagger}_{release}$ represent the energy barrier heights to elongation or release at template position *I*. $\Delta G^{0}_{forward}$ indicates the thermodynamic free energy of the elongation step from position *I* to *I* + 1. $\Delta G^{0}_{f,complex}$ represents the standard free energy of formation of the elongation complex at any template position. It consists of three components: the standard free energy of formation of the open complex (melting of the duplex DNA), the formation of the RNA:DNA heteroduplex and binding of the RNA polymerase to the nucleic acid components of the elongation complex. The figure is taken from von Hippel and Yager (1991, 1992).

the mRNA transcripts. All exonucleases isolated in *E. coli* to date hydrolyse RNA in a 3′ to 5′ direction, and no 5′ to 3′ exonuclease has been identified. This could be one reason why bacterial mRNAs do not require a special 5′ Cap structure is as it is present at the 5′ ends of eukaryotic mRNAs (Box 5.1 and Fig. 5.4).

Degradation of prokaryotic mRNAs is therefore very likely to be initialized by endonucleolytic cleavage reactions. In fact it has been observed in some cases that strong factor-independent terminators, linked to upstream genes, can enhance the expression of these upstream genes by stabilizing the mRNA sequences. This observation has been confirmed by a systematic study employing different mutations in the intrinsic terminator region of the *crp* gene (the gene for the catabolite regulator protein; see Section 7.3.2). It can be

Box 5.1 Differences between prokaryotic and eukaryotic mRNAS

There are several fundamental differences between eukaryotic and prokaryotic transcription products. Besides the fact that in eukaryotes different RNA polymerases are responsible for the synthesis of mRNAs, rRNAs and small stable RNAs, the structures of the products differ in several respects. Eukaryotic transcripts are normally synthesized as precursors which contain **intron** and **exon** sequences. Exons are the fragmented structural genes, and are interrupted by intron sequences. The intron sequences are removed and the exon sequences are ligated (usually in a colinear way) by a process termed **splicing** before the mature mRNAs are translated in the cytoplasm or before stable RNAs are assembled into functional organelles. Splicing normally involves a complex ribonucleoprotein (RNP) organelle called the **spliceosome**, which consists of a number of different proteins and several small U-rich RNA molecules (small nuclear RNPs (snRNPs)). Some transcripts, for example the precursor rRNAs of lower eukaryotes, are spliced autocatalytically by a mechanism which has been described as self-splicing. Well known examples of self-splicing are the precursor rRNA molecules of the ciliate *Tetrahymena thermophila,* where the phenomenon was first discovered. Interestingly, the bacteriophage T4 genome contains some introns which are removed by a self-splicing mechanism. In contrast, prokaryotic RNA transcripts do *not* normally contain introns.

A second difference between eukaryotic and prokaryotic RNAs is related to the number of genes present in a single mRNA molecule. Whereas most prokaryotic mRNAs contain more than one structural gene eukaryotic mRNAs generally encode only one gene. In other words, prokaryotic mRNAs are usually **polycistronic** while eukaryotic mRNAs are generally **monocistronic**.

RNAs transcribed from eukaryotes and prokaryotes differ furthermore at their 5′ and 3′ ends. Eukaryotic mRNAs contain a modified guanosine residue (7-methylguanosine) linked by a 5′–5′ phosphotriester bridge to the 5′ most nucleotide of the transcript (see Fig. 5.4). This unusual structure is added post-transcriptionally and is termed **Cap**. The Cap structure serves several functions. It protects the transcript from 5′ exonucleolytic degradation and, furthermore, the Cap is a target for binding of specific proteins which recognize 5′ capped mRNAs and which are very likely involved in directing ribosomes to the translational start site of the transcript.

Transcripts from eukaryotes and prokaryotes differ also at their 3′ ends. Normally, during the 3′ end formation of eukaryotic mRNAs, the primary transcripts are cleaved 20–30 nucleotides downstream of a specific signal sequence (AAUAAA). The cleaved transcript is subsequently polyadenylated at the 3′ end in a complex reaction involving cleavage specificity factors and poly(A) polymerase. The length of the poly(A) tail ranges from about 80 to 200 adenylate residues. The poly(A) tail probably functions in regulation of mRNA stability. It also affects the translational efficiency of the respective mRNAs (specific poly(A)-binding proteins are involved in this process). The mRNA turnover clearly depends on the length of the poly(A) tail (and on the presence of the Cap structure). The lifetime of eukaryotic transcripts is generally much longer (up to several hours) than that of prokaryotic mRNAs (between several seconds and minutes).

While it has long been believed that polyadenylation is specific to eukaryotes it is now known that prokaryotic transcripts may also be polyadenylated. The length of the 3′ poly(A) tails of prokaryotic RNAs is considerably shorter, however, (between 10 and 60 residues), and only a small fraction (10–40%) of the transcripts are found in a polyadenylated state. Moreover, whereas the presence of poly(A) increases the lifetime of eukaryotic mRNAs the same signal seems to initiate degradation in the case of prokaryotic transcripts. It is likely, however, that polyadenylation in prokaryotes also affects the efficiency of translation, most probably *via* the RNA-binding ribosomal protein S1, which seems to recognize polyadenylated mRNAs. Interestingly, recent findings indicate that poly(A) residues are also found in some conditions (e.g. deficiency of processing enzymes) at the ends of stable prokaryotic RNAs. Although exceptions are known the general differences between prokaryotic and eukaryotic RNAs may be summarized as in Table B5.1.

Table B5.1 General differences between prokaryotic and eukaryotic transcripts

Prokaryotic transcripts	**Eukaryotic transcripts**
No introns	Introns
*Poly*cistronic	Monocistronic
5′ end: *No* Cap structure	5′ end: Cap structure
Low degree of polyadenylation (10 to 60 residues)	Ubiquitous polyadenylation (~200 residues)

demonstrated that the GC-rich stem structure contributes both to the efficiency of termination and mRNA stability. The length of the oligo-U tract flanking the hairpin region has an influence on the efficiency of termination but not on mRNA stability. When different terminator mutants are fused to unrelated upstream RNAs the lifetime of these RNAs are enhanced in a similar way. The study proves that the 3′ hairpin structure of intrinsic terminators does not only provides the signal for transcription termination but also serves as an important element for the stabilization of mRNA transcripts.

A similar stabilizing effect by special RNA structures at the 3′ end has been described for a family of mRNAs. In this case repetitive extragenic palindromic (REP) sequences have been found to act as barriers to 3′ exonucleolytic digestion (see also Box 5.1 for the effect of 3′ polyadenylation on the lifetime of mRNAs in prokaryotes and eukaryotes).

5.2 Factor-dependent termination

Intrinsic termination reactions can be observed *in vitro* without the addition of components other than those needed for transcription elongation. In contrast, factor-dependent termination requires the addition of a special protein factor. The bacterial **Rho** protein is the universal factor for this type of termination. Compared to intrinsic terminator sequences the Rho-dependent terminators are difficult to recognize from the sequence context of the template. It is shown below that no clear consensus sequence for Rho-dependent terminators can yet

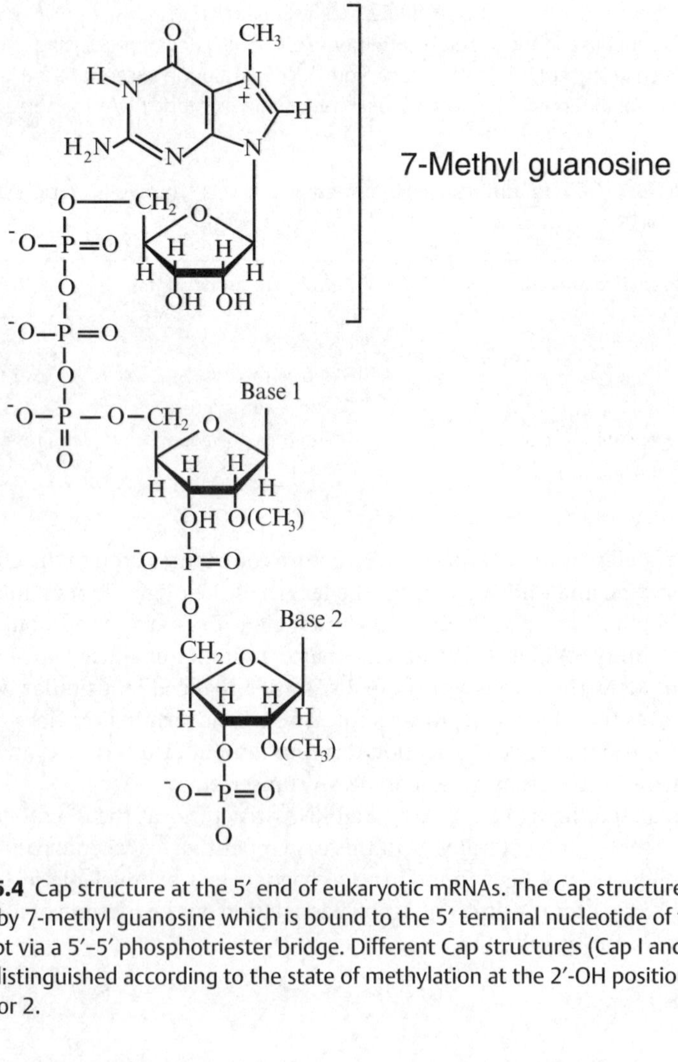

7-Methyl guanosine

Figure 5.4 Cap structure at the 5′ end of eukaryotic mRNAs. The Cap structure is formed by 7-methyl guanosine which is bound to the 5′ terminal nucleotide of the transcript via a 5′–5′ phosphotriester bridge. Different Cap structures (Cap I and Cap II) can be distinguished according to the state of methylation at the 2′-OH positions at bases 1 or 2.

be defined. Moreover, the endpoints of a Rho-dependent termination event are rather heterogeneous. In contrast to intrinsic termination sites, which are heterogeneous by only a few nucleotides, transcripts released by Rho-dependent termination often deviate in length by more than 50 nucleotides. The distribution of endpoints within this range is not random, however. Within the termination region some sites are represented at a higher intensity. It has been shown that these sites correlate exactly with positions where the translocation of the elongation complex is discontinuous and RNA polymerase pauses briefly before it resumes a forward movement. It can be concluded from this observation that the transcription endpoints for Rho-dependent termination are determined by the RNA polymerase step times and the pausing of the translocating elongation complex.

To understand the mechanism of Rho-dependent termination, details of the structure and activity of the termination factor Rho must be known, as must information on the features that comprise a Rho-dependent termination site. These issues are summarized in the next section.

5.2.1 Structure and function of the Rho factor

The functional Rho protein exists as a hexamer of identical subunits with a molecular weight of 46 kDa (419 amino acids) each. According to electron micrographs and low angle X-ray studies the structure of the Rho hexamer can be described as a flat ring-shaped molecule composed of six globular subunits. The ring-shaped structure has a diameter of 120 Å and a central hole with a 30 Å diameter. Each of the globular subunits has a dimension of about 42 Å. The protein structure of a single Rho subunit can be divided into three structural domains. The three domains are readily obtained by mild proteolytic digestion with trypsin. The N-terminal domain (NTD) comprises the first 130 amino acids. Within that domain typical sequence motifs characteristic for RNA-binding proteins are located (amino acids 20–90). In addition, a sequence similar to a region of the regulatory subunit of aspartate transcarbamoylase, known to bind cytidine, is present in the NTD. The second domain extends from amino acid position 131–283. It contains a sequence homologous to functional ATP-binding sites. The third domain, covering the C-terminal part of the protein, contains some conserved elements for NTP binding. Cross-linking and affinity labelling studies have shown no binding of RNA or NTPs to this domain, however.

In accordance with the functional domains identified in the isolated Rho subunits the Rho hexamer exhibits RNA binding and **nucleoside triphosphate phosphohydrolase** (**NTPase**) activities. The NTPase activity cleaves nucleotide triphosphates (NTPs) into nucleotide diphosphates (NDPs) and inorganic phosphate (Pi). The NTPase of Rho is dependent on RNA-binding. This means that binding of Rho to RNA is a prerequisite for NTPase activity. Since ATP is the preferred substrate this activity will be referred to as ATPase.

5.2.2 The role of RNA structure in Rho factor function

It is clear that Rho factor is an RNA-binding protein. Binding of Rho to various RNA molecules has been studied intensively. However, the rules which determine a specific Rho factor binding site have been elusive for many years and a satisfactory explanation cannot yet be given. Early studies with synthetic polynucleotides revealed that poly(C) or RNAs with a high C content were stably bound by Rho. A minimal RNA length of about 100 nucleotides was necessary for binding. Based on nuclease protection studies a length of 70–80 nucleotides has been determined to be in close contact with the Rho factor. This corresponds to binding about 12 nucleotides to each Rho monomer. Although C residues seem to be essential for Rho binding no clear sequence consensus has been established. However, comparison of the known sequences upstream of natural Rho factor termination sites has revealed an interesting spacing of C residues. Within natural RNAs to which Rho factor strongly binds C residues can generally be found at a spacing of 12 ± 1 nucleotides. This spacing corresponds exactly to the binding site size of one Rho monomer. Whether or not this spacing has a particular importance for the activity of the Rho factor is not known, however.

Other more complex sequence elements from natural RNA sites have been characterized as specific interaction sites for Rho. These sites are termed ***rut* sites** (Rho utilization sites). They are believed to act as entry sites where the factor first contacts the transcript. *Rut* sites are always found upstream of the Rho-dependent transcription stop positions. The sequence of the Rho-dependent λ tR1 terminator, which terminates transcription of the λ *cro* RNA, can be taken as an example for the general structure of a *rut* site (Fig. 5.5).

It consists of two segments of unstructured RNA, about 20 nucleotides in length each, termed *rut* A and *rut* B, respectively. The two RNA segments have a high C and a very low G content (formation of stable GC base pairing is thus strongly restricted). Both segments are separated by a five base-pair stem structure with a loop of five nucleotide residues. Binding of the λ *cro* RNA to Rho has been analysed in detail. It can be shown that both ionic and non-ionic bonds contribute to the interaction. The measured change in the standard free energy of the binding reaction ($\Delta G^{0'}$) is 12.6 kcal/mol. This value is strongly affected if base changes are introduced into the *rut* site. However, no systematic rules for the necessity of a certain sequence have been recognized. Enzymatic protection studies show that many of the C residues and a single unpaired G residue are less accessible when the λ *cro* RNA *rut* site is bound to the Rho factor. The direct interaction of Rho with the *rut* site of the *cro* RNA, which finally causes heterogeneous termination many base pairs downstream, is consistent with the view that Rho first initiates its interaction at the 5′ end of a termination structure, and then translocates along the RNA into the 3′ direction. Subsequently, when Rho reaches the transcription complex, termination takes place many nucleotides distal to the entry site.

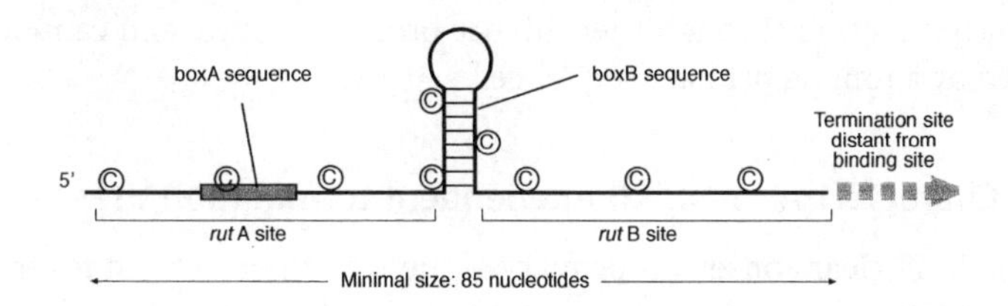

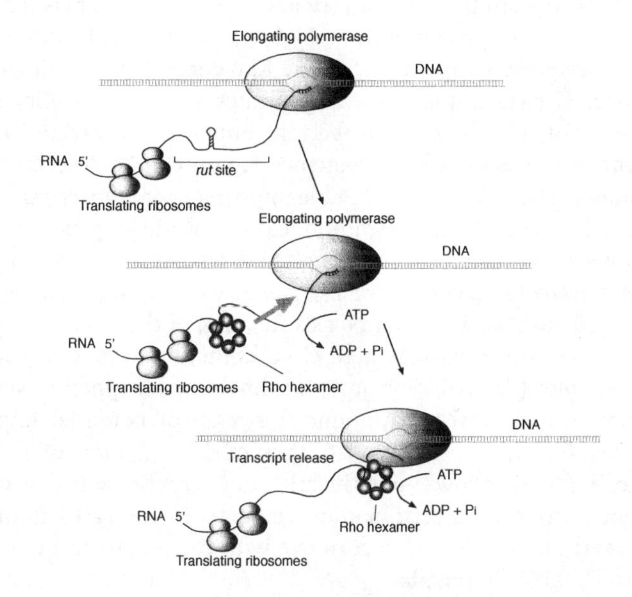

Figure 5.5 Structure of Rho factor entry sites and mechanism of Rho-dependent termination. (a) Typical structures of Rho utilization sites (*rut* sites) have a minimum length of 85 nucleotides, which corresponds to roughly 12 nucleotides per Rho monomer. Such entry sites can often be dissected into two parts, an upstream region (*rut* A) containing a boxA sequence (see Fig. 5.11) which is separated by a boxB-like hairpin from the downstream (*rut* B site) region. Except for the boxB hairpin *rut* sites do not contain stable secondary structural elements. A curious spacing of C residues (indicated by circled letters) at a distance of 11±1 nucleotide can often be found. These residues are considered to make contact with the Rho monomers. (b) The mechanism of Rho-dependent termination is summarized as a cartoon. An elongating RNA polymerase which transcribes an mRNA is schematically depicted. Translating ribosomes are bound to the growing transcript. If a *rut* site is transcribed which is not covered by ribosome, Rho may bind as a hexamer to the site. The bound Rho approaches the transcribing RNA polymerase under hydrolysis of ATP. If RNA polymerase enters a pause site, Rho factor catches up with the elongating complex and causes transcript release from the stalled or paused complex. This release involves an intrinsic Rho factor RNA–DNA helicase activity which is also ATP-dependent.

Comparison with many known sites of Rho-dependent termination has not revealed a high degree of sequence conservation for the *rut* site. Thus *rut* sites obviously function through their higher order structure, and cannot easily be defined by a unique primary sequence.

5.2.3 Characterization of Rho-dependent termination sites

Although no clear consensus primary sequence can be defined for the localization of a Rho factor-dependent termination site some sequence rules have proven to be very useful for the prediction and identification of Rho-dependent terminators. Of particular importance for these predictions is the high content of C residues concomitant with a very low G content. This bias for C against G nucleotides has been recognized at high statistical significance upstream to the 3′ transcription endpoints of many Rho-dependent terminators. Thus, a high C over G base content, which indicates a low probability for a stable secondary structure, seems to be an important characteristic of Rho-dependent terminators. Such a sequence region can be described as a C > G-rich 'bubble'. These C > G-rich bubbles upstream of transcription endpoints have proven to be a useful recognition feature for Rho-dependent terminators. Such bubbles structures can be sought by relatively simple computer programs. When the C > G content of a sequence window, large enough to fit a *rut* site (about 80 base pairs), is plotted as a function of the nucleotide positions of a cistron, sites with relatively high C > G content can be easily identified as bubbles the plot (Fig. 5.6). For quite a number of termination sites analysed these bubbles are consistently found upstream of potential Rho-dependent terminators. In addition to the verification of the endpoints of a transcription unit the method allows identification of cryptic intracistronic factor-dependent transcription termination elements. In the cases studied the efficiency of termination seems to correlate with the length and the relative C:G ratio of the bubbles in the plot.

5.2.4 Rho factor is an RNA:DNA helicase

The Rho factor has yet another interesting activity which may provide an important basis for mechanistic models of its termination function. This activity is an **RNA:DNA helicase**. Helicases are uniquely found in all living organisms and can be recognized by conserved amino acid sequence elements. These elements are characterized in the 'one-letter-code' for protein sequences as **DEAD boxes**. The DEAD box structure of helicases is conserved between bacteria and eukaryotes. Helicases are capable of unwinding double helical nucleic acids into single strands. The RNA-dependent ATPase of Rho is essential for helicase activity. It has been shown *in vitro* that short duplexes between DNA and complementary RNA can be denatured by Rho in an ATPase-dependent reaction. The ATP-dependent RNA:DNA helicase activity of Rho is a directional

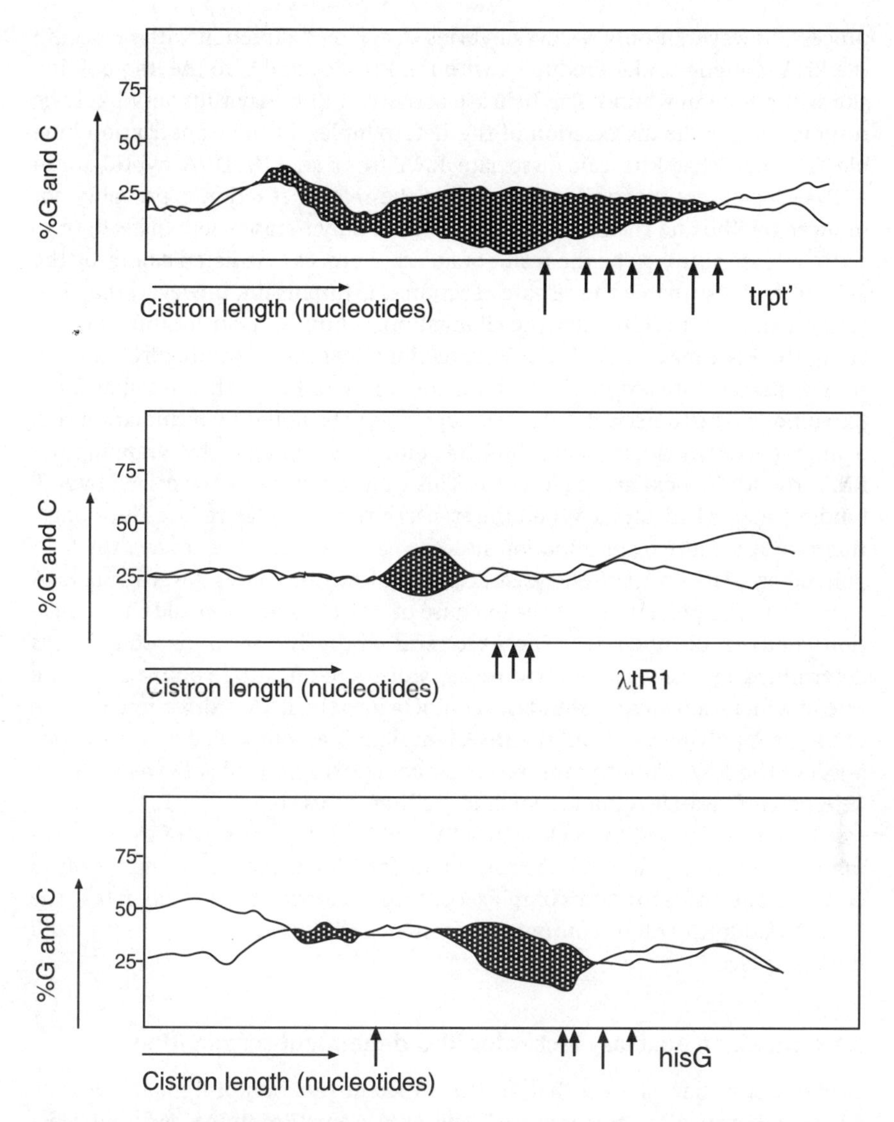

Figure 5.6 Characterization of Rho-dependent termination sites by C>G-rich bubbles. The figure shows the schematic results of a computer analysis of the trpt' terminator (upper panel), the λtRl region (middle panel) and the hisG terminator (lower panel). The cytosine and guanosine percentages for widows of 78 nucleotides are plotted as a function of the nucleotide length in each case. Arrows point to the observed termination sites, and the length of the arrow corresponds roughly to the efficiency of termination. The areas with high C over G content (C > G-rich bubbles) are shaded. The data is taken from Alifano *et al.* (1991) and Rivellini *et al.* (1991).

process, however. It only resolves hybrids in a 5′ to 3′ direction with respect to the RNA sequence. A heteroduplex with the RNA located 5′ to the Rho binding site will not be unwound. The helicase activity of Rho may thus very likely be responsible for the dissociation of the heteroduplex in the transcription bubble. The fact that Rho can dissociate RNA from an RNA:DNA hybrid in an ATPase-dependent reaction indicates that the release reaction is probably not induced by Rho and then executed by the RNA polymerase itself. Instead, transcript release seems to be the result of an *active* process mediated *directly* by the Rho protein. To enable the release reaction it is important, however, that Rho factor manages to approach the elongation complex. Translocation of Rho along the RNA may be achieved by directional extension of Rho:RNA contact sites. A straightforward mechanism for the action of Rho is then feasible. First, Rho binds to the transcript at a *rut* site upstream of a potential termination site. It migrates downstream along the RNA, either by sliding or by wrapping the RNA around the hexameric structure. This translocation may be driven by ATP binding and/or hydrolysis. When the transcription complex reduces its elongation rate at a pause or termination site Rho can approach and contact the RNA polymerase. The RNA:DNA helicase activity of Rho dissociates the growing RNA chain from the polymerase at the expense of ATP consumption and the elongation complex disintegrates. The exact site of the transcription endpoint is determined by the relative forward rates of the elongating complex and the rate at which Rho factor extends its contacts into the RNA 3′ direction. Release of the RNA polymerase from the DNA template is accompanied by the dissociation of the RNA from the transcription complex. Rho probably weakens the RNA polymerase–DNA interaction and facilitates this step.

It should be immediately clear, from the above, that pausing of the elongating complex is an important requirement for Rho-dependent termination to occur. If the transcription complex continues elongation at high rates, the RNA:DNA duplex will be continuously shifted and release of the transcript will not take place.

5.2.5 NusG, an auxiliary factor for Rho-dependent termination

Studies *in vivo* have shown that, for the efficient function of Rho, at least one additional protein factor is required. The 21-kDa **NusG** protein was found to act as such an auxiliary factor. Mutations in NusG have been shown to be able to suppress mutations in Rho, for instance. Furthermore, cells depleted of NusG are defective in factor-dependent termination. It has been shown for λ tR1 and *trp*t' termination sites that Rho-dependent termination decreases significantly in the absence of the NusG protein. NusG has first been characterized as a component of the phage λ transcription antitermination system (see Section 5.4). It is known that NusG, in concert with several other protein factors, modifies the transcription elongation complex and renders it resistant to termination. From such a function it would be expected that NusG is able to suppress

Rho-dependent termination. However, the contrary is the case. Probably NusG stabilizes the antitermination complex but does not directly function itself as an antiterminator. NusG does not only stimulate termination at Rho-dependent termination sites, it is also able to shift the termination endpoints in a 5′ direction. Most remarkably, NusG on its own does not cause or enhance termination. It also does not stimulate or otherwise affect Rho-independent termination. That means NusG specifically supports the function of Rho. In line with this it has been shown that Rho and NusG are able to bind to each other. This binding is weak, however, and the dissociation constant for this interaction is in the range of 10 μM. In addition, it is known that NusG binds directly, though weakly, to RNA polymerase. Binding studies with transcription complexes have shown that NusG strongly associates with stalled elongation complexes when Rho is bound to the nascent RNA. Moreover, NusG enhances the productive binding of Rho to a site 5′ to high affinity *rut* sites. Together, these observations indicate the existence of a quaternary complex in which NusG bridges Rho binding to RNA polymerase and participates cooperatively in the recognition of Rho-dependent termination signals.

The function of NusG seems to be most important for the cell at termination sites where Rho alone is inefficient due to kinetic limitations. NusG therefore has the strongest stimulatory function at sites where termination by Rho alone is hindered due to slow binding or weak interactions with the transcript. At these sites Rho cannot readily approach RNA polymerase. Thus, specifically under conditions of weak Rho binding, NusG helps to overcome the kinetic limitations of Rho, thereby supporting termination.

Transcription elongation studies in the presence and absence of functional NusG have shown that NusG is able to accelerate transcription elongation rates *in vitro* and *in vivo*. The effect appears to result from the suppression of specific RNA polymerase pausing by NusG. It is not easy to reconcile how enhanced elongation rates will cause an increase in termination efficiency, since the extent of pauses has been found to correlate with termination. Probably NusG interaction with the elongation complex weakens the RNA:DNA hybrid or the binding of the nascent transcript to the RNA polymerase. This could explain the observed increase in elongation rates and the facilitated release of RNA at Rho-dependent terminators. Interestingly, NusA, another transcription factor, already described as an important component for the regulation of specific transcriptional pauses (see Section 4.3.4), acts by slowing down transcription elongation. Thus the two factors NusA and NusG, by binding to RNA polymerase, seem to affect a common step in transcription elongation; however, they do this in opposing ways.

It should be noted that NusG does not only affect the activity of the termination factor Rho. A different termination factor, namely the **Nun protein**, is also dependent on the presence of NusG. The Nun protein is expressed by the phage HK022, a relative of phage λ. It acts as a separate termination factor specific for the phage λ *nutR* site (see below).

5.2.6 Alternative termination factors

Whether or not transcription termination is a factor-dependent or factor-independent process has largely been defined by studies *in vitro*. It remains questionable, however, if transcription termination inside the cell, at sites defined operationally as factor-independent *in vitro*, is actually taking place without the participation of proteins. The fact that in the test tube termination can occur at a specific site without the addition of a factor does not rule out the possibility that inside the cell the termination event is supported or facilitated by termination factors. Only if the efficiency or the site of termination differs between *in vivo* studies or studies with purified RNA polymerase can the participation of a termination factor be recognized. Such a situation has been observed when the early termination events of bacteriophage T3 or T7 were compared *in vivo* and *in vitro*. For instance, termination at the T3Te site, which has a sequence typical of Rho-independent terminators, is very inefficient *in vitro*. In addition, transcripts obtained *in vivo* are several nucleotides shorter than transcripts terminated at T3Te *in vitro* with purified RNA polymerase. A protein factor was identified and partly purified from *E. coli* cells, which affected termination at T7Te and T3Te. This protein was designated **Tau**. In the presence of Tau termination at the early bacteriophage T7Te and T3Te terminators occurs at the same sites used *in vivo*. The Tau protein, therefore, almost certainly represents a transcription termination factor, which seems to act at sites which show all the criteria of factor-independent terminators.

Several other transcription factors which can affect termination have already been mentioned in this chapter, and hence, they may be defined as auxiliary termination factors. One of these proteins is NusA. NusA can cause termination at some sites but not at others. NusA is able to promote pausing of RNA polymerase at specific sites. Its activity as a termination factor is probably related to this function. Reducing the forward rate of the elongation complex in a NusA-dependent reaction may allow more time for a conformational transition that triggers the release of the nascent transcript. It should be noted that NusA does not promote termination at Rho-dependent terminators. Instead, at some specific sites, NusA, together with several other components, plays an important role in the suppression of termination events (antitermination complexes, see below).

The phage-specific protein **Nun** has also been identified as a separate termination factor (see above). The Nun protein is encoded by the phage HK022, a relative of λ. This protein is dispensable for HK022 growth but has an inhibitory effect on the growth of λ phages in *E. coli* strains that carry an HK022 prophage. Obviously, the HK022 Nun protein interferes with λ phage development. It is known that the Nun protein interferes with the λ antitermination system. The phage λ uses a transcription antitermination mechanism for the expression of the delayed early genes (see below). Readthrough of termination sites (e.g. λ tR1) is essential for phage development. The Nun protein inhibits the λ antitermination system. It acts as a terminator at tR1, inhibiting the formation of any readthrough products. Obviously, Nun blocks λ infection in a cell which contains both phage genomes, providing a selective advantage for the HK022 phage.

Nun is an RNA-binding protein containing arginine-rich sequence motifs. It is able to interact with a site of the transcript which is normally used to assemble the antitermination complex. Nun can induce termination of transcription clustered several hundred base pairs downstream from the site of interaction with the λ transcript. Termination by Nun is independent of the action of Rho.

5.2.7 The role of Rho factor in transcriptional polarity

An important function as a *coupling factor* between *transcription* and *translation* has been assigned to Rho. Through binding to *rut* sites Rho factor can distinguish between translated or untranslated RNAs. Efficiently translated mRNAs are covered by ribosomes (polysomes). Thus, potential *rut* sites are not accessible for Rho, and the termination factor is unable to find an entry site. Many Rho-dependent termination sites are intragenic and appear to be cryptic, as long as translation efficiently masks the entry sites for Rho. When translation ceases, a phenomenon termed **polarity** can be observed. Polarity can be defined as a gradient of decreasing transcription efficiency along a transcription unit. The same number of transcripts that have been initiated at the promoter site do not reach the end of the transcription unit because of premature termination. Polar transcription is directly related to the efficiency of translation, and Rho represents the coupling unit between the two reactions. The phenomenon is often linked to *polar mutations*. Polar mutations change a sense codon for a specific amino acid within a gene into a nonsense or translational termination codon. As a result, translating ribosomes that reach this position will terminate protein synthesis instead of incorporating the correct amino acid at this site. Many genes contain cryptic Rho-dependent termination sites which are usually masked by ribosomes when the gene is actively translated. A premature block in translation exposes such sites. As a consequence, transcription is terminated at all potential termination sites downstream from the point where translation has been stopped. In this way, continuous transcription of a gene is linked to uninterrupted translation. Polarity in transcription may be seen as a device of the cell to save the high energy costs of mRNA synthesis when the resulting transcription product cannot be used for successful translation. Termination of transcription downstream of inefficiently translated mRNAs may not only be caused by polar mutations. The rate and efficiency of translating ribosomes may also be influenced by nutritional deprivation. For instance, amino acid limitation slows down or arrests translation. In line with this, it has been observed that under conditions of amino acid starvation numerous truncated mRNA transcripts occur in the cell owing to premature Rho-dependent termination events. By this mechanism Rho factor protects the cell from wasteful mRNA synthesis under conditions of starvation. As will be seen later, a similar system, termed **attenuation**, is used to regulate the transcription of many biosynthetic operons. In contrast to the polar transcription effects described here, attenuation generally involves Rho-independent termination, however.

5.2.8 An alternative function of Rho: stabilization of mRNAs

The analysis of several *rho* mutants has revealed that Rho protein is capable of carrying out functions separate from transcription termination. One of the functions that became apparent was the capacity of Rho to affect the lifetime of bulk mRNA. Rho factor has a general tendency to bind to RNA. It has been shown that specific mutations in *rho* do not affect the termination properties of the factor but the ability to bind to bulk mRNA. The reduced affinity of Rho mutants for mRNA-binding causes a higher susceptibility of the mRNA fraction to enzymatic degradation. The finding suggests that Rho factor plays a role in stabilizing mRNAs by specific protein-RNA interactions. The stabilization of mRNAs through interaction with an RNA-binding protein is not restricted to the Rho factor. A similar protective binding effect has also been shown for some other RNA-binding proteins, for instance for the pBR322 plasmid copy number control protein Rop.

5.3 Attenuation of transcription

Attenuation describes the control of transcription of distal genes by regulated termination. Usually (but not exclusively, as seen below) intrinsic (Rho-independent) termination structures are involved in attenuation. The decision between termination or readthrough of transcription is triggered by a switch in the secondary structure of the growing RNA chain, which either folds into a terminator hairpin or an antiterminator structure. The two structures are mutually exclusive alternatives, of which only one is termination-proficient. The change is induced by differences in the translation or transcription rates or, in some special cases, by binding of regulatory proteins.

Usually most of the regulatory determinants characterized today act as DNA structural elements. The fact that regulation of gene expression involves RNA secondary structures rather than DNA determinants may be taken as an indication that a primordial 'RNA world' may have preceded the 'DNA world' of regulation. Transcriptional attenuation therefore most likely belongs to those regulatory mechanisms that originated very early in evolution.

Attenuation of transcription has been most thoroughly investigated for a number of amino acid biosynthetic operons. These operons are characterized by a **leader region**, consisting of 100–200 nucleotides. These leader regions are localized upstream of the initiation codons for the first structural gene. In each case the leader regions contain an open reading frame which codes for a polypeptide of about 20 amino acids. These leader peptides have no apparent function but contain a cluster of those amino acids which are the product of the enzymes encoded in the respective transcription unit (Fig. 5.7).

A second peculiar characteristic of the leader transcripts is the capacity to form alternative secondary structures, which are mutually exclusive. An example is given for the leader transcript of the *trp* operon (Fig. 5.8).

The leader transcript of the *E. coli trp* operon is 162 nucleotides long, and has four sequence segments designated 1, 2, 3 and 4. The sequence segments can form pairwise stem structures, yielding either one or two hairpins, e.g. between segments 1:2 and 3:4, or alternatively 2:3. The 1:2 hairpin has a typical structure of an RNA polymerase pausing site. The hairpin formed by the stem elements 3:4 is a typical Rho-independent transcription terminator, which is followed by a run of U nucleotides. Hence this structure is called the terminator structure or **attenuator**. When this structure is formed transcription is terminated, yielding a transcript of 141 nucleotides. The alternative structure formed by the stem elements 2 and 3, which is mutually exclusive to the attenuator stem structure

Figure 5.7 Leader peptides of amino acid biosynthetic operons. The nucleotide sequences of four representative amino acid biosynthesis operons (*his*, *phe*, *thr* and *trp*) are shown. The sequences start with the AUG initiation codon and the derived amino acid sequences are indicated underneath. The striking clusters of amino acid codons specific for the respective operons are highlighted.

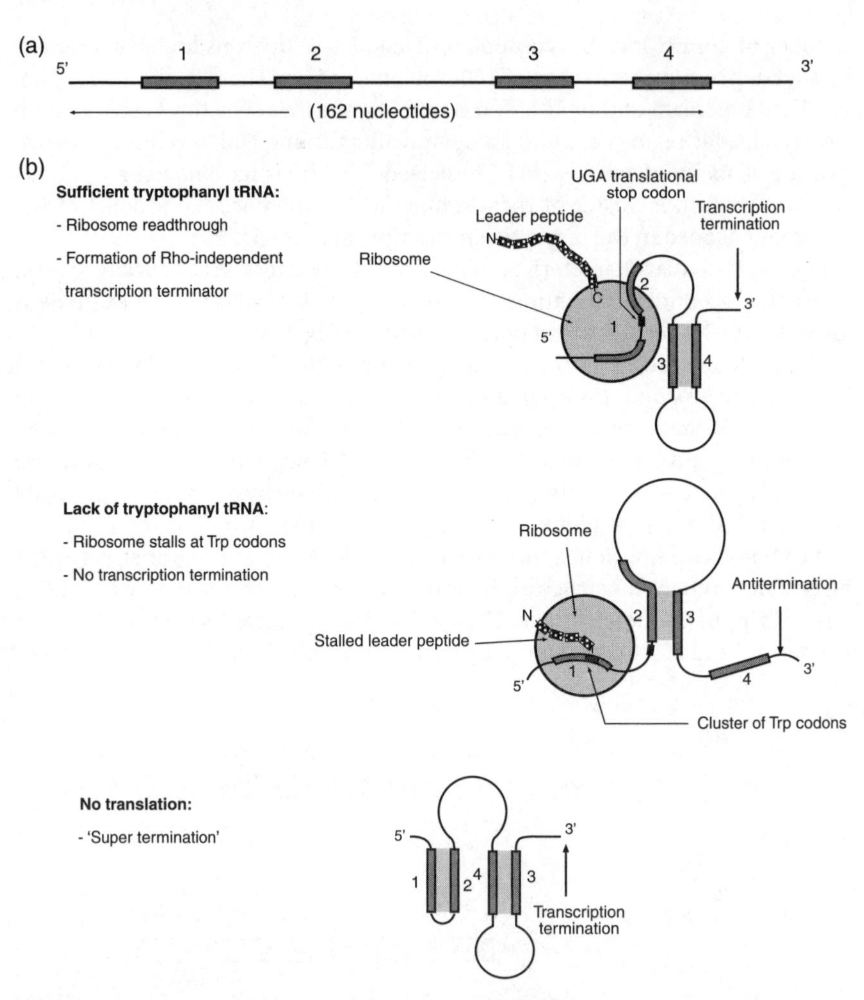

Figure 5.8 Mutually exclusive secondary structures of the *trp* leader. (a) The Trp leader transcript is shown with four potential regions (shaded boxes 1–4) capable of forming mutually exclusive base pairs. (b) Alternative base pair formations of regions 1–4 are presented during translation of the leader transcript. When there is an ample tryptophanyl supply, ribosomes (shaded circle) are able to readthrough the leader region and reach the UGA stop codon. Formation of the Rho-independent terminator structure 3:4 is supported and transcription is terminated. When tryptophanyl tRNA is lacking ribosomes stall at the cluster of Trp codons. The alternative structure involving base pair regions 2:3 is favoured and transcription termination does not occur. In the absence of any translation a double terminator structure is formed and transcription is efficiently terminated. The figure is arranged according to Yanofsky (1981).

3:4, does not cause transcription termination. It is called the **antiterminator structure**, therefore.

To explain regulation based on the structural elements described above a mechanism must exist which enables interconversion between the mutually exclusive RNA structures. The answer to how this interconversion might be brought about is provided by the phenomenon of **transcriptional-translational coupling**. In other words, an essential element for regulation by attenuation is the translation of the leader peptide by ribosomes. In case of the *trp* leader RNA translation yields a peptide of 14 amino acids, which contains two tryptophan residues at position 10 and 11 (see Fig. 5.7). Once transcription from the *trp* promoter has reached a position downstream from structure 1:2, RNA polymerase will pause and allow initiation of the leader peptide translation. RNA polymerase pausing is released when the translating ribosome approaches the 1:2 stem structure. Transcription continues into the attenuator region, where the decision for readthrough or termination will be made. If for some reason (ribosomes or other important components for translation are limiting) no translation of the leader peptide takes place, RNA polymerase will spontaneously leave the pause site after some time and transcribe into the attenuator structure. The termination hairpin 3:4 will form and transcription is very efficiently stopped (*no translation—therefore no need for mRNA*). If translation is unrestricted and aminoacylated tRNAs are abundant and readily available for translation (including the tRNAs charged with tryptophan), ribosomes and transcribing RNA polymerase move in a synchronized way. Translating ribosomes will reach the stop codon. The formation of the 1:2 stem structure is prevented by the translating ribosome, but the attenuator 3:4 will form and cause termination of transcription (*sufficient aminoacyl tRNAs present—therefore no need for transcription of amino acid biosynthesis genes*). If the number of tRNAs charged with tryptophan is limited, ribosomes will stall at the double Trp codon, waiting for a charged tryptophanyl tRNA to bind. The stalling ribosomes at codons 10 and 11 will prevent secondary structure formation about 13 nucleotides on either side of the '*hungry*' codons. Formation of the 1:2 stem is impossible and the antiterminator stem 2:3 will form. This precludes formation of the terminator 3:4 and transcription continues into the structural genes of the *trp* operon (*limiting amounts of tryptophanyl tRNA—therefore transcriptional readthrough*).

The above considerations show that the essence of attenuation control resides in the kinetic relationship between the rate of RNA polymerase translocation, the movement of ribosomes along the leader transcript and the rate of RNA secondary structure folding. Attenuation thus represents a device for transcription–translation coupling. It is also important to note that the process of leader peptide translation, not the peptide itself, is responsible for attenuation control. This has been shown, for instance, by frame shift mutations in the leader peptide or by providing the peptide in *trans*. On the other hand, the importance of the secondary structures of the leader transcript is supported entirely by experimental data. For instance, studies of the nuclease sensitivity of the leader transcript have revealed a pattern of protection and accessibilities consistent with the existence of the proposed secondary structures. Moreover, the secondary structure of the *trp* leader region has been confirmed by NMR studies. Base substitutions which weaken the stability of the antiterminator stem increase termination, whereas compensatory mutations which re-establish base pairing restore the original function (see Fig. 5.2). Deletions of the terminator stem prevent attenuation control. For some of the mutations it has been shown that they are effective only when present in the transcribed DNA strand. No effect has been observed when base substitutions have been placed only in the non-transcribed strand of the heteroduplex template, again ruling out that DNA structure is responsible for attenuation control.

Another point is noteworthy. Often the end products of biosynthetic operons act as feedback regulators. In this case the end products are amino acids. However, in the case of the attenuation mechanism, it is not the end products (amino acids) that are sensed as regulatory compounds. Rather the translational capacity, given by the abundance of charged tRNAs and the general activity of ribosomes, determines regulation, and not the actual level of the free amino acid pool. This is why attenuation can also be influenced by defects in the tRNAs or the corresponding aminoacyl tRNA synthetases, responsible for charging.

A somewhat different attenuation control mechanism from that described for the amino acid biosynthetic operons above operates during the control of the *pyrBI* and *pyrE* operon expression. The *pyrBI* operon, for instance, encodes the pyrimidine biosynthetic enzyme aspartate transcarbamoylase which is not directly involved in amino acid metabolism. Similar elements are found in the attenuation of *pyrBI* or *pyrE* transcription as those found in the case of the *trp* operon. For instance, a 158-nucleotide leader precedes the first structural gene of the *pyrBI* operon. The leader transcript contains an open reading frame which encodes a polypeptide of 44 amino acids. A very efficient Rho-independent terminator structure is located near the downstream end of the leader open reading frame, and 23 nucleotides upstream of the initiation codon of the first structural gene. In addition, the leader contains a UTP-sensitive transcription pause site. This site consists of a run of uridines and precedes the terminator in the leader open reading frame. Transcription termination within the *pyrBI* leader region is directly controlled by the extent of coupling translating ribosomes to transcribing RNA polymerase. Initiated RNA polymerase starts transcribing through the leader open reading frame. At low intracellular UTP levels transcription is slowed down and RNA polymerase stalls at the UTP-sensitive pause site. Ribosomes that have started translation move up to the stalled RNA polymerase. The formation of a potential Rho-independent terminator hairpin is physically blocked by the presence of the translating ribosomes. RNA transcription can thus continue into the downstream genes. At high intracellular UTP concentrations, however, RNA polymerase will not slow down and stall at the UTP-sensitive pause site. Instead, it will transcribe away from the

(a) *pyrBI* operon leader region

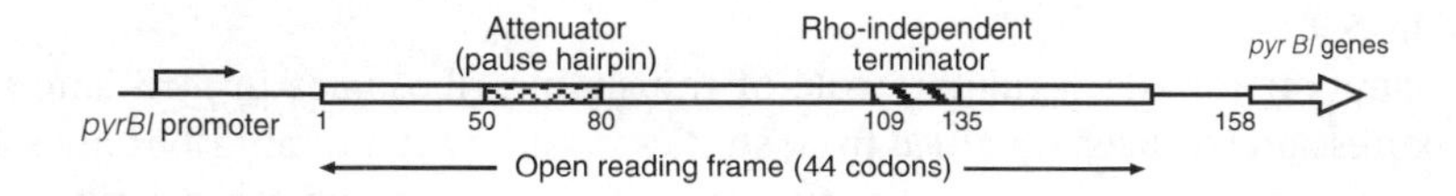

(b) *pyrBI* transcript

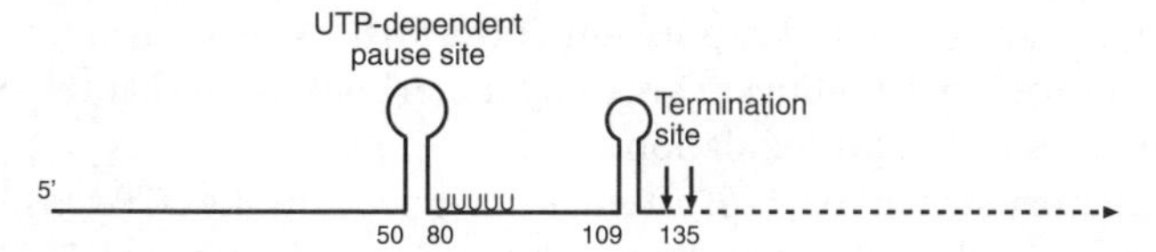

(c) Low UTP concentration - strong transcription pausing - antitermination

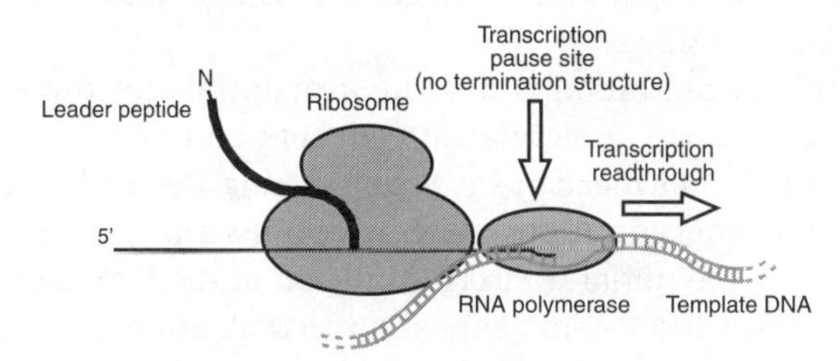

(d) High UTP concentration - weak transcription pausing - termination

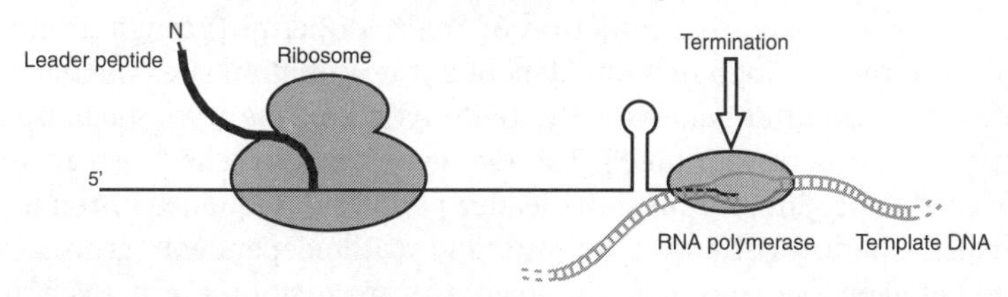

Figure 5.9 Structure of the *B. subtilis pyrBI* attenuator. (a) The leader region of the *pyrBI* operon is schematically drawn. The 158-nucleotide transcript codes for an open reading frame of 44 amino acids. Within the leader transcript there are two sequence regions with the potential to form hairpin structures. A site between nucleotides 50 and 80 has the capacity to fold into a pausing hairpin which is flanked by a run of pyrimidines (attenuator). The sequence 109–135 contains a Rho-independent termination structure. (b and c). At low cellular UTP concentration RNA polymerase stalls at the attenuator structure. Translating ribosomes keep up with the transcription complex and obstruct any Rho-independent terminator structure. Transcription proceeds into the *pyrBI* genes. (d) At high UTP concentration the rate of transcription is not hindered at the attenuator site. The transcription complex runs well ahead of the translating ribosomes. This causes the transcript to fold into a termination hairpin. Readthrough into the *pyrBI* genes is stopped.

ribosomes. This enables the formation of the terminator hairpin and transcription will be terminated before RNA polymerase reaches the structural genes (Fig. 5.9).

Support for the regulatory role of translating ribosomes in *pyrBI* and *pyrE* expression has been obtained from studies in cells with mutant ribosomes. It is known that mutations in the *rpsL* gene (the gene coding for the ribosomal protein S12) result in ribosomes with a significantly reduced translation rate (about one-third of the normal rate). Strains containing such 'slow' ribosomes exhibit significantly reduced expression of the *pyrBI* gene, even when grown in media with excess pyrimidines. This study underlines again that the process of translation is essential for regulation.

The regulatory switch for *pyrBI* attenuation is provided by UTP-sensitive pausing, which responds to different intercellular concentrations of UTP. The cellular UTP concentration can be regarded as the *signal* and RNA polymerase as the *sensor*. There is quite a large natural variation in the UTP content in *E. coli* which can affect the rate of UTP incorporation. Concentrations may range from less than 50 μM to more than 1 mM.

The intracellular UTP concentration is not only responsible for the attenuation control of *pyrBI* expression. A significant contribution to overall regulation is caused by an additional mechanism acting during the early steps of transcription. At high UTP concentrations, reiterative transcription occurs at a cluster of AT base pairs in the initial transcribed region of the template (transcript slippage; see Section 3.4.2). This slippage reaction inhibits promoter clearance, thereby reducing *pyrBI* expression by about sevenfold.

Not all attenuation mechanisms are effective at intrinsic Rho-independent terminators. For instance, regulation of the *tna* operon through attenuation involves the readthrough of a Rho-dependent termination site. The *tna* operon encodes the structural genes for the hydrolytic enzyme tryptophanase (*tnaA*) and tryptophan permease (*tnaB*). The two genes are preceded by a regulatory region (*tnaC*) encoding a 24-residue leader peptide. A sequence of roughly 200 base pairs, characteristic of a *rut* site, and a Rho-dependent terminator are located between *tnaC* and *tnaA*. Details of the attenuation mechanism are not known. However, the availability of tryptophan plays a central role. Rapid translation of the leader peptide at high levels of tryptophan could block Rho binding to the *rut* site. In the absence of tryptophan an out-of-frame stop codon in the *tnaC* leader seems to cause ribosome release. This may allow Rho factor to associate with the transcript and to cause termination.

Another interesting example of attenuation control involving Rho-dependent attenuators is known for the expression of the *rho* gene itself. The structural gene for Rho is preceded by a series of Rho-dependent terminators. The degree of termination at these sites depends on the cellular concentration of Rho. Thus, the amount of Rho in the cell is autogenously regulated by the gene product through attenuation.

Attenuation also plays a specific role in the complex regulation of ribosomal proteins, the constituents of the translational apparatus. Many of the ribosomal proteins are clustered in transcription units where proteins for the large and small ribosomal subunits are often encoded together with other proteins essential for translation or transcription. One example is the S10 operon, which contains the genes for 11 ribosomal proteins, seven for the large and four for the small ribosomal subunits. Transcription of this operon is regulated by attenuation involving the leader region of the transcription unit. Attenuation is mediated by one of the products of the operon, namely the ribosomal protein L4. Hence, L4 is the central regulator for the S10 operon. First it is involved in translational feedback regulation, which is common for many ribosomal proteins. Translational feedback is determined by a reversible block in translation resulting from binding of excess ribosomal proteins (in this case L4) to their own mRNA. L4, which is an RNA-binding protein, normally interacts directly with the ribosomal 23S rRNA. It also binds to an S10 mRNA sequence, which has a homologous secondary structure to the 23S rRNA-binding site; however, it has a somewhat weaker affinity. When 23S rRNA is limiting, excess L4 will bind to the S10 mRNA and thereby inhibit translation of the S10 operon (for a schematic summary of the regulation of the S10 operon see also Fig. 8.13). Apart from this function in translational feedback control, L4 binding to the S10 leader RNA also inhibits transcription of the distal portions of the mRNA. Rho-independent termination occurs in the presence of excess L4 protein, about 140 nucleotides downstream of the start site of the S10 operon. Translation of the S10 mRNA does not seem to be involved in this type of regulation. However, NusA is required for efficient transcription termination at the S10 attenuator. The precise mechanism for this type of attenuation is still unclear. Probably, S10 promotes pausing of the transcription elongation complex, which may allow L4 to associate. Interaction of L4 with the paused transcription complex propably triggers RNA release at the attenuator site.

Many of the *E. coli* attenuation mechanisms described here involve translating ribosomes. Often the corresponding genes in *B. subtilis* are also regulated through attenuation. In contrast to the *E. coli* genes, however, attenuation in *B. subtilis* does not involve translating ribosomes but specific RNA-binding proteins. Characteristic examples are the biosynthetic operons for tryptophan or pyrimidine nucleotides specified in more detail below.

The *trp* operon of *B. subtilis* contains a regulatory leader region of 204 nucleotides preceding the first six structural genes (*trpE, D, C, F, B, A*). In accordance with the general attenuation mechanism the *B. subtilis trp* operon leader transcript is able to form alternative hairpin structures. One of the hairpin structures resembles an intrinsic terminator, which efficiently blocks transcription. The terminator is only active when sufficient amounts of tryptophan are available. In the absence of tryptophan, readthrough of the terminator into the structural genes is observed. There is no translatable open reading frame in the leader, hence, translation is not involved in this attenuation control. Instead, an RNA-binding protein has been identified which interacts with the leader of the *trpEDCFBA* mRNA. This protein is the *mtrB* gene product, and it is termed **TRAP** (*trp* RNA-binding attenuation protein). The 8- kDa TRAP protein (75 amino acids) exists as a toroidal-shaped multimer composed of 11 subunits. The structure is comparable with the shape of a doughnut. Each subunit is able to bind one molecule of tryptophan. Tryptophan binding enhances the affinity of the protein for RNA interaction. Interestingly, the *trp* leader RNA contains 11 repeats of a GAG or UAG sequence motif, separated by two nucleotides each. The multimeric TRAP protein seems to bind preferentially to these sequences, wrapping the leader transcript around its toroidal structure. Binding is very likely to involve three basic amino acids of each TRAP monomer (Lys37, Lys56 and Arg58), which are regularly spaced within the structure of the multimer. The Lys–Lys–Arg (KKR) sequence motif seems to be highly specific for binding of the unstructured repeat of GAG and UAG sequences.

In the presence of tryptophan, binding of TRAP to the leader RNA blocks formation of the alternative secondary structure, allowing the terminator hairpin to form. Hence, TRAP binding causes efficient termination. No other factor seems to be required for the TRAP-directed attenuation mechanism.

In a similar way transcription of the *B. subtilis pyr* operon is regulated by an RNA-binding protein. The *pyr* operon encodes eight enzymes necessary for the biosynthesis of pyrimidine nucleotides. In addition, the regulatory protein PyrR and the uracil permease PyrP are encoded by the same operon. The transcription of the operon is coordinately repressed in the presence of excess pyrimidine nucleotides. This repression involves three intrinsic transcription terminators. They are located in the leader region upstream of *pyrR* and in the intercistronic regions between *pyrR* and *pyrP* and *pyrP* and *pyrB*. Each of the terminators is preceded by a sequence capable of folding into an alternative secondary structure (antiterminator hairpin). The formation of the terminator and the antiterminator structures are mutually exclusive. The 20-kDa protein PyrR binds to the *pyr* transcript. Binding of PyrR prevents folding of the antiterminator structure. Apparently, binding of PyrR is allosterically regulated by uridine monophosphate (UMP), the end product of the *pyr* operon. Thus, UMP activates PyrR for binding to the early region of the *pyr* mRNA. PyrR binding in turn disrupts the structure of the antiterminator hairpin. The alternative terminator structure is formed and transcription into the distal portions of the *pyr* operon is terminated.

5.4 Antitermination

The regulatory principles of attenuation and antitermination have many common feature, and often the borders of the two regulatory systems are not precisely defined. However, according to the classical subdivision, the mechanisms of attenuation and antitermination are very different. The subdivision that is undertaken here follows this classical definition. Some of the related systems described further below have originally been subdivided rather arbitrarily, and a strict distinction between the two regulatory mechanisms is of rather formal nature.

The typical mechanism of antitermination results from the modification of RNA polymerase early during transcription elongation. The modified RNA polymerase resists terminators which are encountered during the transcription of the downstream operon. The property of the modified RNA polymerase to ignore termination signals may last over time and over transcription distances of several thousands of base pairs. The sites where termination readthrough occurs are normally far away from the site where the antitermination-competent polymerase is assembled. Therefore, antitermination is often described as 'action at a distance'.

5.4.1 Antitermination in phage λ gene expression

Antitermination was first discovered as a way in which the *E. coli* phage λ regulates its late gene expression. Most of our knowledge about this regulatory mechanism stems from studies with this system. Gene expression in λ may, therefore, be considered as the hallmark for antitermination.

5.4.2 The phage λ N protein system

Two different antitermination systems operate during phage λ gene expression. They depend on the participation of either of the phage-specific proteins N or Q. The N-dependent antitermination system, which is of higher complexity and for which more information is available, is described first.

Early transcription of the λ genome starts from two divergent promoters, P_L and P_R, respectively. The first genes transcribed are (leftward) the λ *N* and (rightward) the *cro* genes. Transcription in both the left and the rightward direction is then stopped at the Rho-dependent terminators t_{L1} or t_{R1}. The *N* gene codes for a basic protein of 12 kDa that has no apparent homologue in uninfected *E. coli* cells. When sufficient **N protein** has been synthesized, termination at these two terminators at the end of the early operons is prevented. RNA polymerase is now able to transcribe beyond the terminators t_{L1}, and all other terminators that follow in the leftward direction of transcription (t_{L2} and t_{L3}). At the same time the terminators t_{R1}, t_{R2}, t_{R3} and t_{R4} in the rightward direction of transcription are also ignored (Fig. 5.10).

The change from the termination-proficient to the termination-resistant form of RNA polymerase is not only dependent on the presence of the phage N protein. A number of additional host factors is required to render RNA polymerase as an antitermination-proficient form. These are the **Nus proteins** (N utilization substances) NusA, NusB, NusE and NusG (Table 5.1).

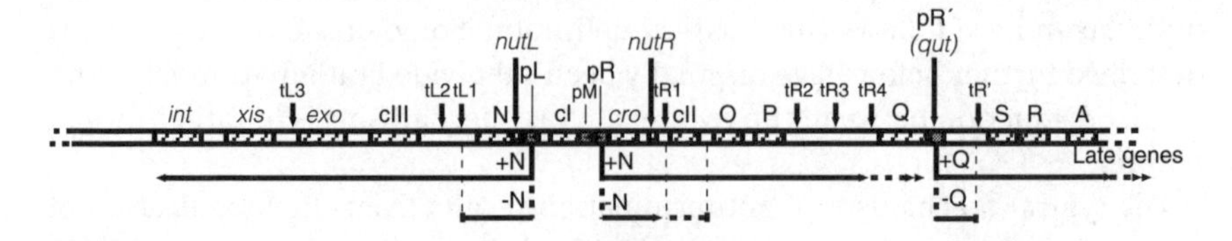

Figure 5.10 Organization of the phage λ. A physical map of the phage λ genes is schematically shown. Promoters are indicated by dark boxes marked with an arrow head. The phage genes are drawn as shaded boxes. Termination sites are marked by short vertical arrows. Transcription starts in the leftward or rightward direction at the two promoters pL and pR, respectively. The cI gene is transcribed from the minor promoter pM in the leftward direction. The late genes are transcribed from pR′ The *nut* site regions *nutR*, *nutL* and *qut* are indicated by long vertical arrows. In the absence of the N protein, transcription from pL and pR is terminated at tL1 or tR1, respectively. In the presence of N protein, antitermination complexes are formed at the *nutR* and *nutL* regions and transcription runs through the terminators tR1–tR4 and tL1–tL3. Similarly, in the absence of Q protein late transcription is terminated at tR′ while in the presence of Q an antitermination complex is constituted at the *qut* site and transcription continues into the late genes.

NusE is in fact identical to the small subunit ribosomal protein S10. RNA polymerase provides a platform at which the antitermination factors (Nus proteins) assemble. To constitute a termination-resistant polymerase it is essential, however, that the enzyme transcribes through a sequence region close to the promoter. This sequence is termed ***nut* site** (N utilization site). While the Nus proteins are acting in *trans* to form an antitermination complex the *nut* site can be considered to be a necessary *cis*-acting sequence for the formation of a termination resistant elongation complex. Two *nut* sites are found in the λ genome, one in the rightward and one in the leftward transcription unit, distal to the promoters P_R and P_L. They are designated *nutR* and *nutL*, respectively. The *nut* sites are composed of different sequence elements. They each contain a boxA and a boxB sequence (Fig. 5.11). In the early literature a boxC sequence was additionally assigned as a *nut* element. It is now known, however, that boxC, a conserved UGUGUGGG sequence, probably does not function in antitermination but rather seems to act as a processing signal for RNase III. Comparison of different lambdoid phages revealed that boxA has a conserved sequence, whereas boxB is an interrupted palindrome able to form a hairpin structure. From nuclear magnetic resonance structure determinations it is known that the hairpin has a pentaloop. Four nucleotides of the pentaloop form a GNRA-like **tetraloop** structure (see Box 4.1). The looped-out fifth nucleotide makes extensive hydrophobic interactions with the arginine-rich structure of the phage-specific N protein, which is tightly packed against the major groove of the boxB hairpin. Studies with mutants have shown that bases in the loop of the hairpin structure are of particular importance for the specificity of N pro-

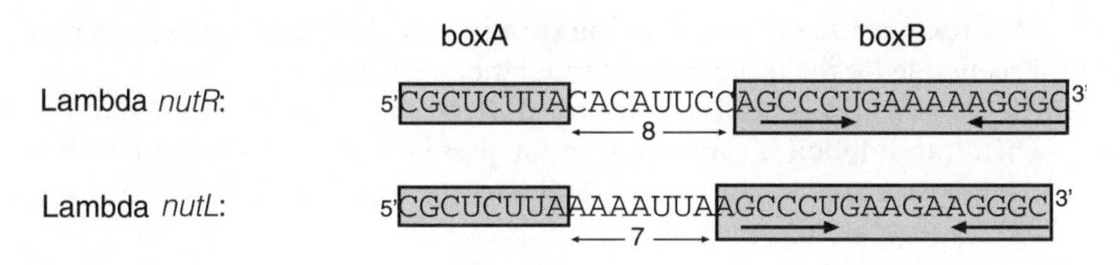

Figure 5.11 Phage λ *nut* site sequences. The sequences of the phage λ *nut* site sequences *nutR* and *nutL* are shown. BoxA and boxB regions are shaded, and their length in nucleotides given. The inverted repeat structure of boxB is indicated by arrows. A consensus boxA sequence derived from various phage sequences is also shown.

Table 5.1 Nus factors for phage λ antitermination

Protein	Molecular weight (kDa)	Origin	Properties
NusA	55	Host-encoded	Acidic
NusB	16	Host-encoded	Neutral
NusE *	12.5	Host-encoded	Basic
NusG	21	Host-encoded	Neutral
N	12	Phage-specific	Basic
Q	22.5	Phage-specific	Basic

* NusE is identical to the ribosomal protein S10.

tein binding. The primary sequence of the stem nucleotides of the hairpin does not show particular sequence conservation, however. Interestingly, related lambdoid phages, which have slightly different N proteins, also require their own specific *nut* sites. If the *nut* sites are changed correspondingly the different N proteins can be functionally replaced.

There is no doubt today that the functional component of the *nut* element is RNA and not DNA. The evidence for this notion can be summarized as follows:

1. RNA polymerase has to transcribe through the *nut* region to become termination resistant.
2. If ribosomes translate into the *nut* region, owing to reading frame mutations, they mask the *nut* site function.
3. N protein binding requires transcription of the boxB sequence.
4. Anti-boxB oligonucleotides (complementary to the RNA) specifically interfere with antitermination.

5. After transcription through *nut* the corresponding DNA sequence becomes dispensable for the formation of termination-resistant polymerase.
6. The *nut* site RNA is protected against nucleases and chemical modification when transcription is carried out in the presence of Nus factors. However, extensive treatment with RNase T1 releases the N protein from the complex and prevents antitermination.

Antitermination in λ depends on the stable interaction of the N protein and the host Nus factors with RNA polymerase. Although some of the Nus proteins are able to bind RNA polymerase directly, stable binding is only observed after RNA polymerase has transcribed through the *nut* region. Obviously both RNA polymerase and sequences of the *nut* region contribute to stable antitermination complex formation. The RNA polymerase provides the platform for a series of complex interactions which constitutes a ribonucleoprotein complex that is resistant to transcription termination. This complex has been termed the **elongation control particle** (**ECP**). Details of the interactions, which stabilize the antitermination complex, are depicted schematically in Fig. 5.12.

The complex pattern of interactions has been analysed by gel mobility shifts (see Box 6.2) and by affinity chromatography with immobilized RNA polymerase. The different components of the complex are held together by combined protein–protein and protein–RNA interactions, which can be summarized as follows: NusA, NusE (S10) and NusG are able to bind directly to the

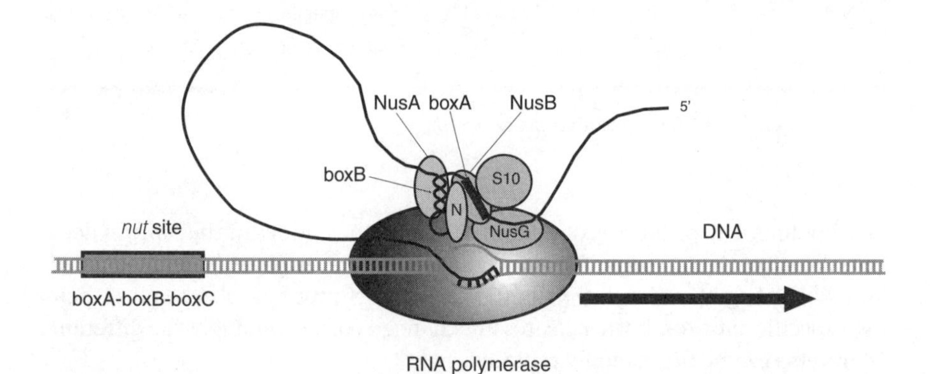

Figure 5.12 Structure of a phage λ antitermination complex. An elongating RNA polymerase that has transcribed over a *nut* region consisting of boxA, boxB and boxC elements is schematically depicted. The growing transcript, indicated as a dark line, extends from the transcription bubble (the active centre) of the elongating complex and forms a closed structure by binding of the *nut* site RNA sequences boxB (indicated as a helical stem-loop structure) and boxA (indicated as a thick dark line) to RNA polymerase. The phage-specific protein N and the host proteins NusA, NusB, S10 and NusG participate in antitermination complex formation and stabilize the complex through cooperative interactions. This complex is able to read through terminators hundreds of base pairs downstream.

RNA polymerase. Binding is stabilized by cooperative interactions, and single isolated proteins bind only weakly. There is evidence for two molecules of NusA in the complex. NusA has already been shown to be an efficient modulator of transcription. It reduces the elongation rate, enhances specific pauses, and promotes termination at specific sites. Binding of one NusA molecule occurs after the specificity factor σ is released from the transcription complex. Binding of NusA and σ to the RNA polymerase core are mutually exclusive. The binding of the first NusA molecule is sequence-independent. Binding of the second NusA protein requires that the RNA polymerase has passed through the *nut* site. NusE (S10) binds as a dimer together with NusB. The two proteins interact weakly with each other (K_D is approximately 10^{-6} M). In a similar way, NusA and the N protein are able to interact in solution (K_D is approximately 10^{-7} M). Thus, NusA seems to be the adapter which couples N to the RNA polymerase. In addition, N binds strongly to the *nut* site boxB sequence (K_D is roughly 10–20 nM). The N protein contains a cluster of arginines, which is a typical sequence motif found in many RNA-binding proteins. In addition, there is some evidence that NusB and NusG bind to boxA, whereas NusA binds to boxB. Although each individual interaction appears to be very weak, the complete assembly of the complex is a highly cooperative process. All components are needed in concert to form a complex that lives long enough to suppress termination signals thousands of base pairs downstream of the site of assembly.

Within the antitermination complex, protein N and the Nus proteins are tethered to the surface of the RNA polymerase by the RNA *nut* site. This physical link to the RNA polymerase is important. When transcripts containing *nut* elements are provided in *trans* they do not support antitermination of an operon without *nut*. The fact that the *nut* site close to the 5′ end of the transcript is linked to the RNA polymerase, bridged by several Nus proteins, implies that during elongation of the complex the growing transcript will form a steadily increasing loop of RNA.

It has been noted that the readthrough of close-by terminators (close to the *nut* site) only requires a subset of the components needed for the complete antitermination complex. These include the boxB sequence and the proteins N and NusA. This minimal antitermination system is able to readthrough both Rho-dependent and intrinsic terminators. To suppress termination signals more distant from the *nut* site a more processive antitermination complex is required. Hence, the complete set of Nus factors, including NusB, NusE and NusG, must be present. The complete antitermination system not only overrides factor-independent and factor-dependent terminators, it also has the capacity to transcribe through tandem terminators. Because of this property the complex has vividly been termed a 'juggernaut'.

N-dependent antitermination is not strictly linked to a particular promoter. It has been shown that λ *nut* sites can be functionally separated from their promoters P_L or P_R. Furthermore, the distance of the *nut* site to the promoter is not a stringent requirement, provided that RNA polymerase passes through the *nut* site. Antitermination requires the bacterial RNA polymerase, however. If transcription units containing a functional *nut* site are under the control of the phage T7 RNA polymerase, for instance, no functional antitermination complex is assembled.

The N protein is known to inhibit pausing at both Rho-dependent and Rho-independent terminators. This property may provide a mechanistic explanation for antitermination, namely to *enhance* the speed of transcription. It is easily feasible that termination is impeded if RNA polymerase passes too quickly through the terminators. It is interesting in this regard that the presence of a consensus boxA sequence fused to a synthetic promoter enhances the transcription elongation rate of several genes *in vivo* (see Section 8.5.5). The contributions of the transcription factors NusA and NusG are puzzling. NusA by itself is known to reduce the elongation rate and to enhance specific pauses. In a similar way, NusG has been shown to support termination. How can these apparently conflicting properties be reconciled with the antitermination function? As a solution to the puzzle it has been supposed that the major function of the two proteins resides in the stabilization of the antitermination complex, facilitating the contact of other Nus factors with the *nut* site and with the RNA polymerase.

The Nun protein of phage HK022, mentioned earlier (see Section 5.2.6 above), competes with N for binding to boxB. It has a similar RNA-binding motif (arginine cluster) as the N protein. Binding of Nun to the antitermination complex instead of the N protein induces termination rather than antitermination. Although the two proteins use the same assembly route they trigger opposing functions of the RNA polymerase.

Finally, it should be mentioned that transcription antitermination is not restricted to the regulation of prokaryotic gene expression. Suppression of pausing and termination sites is also observed in eukaryotic transcription units.

An interesting analogy to λ N-dependent transcription antitermination is apparent in the transcription of the human immunodeficiency virus (HIV). Here, the transcriptional activator protein Tat binds to a specific site of the transcribed RNA (Tar) and prevents premature termination. Tat, like the phage-specific N protein, contains the same arginine cluster, which is characteristic of many RNA-binding proteins.

5.4.3 The phage λ Q protein system

The second antitermination system acting in phage λ relies on the 22.5-kDa basic phage-specific protein Q. The presence of this protein prevents termination at the intrinsic tR′ terminator downstream from the late promoter $P_{R'}$ and allows transcription into the late phage genes. Therefore, expression of the late phage genes depends on the presence of Q. In contrast to the N protein, which recognizes an RNA stem-loop structure, Q is a *DNA-binding* protein. It interacts with a site that spans the promoter $P_{R'}$. This site is termed *qut* (Q utilization). The Q-dependent antitermination system differs in a second respect from the N system. The *qut* site and the $P_{R'}$ promoter form a single functional unit and cannot be dissected or replaced.

RNA polymerase, which starts transcription at $P_{R'}$, pauses for several minutes immediately downstream of the promoter at position +16/17. It has been shown that the DNA sequence in the non-template strand is responsible for this pause. It contains a sequence motif which is similar to the –10 region of σ^{70} promoters and the presence of σ^{70} has been reported in the paused complex. When Q binds to the *qut* site it causes RNA polymerase to escape from the pause. After resuming transcription, the RNA polymerase is able to transcribe through the tR′ terminator into the late λ genes. The presence of NusA greatly supports the efficiency of Q to chase RNA polymerase from the +16/17 pause site and to readthrough the tR′ terminator. With the exception of NusA no other host factor seems to be involved in the Q-dependent antitermination system. It is assumed that Q interacts directly with the RNA polymerase. The interaction probably takes place at the same site used by the N protein (Fig. 5.13).

It is not known exactly how Q suppresses termination at sites far downstream from the *qut* site because Q binding to *qut* is essential for termination resistance. Models explaining antitermination by Q have to take into account that perhaps the *qut* site remains linked to the RNA polymerase while it elongates through the late λ genes. This could be envisaged if the DNA forms a large loop while the *qut* site and Q are carried along with the RNA polymerase. An alternative explanation might be that the polymerase undergoes a long time conformational change, which is mediated by the transient contact between *qut*-bound Q. This conformation might be frozen by the presence of Q and probably enables RNA polymerase to readthrough terminators, even far distal from *qut*.

5.4.4 rRNA transcription and antitermination

Antitermination is by no means restricted to the transcription of λ genes. Many examples exist for transcription antitermination in bacteria and also in eukaryotes. The mechanism by which terminator readthrough occurs can be completely different, however, in the various cases. A mechanism involving many elements in common with phage λ antitermination has been described for the transcription of rRNA operons in bacteria. In *E. coli* rRNAs are transcribed from seven operons which are highly homologous (see Section 8.7). Since rRNAs are not translated, putative termination sites are not protected during transcription by translating ribosomes, and therefore may be prone to termination-factor Rho binding. Hence, it might be expected that, during rRNA transcription, premature termination could frequently occur at cryptic termination sites, if not prevented by a specific antitermination system. This conjecture was confirmed when Rho-dependent termination signals were placed within rRNA transcription units. These terminators did not cause efficient termination of transcription but RNA polymerase readthrough was observed

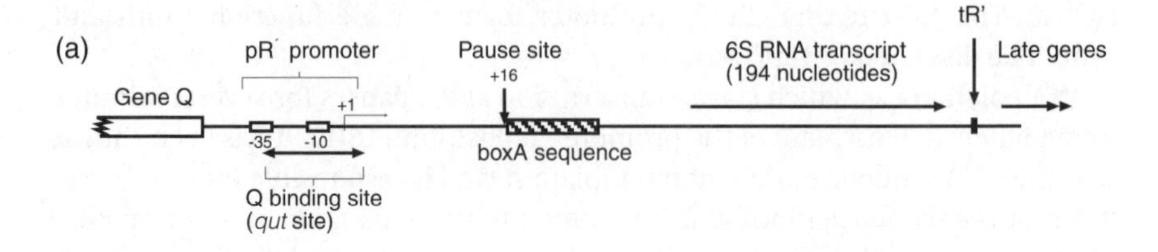

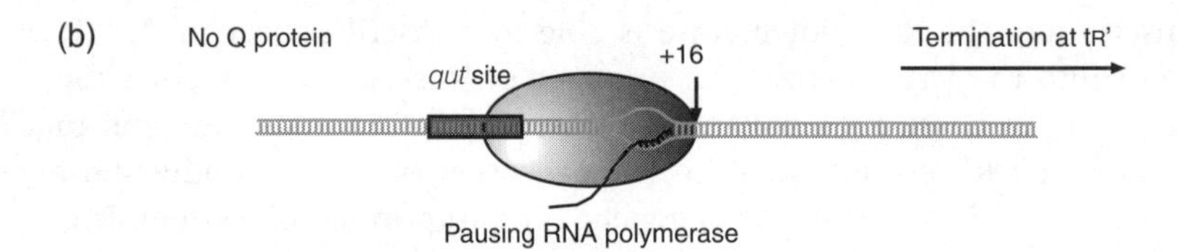

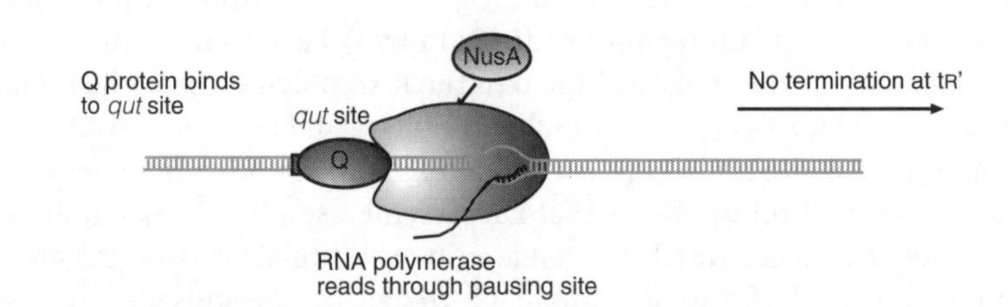

Figure 5.13 Structure of the λ Q antiterminator region. (a) A map of the phage λ late gene regulatory region is shown. The pR′ promoter and the overlapping Q binding site (*qut* site) are indicated. A conserved boxA sequence is found flanking a strong pausing site at position + 16 of the λ pR′ start. A Rho-dependent transcription terminator (tR′) is found 194 bases downstream. Termination at tR′ results in a 6S RNA transcript. (b) Termination at tR′ is dependent on the availability of the phage Q protein. In the absence of Q, RNA polymerase undergoes pausing at position + 16 without institution of antitermination activity. Transcription is terminated at tR′. In the presence of Q, binding occurs to the *qut* site. Bound Q appears to contact RNA polymerase and facilitates reading through the pause site. Q either stays bound to RNA polymerase or, if still tethered to the *qut* site, may cause a DNA loop during the elongation of RNA polymerase. As a result, the modified RNA polymerase reads through the tR′ terminator into the late phage genes. Although the presence of NusA is not necessary for antitermination at *qut* it is known to enhance the readthrough efficiency.

through these sites. Sequence inspection of bacterial rRNA operons revealed the presence of *nut*-like sequences close to the rRNA promoters P2 upstream of the 16S rRNA gene. A boxA-like sequence with close homology to the λ phage boxA is present in all seven rRNA transcription units. In addition, an interrupted palindrome comparable to the λ boxB sequence is found upstream of the boxA-like sequence in all seven rRNA operons. A second set of *nut*-like sequences can be found in the spacer region between the genes for 16S and 23S rRNA, and a conserved sequence identical to the boxC sequence is also present upstream of all 16S rRNA genes. The arrangement of the *nut*-like sequences upstream of the 16S rRNA occurs in a different order compared to phage λ, however. The order is boxB–boxA–boxC in all rRNA operons.

It transpires that the boxA-like element is the crucial determinant for the suppression of terminators placed within rRNA genes. However, Nus proteins are also required. On the other hand, it is clear that the N protein, which is not present in uninfected *E. coli* cells, cannot be involved in rRNA antitermination. No other equivalent protein which could substitute for phage λ N protein has been identified to date in *E. coli*. In addition to the boxA sequence, NusB seems to play an important role. Together with S10 (NusE) it can bind directly (in the absence of RNA polymerase) to the *rrn nut*-like boxA sequence. Interestingly, transcription elongation of *rrn* operons is impaired in cells with mutations in NusB. The other Nus factors are also required; however, not with the same stringency. To obtain efficient antitermination effects in *in vitro* conditions, the addition of a crude cell extract proved to be necessary. The complete set of known Nus factors NusA, NusB, NusE and NusG alone did not suffice. From this observation it was concluded that one or more unidentified cellular components, in addition to the known Nus factors, are probably required for the complete rRNA antitermination system. In contrast to the N-dependent antitermination system of λ the rRNA antitermination system enables RNA polymerase to transcribe through Rho-dependent but not to intrinsic or factor-independent terminators (note that strong Rho-independent terminators, usually arranged in tandem, are found at the end of rRNA operons). Several additional observations indicate that any rRNA transcriptional antitermination system should be substantially different from the phage λ system. The lack of the N protein and the different order in the arrangement of the *nut*-like sequences boxA, B (and C) has already been mentioned. The response to mutations in the boxA/boxB sequences is different for rRNA and N-directed antitermination. The role of the boxA sequence is of much greater importance in rRNA antitermination and apparently seems to be sufficient to prevent premature transcription termination. Moreover, certain terminators do not respond in the same way in both systems. The Rho-independent strong terminators at the end of the rRNA operons are particulary differentially affected.

Antitermination of rRNA transcription cannot be regarded as an independent function. The leader sequences of rRNA operons are known to participate in a number of additional functions during the synthesis, folding and maturation of rRNAs. The *nut*-like sequences within the rRNA leader region have been shown to assist and facilitate structure formation and assembly of the small ribosomal subunit in a chaperone-like way. Folding of the rRNA secondary structure, assembly of the ribosomal particles and processing of the long primary rRNA transcript occur simultaneously with transcription. Preribosomal particles are formed before RNA polymerase has reached the end of the transcription unit. During these complex steps of ribosome biogenesis, parts of the leader rRNA *nut*-like sequences undergo transient interactions with sequence elements within the first 400 nucleotides from the 5′ end of the mature 16S rRNA. The role of the Nus proteins during these interactions is unclear. It should also be borne in mind that NusE, which directly interacts with the rRNA leader boxA sequence, is an integral ribosomal constituent (S10) found in the mature 30S ribosomal subunit. It is feasible that the complex steps of transient RNA–RNA or RNA–protein interactions which facilitate 30S particle formation may effectively prevent Rho factor from binding to the rRNA transcript. It should be noted that the leader region otherwise provides an ideal *rut* site for Rho factor interaction. If maturation and assembly are considered to be higher order mechanistic aspects, antitermination of rRNA transcription may simply be a byproduct of the complex reactions leading to a mature ribosome. In analogy with the ECP complex mediating N-dependent transcription antitermination in phage λ, the complex responsible for the correct sequence of events during ribosome biogenesis has been termed an **assembly control particle** (ACP). During evolution the phage λ may have adopted parts of this mechanism and refined it for antitermination to regulate delayed early or late gene expression.

5.4.5 *nut* site-independent antitermination mechanisms

A different mechanism of transcriptional control through antitermination involving uncharged tRNA molecules as positive regulators has been identified in *B. subtilis* and several other Gram-positive genera. About 20 aminoacyl tRNA synthetase and amino acid biosynthesis genes which belong to this family have been characterized; they are regulated by a common mechanism. Like the attenuation control described for the *B. subtilis trp* operon or the *B. subtilis pyr* operon, these genes contain intrinsic terminator structures in their mRNA leader sequences upstream of the translational start signal for the structural genes. The conserved mRNA leader can fold into alternative structures which disable the formation of the terminator. The conformational switch, which defines whether transcription termination or readthrough occurs, is mediated by a cognate tRNA molecule that interacts with the highly conserved portions of the mRNA leaders. This mechanism is usually classified as antitermination, although in principle it might as well be regarded as attenuation.

The *tyrS* gene of *Bacillus subtilis* is the best characterized member of this type of control. Therefore it shall be presented here as an example in more detail. In *B. subtilis* the aminoacyl tRNA synthetase genes are regulated in response to the availability of the cognate amino acid and not by general amino acid limitations. In the case of the tRNA synthetase gene *tyrS*, uncharged $tRNA^{Tyr}$ interacts with the nascent *tyrS* leader. This interaction occurs between a UAC sequence within the mRNA leader (which specifies the appropriate amino acid codon for the tyrosyl tRNA) and the anticodon of the $tRNA^{Tyr}$. The sequence of the mRNA leader that contains the specific codon sequence is termed the **specifier sequence**. The codon-like sequence is located in a conserved bulged loop structure of the leader RNA. In case of the *tyrS* leader transcript a UAC triplet coding for tyrosine interacts with the cognate anticodon sequence of $tRNA^{Tyr}$. For other synthetase genes the corresponding cognate codons are found at the same position of their mRNA leaders. Thus, each gene is induced by limitation for the cognate amino acid and not by general amino acid limitation. The tRNA that is bound by codon–anticodon interaction to the specifier sequence simultaneously binds to a conserved sequence of a side bulge of the antiterminator structure, which is mutually exclusive to the terminator. This sequence is called the **T-box**. It has a conserved sequence of **UGGNACC**. Probably the interaction occurs by base pairing between the CCA nucleotides present in the acceptor stem of all tRNAs and the conserved sequence of the T-box. This interaction changes the leader RNA from a termination-proficient to an antiterminator structure. Charged $tRNA^{Tyr}$ does not interact with the *tyrS* leader and transcription is terminated at the downstream Rho-independent terminator structure (Fig. 5.14).

The observed lack of interaction might be because of a different conformation of the charged *versus* the uncharged aminoacyl tRNA. Alternatively, it is explained by steric hindrance caused by the amino acid linked to the CCA terminus.

The same mechanism acts for a number of additional tRNA synthetase genes in *B. subtilis*, such as *trpS*, *thrS*, *leuS* or *pheS* and also for several tRNA synthetase genes from other Gram-positive bacteria. Each tRNA synthetase gene has a similar leader structure containing the appropriate codon at a conserved position which is recognized by the corresponding anticodon of the respective tRNA. In this way a common antitermination mechanism provides specificity for the lack of individual amino acids. Perturbations in the level of a cognate amino acid will induce only the corresponding individual tRNA synthetase.

The type of antitermination control described above uses uncharged tRNA as an effector and exerts its effect by direct interaction between the mRNA leader and the uncharged tRNA. No translation is involved in this mechanism. In contrast, the antitermination control of amino acid biosynthetic operons in *E. coli* involves aminoacylated tRNA as the sensor, and translating ribosomes are required to cause the switch from readthrough to transcriptional termination.

It is possible that protein factors are involved in mediating the tRNA-directed antitermination mechanism. Proteins may, for instance, be important for the stabilization of the antiterminator structure or facilitate the interaction between the anticodon and the specifier sequence. Although the participation of proteins is likely, no evidence for a direct involvement has been established to date.

A different type of regulation by antitermination of transcription is known to act at several catabolic operons. The mechanism is substrate-induced and effected by **antiterminator proteins**. Probably the best characterized example is the regulation of the *bgl* operon of *E. coli*. The *E. coli bgl* operon contains three genes, *bglG*, *bglF* and *bglB*, that are involved in the utilization of aromatic

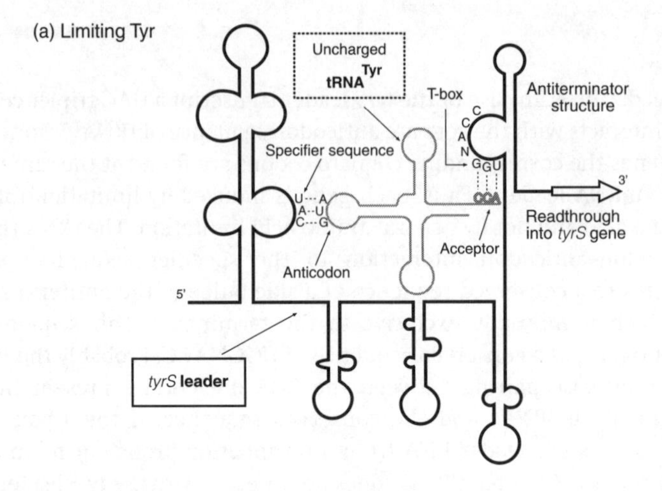

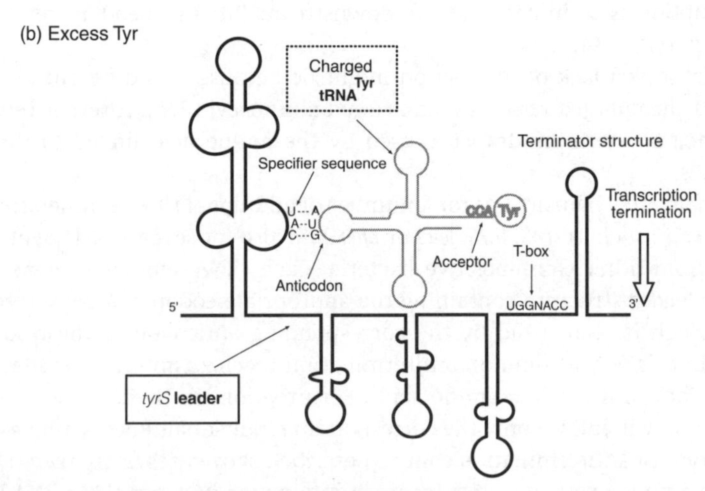

Figure 5.14 Regulation of aminoacyl tRNA synthetase operons in Gram-positive bacteria by transcription antitermination. The leader region of the *tyrS* gene, which confers antitermination activity, is shown. Conserved structural elements located upstream of the translational start site direct the formation of stem-loop structures indicated by a thick dark line. The 3′ end of the leader has the potential to form a Rho-independent terminator hairpin before transcription into the *tyrS* gene occurs. A conserved sequence element, the T-box, forms part of an antiterminator structure with conserved sequences on the 5′ side of the terminator stem. The T-box contains a complementary sequence which can bind to the acceptor CCA sequence of uncharged tRNAs. Within the first stem-loop structure of the leader transcript is a sequence region, termed the specifier sequence, which is complementary to the $tRNA^{tyr}$ anticodon. (a) At limiting Tyr concentrations uncharged $tRNA^{tyr}$ is bound to the leader RNA via complementary base pairing interactions between the anticodon bases and bases of the specifier sequence, and the CCA end and bases within the T-box. (b) At excess tyrosine all $tRNA^{Tyr}$ is aminoacylated. Charged $tRNA^{Tyr}$ is not complexed to the T-box and the 3′ end of the leader forms into a terminator hairpin. Readthrough into the *tyrS* gene is inhibited. The figure is arranged according to Henkin (1994).

β-glucosides. The gene *bglG* codes for a protein that belongs to a family of antiterminator proteins. The *bglF* gene encodes a transporter for β-glucosides. It belongs to the family of phosphotransferases (PTS). The *bglB* gene is responsible for the hydrolysis of phosphorylated β-glucosides. The order of genes within the transcription unit is *bglG–bglF–bglB*. The transcript upstream and downstream from *bglG* contains Rho-independent terminators. Normally, the *bgl* operon in *E. coli* is cryptic, which means it is not transcribed. The inactive promoter can be activated and complete transcripts can be induced by β-glucoside substrates. The antiterminator protein BglG acts as a positive regulator. It can be phosphorylated and dephosphorylated by BglF, the gene product of the *bglF* gene. In its dephosphorylated state BglG exists as a dimer in solution. The BglG dimer can bind to a 32-nucleotide target sequence within the *bgl* mRNA leader, which partially overlaps the first terminator structure. This binding induces an alternative stem-loop structure which precludes formation of the terminator hairpin. Binding of BglG therefore enables RNA polymerase to readthrough the terminators. The *dephosphorylated* dimeric BglG protein represents the active regulator. In its *phosphorylated* form, BglG exists as a monomer and cannot bind to the *bgl* mRNA. Consequently, it can not support transcriptional readthrough. Phosphorylation of BglG by BglF occurs in the absence of β-glucosides. Therefore, at limiting β-glucosides the phosphorylated monomeric BglG has no antitermination activity and transcriptional readthrough of the operon is prevented.

The BglG protein is structurally and functionally homologous to the SacY, SacT and several other proteins of *B. subtilis* involved in the utilization of sucrose. All of these proteins have been shown to belong to a family of regulators able to prevent transcription termination. SacY and SacT can stimulate the expression of the *sacB* or the *sacPA* genes, respectively. Both regulons contain leader sequences with the capacity to fold into alternative terminator/antiterminator structures. Binding of the proteins SacY or SacT stabilizes the antiterminator conformation, enabling readthrough of transcription and expression of the *sacB* or *sacPA* genes. The activity of both proteins is regulated by phosphorylation/dephosphorylation, similar to the case of BglG. In conclusion, many catabolic operons in *B. subtilis* are regulated through antitermination involving specific antiterminator proteins.

Summary

Transcription termination not only determines the end of a transcription unit. The efficiency of termination is not an all or nothing decision and therefore has important regulatory implications for the cell. The expression of many genes is regulated through the modulation of the sites or the efficiency of termination. Generally two different mechanisms that cause transcription termination are known. *Factor-independent termination* is triggered by a specific secondary structure of the nascent transcript and does not require auxiliary proteins. The transcript structure that causes termination consists of a stable *GC-rich stem-loop* followed by a *run of uracil residues*. Formation of the stem-loop structure causes RNA polymerase pausing in the first place. Such pausing always precedes termination. For some transcripts the efficiency of factor-independent termination can be influenced by individual promoters and the early transcribed sequences. Sequences downstream of the termination site can also affect the termination efficiency.

The stable hairpin structures at the 3′ ends of transcripts resulting from factor-independent terminations are of major importance for the stability of the transcripts. They prevent exonucleolytic degradation and thus enhance the lifetime of the respective RNA molecules.

The second mechanism of transcription termination depends on the action of a specific protein, namely the *termination factor Rho*. Functional Rho molecules consist of six identical subunits arranged in a ring-like structure. Rho binds to single-stranded RNA, which lacks strong secondary structure. A minimal binding site of about 100 nucleotides is required and binding occurs preferentially to RNA sequences with a high C and a low G content. RNA-bound Rho protein exhibits ATPase and RNA:DNA helicase activities, both of which are essential for termination. RNA sequences that show specific Rho binding are termed *rut* sites. Usually termination occurs many base pairs downstream from the *rut* site, which Rho uses as an entry site. Termination requires that Rho, after binding to the *rut* site, migrates in a 3′ direction along the transcript, approaching the transcription complex. RNA polymerase pausing allows Rho to catch up with the elongating RNA polymerase. Pausing is thus an important prerequisite for factor-dependent termination.

Several auxiliary proteins are known that affect the efficiency of Rho-dependent termination. *NusG*, which is able to interact with both RNA polymerase and Rho, supports termination at sites of the growing transcript, where Rho by itself cannot bind efficiently. Another Nus protein, namely *NusA*, affects the efficiency of Rho by enhancing RNA polymerase pausing. In addition, several proteins have been characterized which function as phage-specific termination factors. Examples are the *Tau protein*, directing phage T3- and T7-specific termination events, and the *Nun protein* of the lambdoid phage HK022, which supports termination at the phage λ *nutR* region.

The termination factor Rho acts as a *coupling factor* between transcription and translation. Transcription of efficiently translated mRNAs is normally not terminated by Rho, although many cryptic Rho-dependent termination sites may be present. Binding of Rho to the nascent transcripts is inhibited because potential *rut* sites are occupied by polysomes. Conditions that cause ribosome release because of translational arrest or premature translational termination allow the Rho factor to bind to ribosome-free regions of the transcript and to terminate transcription.

The expression of many biosynthetic operons is regulated by a controlled switch mechanism between readthrough and termination at sites termed *attenuators*. The attenuator typically describes an RNA structure which can exist either as a Rho-independent terminator hairpin or an alternative secondary structure (the antiterminator) that allows transcription to proceed. The switch between these two mutually exclusive structures is triggered by the efficiency of translation through the attenuator region. Attenuation thus synchronizes transcription and translation; in case translation ceases transcription will be halted or terminated. Alternatively, under certain limiting conditions, the rate of transcription can be slowed down, causing translating ribosomes to catch up with the transcription complex and change the attenuator structure from termination to readthrough mode. In this case formation of the terminator or antiterminator structure of the transcript depends on transcriptional pausing. Several systems controlled by attenuation which do not involve translation for the switch between the attenuator and antiterminator structure are known. Instead, specific RNA-binding proteins trigger the interconversion. One example is the *TRAP protein* multimer, which binds to the regulatory region of the *B. subtilis trpEDCFBA* operon, causing the formation of a Rho-independent terminator structure. TRAP binding is activated by the amino acid tryptophan. Therefore, efficient transcription of the operon occurs only in the absence of excess tryptophan. In a similar way, the allosterically regulated *PyrR protein* binds to the *B. subtilis pyr* operon leader where it causes a change from the antiterminator to the alternative terminator structure.

Gene expression in lambdoid phages is regulated by a mechanism of termination and *antitermination*. Transcription of genes distal to termination signals is achieved by a modification of RNA polymerase which renders it resistant to termination. In case of the λ *N protein*-specific mechanism the change in termination efficiency is brought about by the formation of an *elongation control particle* consisting of the transcribing RNA polymerase, at which the phage N protein and four different host-specific Nus proteins, together with a region of the growing transcript termed the *nut site*, are assembled. The RNA polymerase–Nus protein–*nut* RNA complex is programmed over time and distance to readthrough terminators. Expression of the late phage proteins is regulated by the *phage Q protein* which, together with the host NusA protein and a DNA determinant called *qut*, renders RNA polymerase to a termination-resistant form.

Premature termination during *rRNA transcription* is prevented by an antitermination mechanism related to the N protein-mediated system, without the requirement of N, however. Only Rho-dependent terminators, not Rho-independent terminators are ignored during rRNA antitermination. The expression of several groups of other genes involves regulation through antitermination. The transcription of certain *B. subtilis* tRNA synthetases is controlled by a common antitermination mechanism involving tRNA molecules as effectors. Moreover, a number of catabolic operons, like the *bgl* operon, are regulated by antitermination through specific RNA binding proteins that trigger the interconversion between terminator and antiterminator structures of the nascent transcripts.

References

Alifano, P., Rivellini, F., Limauro, D., Bruni, C. B. and Carlomagno, M. S. (1991) A consensus motif common to all Rho-dependent prokaryotic transcription terminators. *Cell* **64**: 553–63.

Henkin, T. M. (1994) tRNA-directed transcription antitermination. *Molecular Microbiology* **13**: 381–7.

Rivellini, F., Alifano, P., Piscitelli, C., Blasi, V,. Bruni, C. B. and Carlomaagno, M. S. (1991) A cytosine- over guanosine-rich sequence in RNA activates rho-dependent transcription termination. *Molecular Microbiology* **5**: 3049–54.

von Hippel, P. and Yager, T. D. (1992) The elongation-termination decision in transcription. *Science* **255**: 809–12.

von Hippel, P. and Yager, T. D. (1991) Transcript elongation and termination are competitive kinetic processes. *Proceedings of the National Academy of Sciences USA* **88**: 2307.

Yanofsky, C. (1981) Attenuation in the control of expression of bacterial operons. *Nature* **289**: 751–8.

Further reading

Aksoy, S., Squires, C. L. and Squires, C. (1984) Evidence for antitermination in *Escherichia coli* rRNA transcription. *Journal of Bacteriology* **159**: 260–4.

Alifano, P., Rivellini, F., Limauro, D., Bruni, C. B. and Carlomagno, M. S. (1991) A consensus motif common to all Rho-dependent prokaryotic transcription terminators. *Cell* **64**: 553–63.

Bear, D. G. and Peabody, D. S. (1988) The *E. coli* Rho protein: an ATPase that terminates transcription. *Trends in Biochemical Sciences* **13**: 343–7.

Berg, K. L., Squires, C. and Squires, L. C. (1989) Ribosomal RNA operon antitermination. Function of leader and spacer region *boxB–boxA* sequences and their conservation in diverse micro-organisms. *Journal of Molecular Biology* **209**: 345–58.

Chan, C. L. and Landick, R. (1994) New perspectives on RNA chain elongation and termination by *E. coli* RNA polymerase. In: Conaway, R. C. and Conaway, J. W. (eds.) *Transcription: Mechanisms and Regulation*. New York: Raven Press; pp. 297–321.

Das, A. (1992) How the phage lambda *N* gene product suppresses transcription termination: communication of RNA polymerase with regulatory proteins mediated by signals in nascent RNA. *Journal of Bacteriology* **174**: 6711–16.

Das, A. (1993) Control of transcription termination by RNA binding proteins. *Annual Reviews of Biochemistry* **62**: 893–930.

Friedman, D. I. and Court, D. L. (1995) Transcription antitermination: the λ paradigm updated. *Molecular Microbiology* **18**: 191–200.

Gollnick, P. (1994) Regulation of the *Bacillus subtilis trp* operon by an RNA-binding protein. *Molecular Microbiology* **11**: 991–7.

Henkin, T. M. (1994) tRNA-directed transcription antitermination. *Molecular Microbiology* **13**: 381–7.

Henkin, T. M. (1996) Control of transcription termination in prokaryotes. *Annual Reviews of Genetics* **30**: 35–57.

Landick, R. (1997) RNA polymerase slides home: pause and termination site recognition. *Cell* **88**: 741–4.

Landick, R., Turnbough, C. L. and Yanofsky, C. (1996). Attenuation control. In: Neidhard, F. C., Curtiss III, R., Ingraham, J. L. *et al.* (eds.) *Escherichia coli* and *Salmonella* Cellular and Molecular Biology, Vol.1. Washington DC: ASM Press, pp. 1263–86.

Martin, F. H. and Tinoco, I. J. (1980) DNA–RNA hybrid duplexes containing oligo(dA:rU) sequences are exceptionally unstable and may facilitate termination of transcription. *Nucleic Acids Research* **8**: 2295–9.

Oxender, D., Zurawski, G. and Yanofsky, C. (1979) Attenuation in the *Escherichia coli* tryptophan operon: role of RNA secondary structure involving the tryptophan codon region. *Proceedings of the National Academy of Sciences USA* **76**: 5524–8.

Platt, T. (1986) Transcription termination and the regulation of gene expression. *Annual Reviews of Biochemistry* **55**: 339–72.

Platt, T. (1994) Rho and RNA: models for recognition and response. *Molecular Microbiology* **11**: 983–90.

Richardson, J. P. (1990) Rho-dependent transcription termination. *Biochimica et Biophysica Acta* **1048**: 127–38.

Richardson, J. P. (1991) Preventing the synthesis of unused transcripts by Rho factor. *Cell* **64**: 1047–9.

Richardson, J. P. (1993) Transcription termination. *Critical Reviews in Biochemistry and Molecular Biology* **28**: 1–30.

Richardson, J. P. and Greenblatt, J. (1996) Control of RNA chain elongation and termination. In: Neidhard, F. C., Curtiss III, R., Ingraham, J. L. *et al.* (eds.) *Escherichia coli* and *Salmonella* Cellular and Molecular Biology, Vol.1. Washington DC: ASM Press, pp. 822–48.

Rosenberg, M. and Court, D. (1979) Regulatory sequences involved in the promotion and termination of RNA transcription. *Annual Reviews of Genetics* **13**: 319–53.

Sullivan, S. L. and Gottesman, M. E. (1992) Requirement for *E. coli* NusG protein in factor-dependent transcription termination. *Cell* **68**: 989–94.

Telesnitsky, A. P. W. and Chamberlin, M. J. (1989) Sequences linked to prokaryotic promoters can affect the efficiency of downstream termination sites. *Journal of Molecular Biology* **205**: 315–30.

Vogel, U. and Jensen, K. F. (1997) NusA is required for ribosomal antitermination and for modulation of the transcription elongation rate of both antiterminated RNA and mRNA. *Journal of Biological Chemistry* **272**: 12265–71.

von Hippel, P. H. (1998) An integrated model of the transcription complex in elongation, termination, and editing. *Science* **281**: 660–5.

von Hippel, P. H. and Yager, T. D. (1992) The elongation-termination decision in transcription. *Science* **255**: 809–12.

Whalen, W. A. and Das, A. (1990) Action of an RNA site at a distance: role of the *nut* genetic signal in transcription antitermination by phage-λ *N* gene product. *The New Biologist* **2**: 975–91.

Yager, T. D. and von Hippel, P. H. (1991) A thermodynamic analysis of RNA transcript elongation and termination in *E. coli*. *Biochemistry* **30**: 1097–118.

Yanofsky, C., Konan, K. V. and Sarsero, J. P. (1996) Some novel transcription attenuation mechanisms used by bacteria. *Biochimie* **78**: 1017–24.

6

The role of DNA structure in transcription regulation

This chapter elucidates the role of DNA conformation and topology in the process of transcription. In particular, the information stored in the higher order structure of DNA rather than the direct information residing in the primary sequence is the subject of this section. The chapter begins with a brief discussion on base methylation as a means of transcriptional regulation in prokaryotes. The physical background of DNA curvature and the implications of curved DNA for transcription initiation are summarized next. The special role of curved sequences upstream of strong promoters is highlighted and possible mechanisms as to how such sequences may contribute to promoter efficiency are discussed. There follows a section describing the influence of transcription factors in conjunction with curved upstream DNA sequences on promoter activity. A special section is devoted to the phenomenon of DNA supercoiling and, finally, the tight interrelation between transcription and DNA topology is exemplified by the 'twin supercoiled domain' model.

The informational character of DNA molecules is generally believed to reside in the nucleotide sequence or primary structure. About 20 years ago it became apparent that the information stored in DNA is not only reflected by its primary sequence, however. Clearly, the order of nucleotides in the DNA encodes the primary sequence of proteins or structural RNA molecules. Therefore, the informational character was mainly seen in the propensity to form complementary (Watson–Crick-type) base pairs. A complete new dimension of information residing in the DNA structure was recognized with the discovery that the conformation and spatial structure or topology of the DNA molecule itself contains essential information for the regulation of gene expression. In these cases the primary structure appears to serve no other function than to provide the basis for the spatial structure of the molecule in three dimensions. The architecture or contour of the DNA can either directly serve as a regulatory signal or provide a defined three-dimensional platform which can be recognized by structure-specific proteins with regulatory functions. Implications for transcription resulting from the additional information inherent in changes to DNA conformation or topology are a focus for this chapter. Before a closer look is taken at the architectural aspects determined by DNA conformation and topology, a related question should be addressed. This question concerns the chemical modification of DNA. Given the fact that recognition of DNA can be determined by the nucleotide sequence *and* the higher order structure of the overall DNA molecule, can the chemical modification of DNA bases provide additional information or specificity for this recognition process?

6.1 Is DNA methylation involved in regulation?

The specific methylation at the C5 position of cytidines within DNA sequences (CG sequences) plays an essential role in the regulation of eukaryotic gene expression. Does a similar control mechanism also exist in bacterial transcription regulation? Although DNA methylation occurs in prokaryotes, it does not play a major role in the regulation of bacterial transcription. However, several clear examples are known where bacterial transcription is coupled to the modification of DNA through methylation. It is known that several different methylation systems exist in bacteria. The *E. coli* ***dam*** **gene**, for instance, encodes an adenine-specific methylase which converts adenine residues in the sequence context GATC to 6-methyl adenine. The *dam* methylation functions as an essential component of the **mismatch repair** system. It serves to decorate the newly synthesized daughter strand, enabling its distinction from the parent strand after replication. Because methylation is secondary to replication, newly synthesized DNA transiently exists as hemimethylated molecule. It is the unmethylated DNA strand that is the preferred substrate for the mismatch repair system. There is good evidence, however, that the *dam* methylation system is also involved in the regulation of transcription initiation at some promoters. A number of genes have been identified which contain potential *dam* methylation sites within their promoter sequences. It can be shown that these genes are controlled through methylation of their GATC sequences. Three such examples are summarized in Fig. 6.1. Methylation sites (GATC sequences) are found either in the –10 or –35 regions or in both promoter hexamers. Fig. 6.1 shows the promoter sequences for the genes *dnaA* which codes for the replication initiator protein DnaA, *trpR*, encoding the tryptophan repressor, and *glnS*, which encodes the glutaminyl tRNA synthetase. Other examples, not shown in the figure, are *sulA*, which encodes an inhibitor for cell division and the phage P1 gene *cre*, encoding a DNA ligase responsible for the circularization of the phage chromosome. The transcription efficiency of the genes in question is reduced when the DNA changes from unmethylated to the methylated state. It is therefore assumed that the methyl groups directly interfere with RNA polymerase recognition. What might be the purpose of linking transcription to *dam* methylation? A possible answer may be given by the fact that *dam* methylation is able to link expression to the cell cycle. The logic of such a notion is immedi-

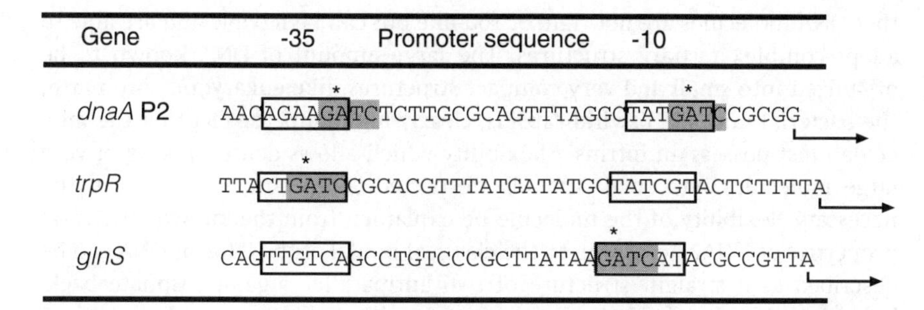

Figure 6.1 *dam* methylation sites within promoters. The *dnaA* P2, the *trpR* and the *glnS* promoter regions are shown. The −35 and −10 regions are boxed. N6 methyl adenosines are indicated by asterisks. The coding strand sequence is given, and GATC sequences are shaded. The transcription start site is indicated by an arrow. The information is taken from Plumbridge (1987).

ately evident for genes like *sulA* or phage P1 *cre*, which are required only once during a cell cycle. It is more difficult to understand why *trpR* or *glnS* should be linked to the cell cycle, however. The presence of *dam* sites within these promoter sequences might therefore be accidental. It remains to be seen whether these genes have an additional function related to the cell cycle that is not yet known or whether their promoters contain *dam* methylation sites by chance. The methylation of GATC sequences by the *dam* methylase is known to play a role in another rather complex regulatory system, namely during the expression of **fimbriae**. The expression of fimbriae is subject to **phase variation**, which is characterized by a reversible switch between alternative expression states: ON or OFF. Two GATC sequences are located in the regulatory region of the fimbrial *papAB* and *papI* operons. The GATC sites overlap binding sites for the regulatory protein Lrp (see Section 8.4) which, together with the PapI protein, controls *pap* gene expression. Methylation of both GATC sequences blocks Lrp binding. On the other hand, binding of Lrp to the unmethylated site allows PapI to bind to the complex. *dam* methylation is inhibited and the system is switched from the OFF to the ON state. It is important to note that this complex regulatory system requires DNA replication to alter the methylation state of the GATC sites involved (see Fig. 8.6).

6.2 Effects of DNA curvature on transcription initiation

From the exact parameters that determine the three-dimensional structure of the double helical form of B-DNA it might be implied that B-DNA can be viewed as a continuous rod-like molecule. It is clear, however, from early DNA studies that DNA molecules are not rigid or rod-like but can be flexible and are able to adopt complex tertiary structures. The large amount of DNA known to be organized into small and very compact structures, like eukaryotic chromatin, the bacterial nucleoid or viral capsids, clearly highlights the fact that the molecule must possess an intrinsic flexibility which allows dense packing of very large molecules into small spatially restricted compartments. How can the necessary flexibility of the molecule be explained from the known structural properties of DNA? According to the classical model of the B-form, DNA can be described as a straight structure of two antiparallel sugar–phosphate backbones in a right-handed helical arrangement with complementary bases paired perpendicular to the helix axis. The B-form DNA helix is further characterized by a **major** and a **minor groove**, which are located on the exterior surface of the two interwound DNA strands. The two grooves are characterized by the location of the C5 positions of pyrimidines and the N7 position of the purines, which point into the major groove, while the C2 keto group of pyrimidines and the C2 position of purines face the minor groove (Fig. 6.2). Each base pair is rotated relative to the next around the helix axis by about 34–36°. A complete helical turn, therefore, involves 10–10.5 base pairs. This rotation is termed **twist**. The planes of the base pairs are usually at right angles with respect to the helical axis. In the A form of DNA the plane of the base pairs is rotated around its small axis by about 20°. This rotation is designated **tilt**. Changes in the tilt angle will deform the helix axis in the direction of the sugar–phosphate backbone. The plane of a base pair can also be rotated around its long axis. This

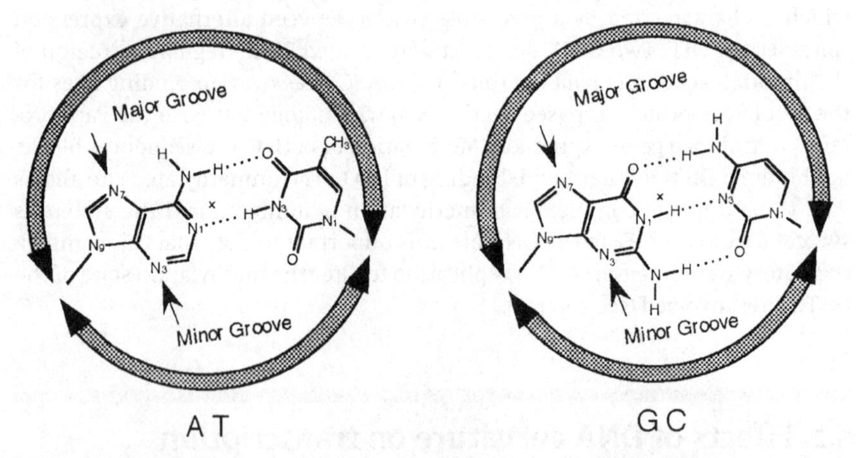

Figure 6.2 Location of the minor and major grooves in helical DNA base pairs. An AT and a GC base pair are shown in the base pair plane of helical DNA. The circular sphere of the DNA is outlined as thick dark lines with arrow heads. The larger segment represents that face of the helix where the major groove is located. The N7 and N3 positions of purines which are accessible in the major or minor grooves, respectively, are marked with arrows.

rotation is given by the **roll** angle. Changes of the roll angle result in deflections of the helix axis towards the major or minor grooves (Box 6.1).

When DNA is free in solution, it shows significant structural polymorphism. It forms a spectrum of structures with alternating B-form or non-B or A-like structures. Even rare conformers such as Z-DNA or cruciform structures may exist (Table 6.1).

In addition, DNA in solution also shows considerable conformational flexibility. Substantial alterations of the angles between adjacent base pair planes may be the result of thermal motions. For instance, increasing or decreasing the twist results in **torsional flexibility**. Alternatively, changes in tilt or roll angles may change the direction of the helix axis, giving rise to **axial flexibility**. Axial flexibility causes changes in the width of the major and the minor grooves, respectively, resulting in a curved shape of the molecule. The flexibility of a given DNA is not always **isotropic**. This means that there are preferred directions of deflection from the straight helix axis. The resulting DNA will then, on time-average, exist in a form which appears to be permanently bent. If this conformation remains without external constraints the DNA is termed to be **intrinsically curved**.

Intrinsic curvature depends on the sequence context of the DNA. Several theories have been derived to explain and predict sequence-dependent intrinsic curvature of a given DNA. Clearly, the main contribution to intrinsic curvature in natural DNA can be attributed to **A-tracts**. A-tracts are runs of AT base pairs with between two and six consecutive adenosines in one strand. They can be found in many natural DNAs where they are associated with strong intrinsic curvature. Notable examples are the fragments which can be isolated from the mitochondrial minicircle DNA of trypanosomatides (**kinetoplast DNA**). In addition, the DNA upstream sequences from several strong promoters belong to the same category of DNAs with strong intrinsic curvature. In each case the curved DNAs normally contain several A-tracts within their sequence. These A-tracts are not distributed at random, but appear to be spaced in a 10–11 base pair periodicity. Because of this periodicity the centres of the A-tracts are located at one side of the helix axis (roughly 10.5 base pairs per turn). Based on ligase catalysed rates of circularization, estimates of 17–23° have been made on the magnitude of DNA curvature resulting from a single A-tract of six consecutive adenosines (Boxes 6.2 and 6.3). Since both A- and GGCC-tracts may contribute to the intrinsic bend this value may be slightly overestimated. Different studies find a bending angle of about 12° for such an A-tract.

How can the curvature of a DNA fragment be determined? The most versatile method, still applied very successfully today, is polyacrylamide gel electrophoresis. DNA fragments containing a bent region migrate more slowly on polyacrylamide gels than straight (reference) DNA fragments of identical length (see Box 6.2).

At temperatures above the melting transition of the curved region (usually above 40°C), the effect of the curvature is lost and the corresponding DNA

Box 6.1 DNA structural parameters

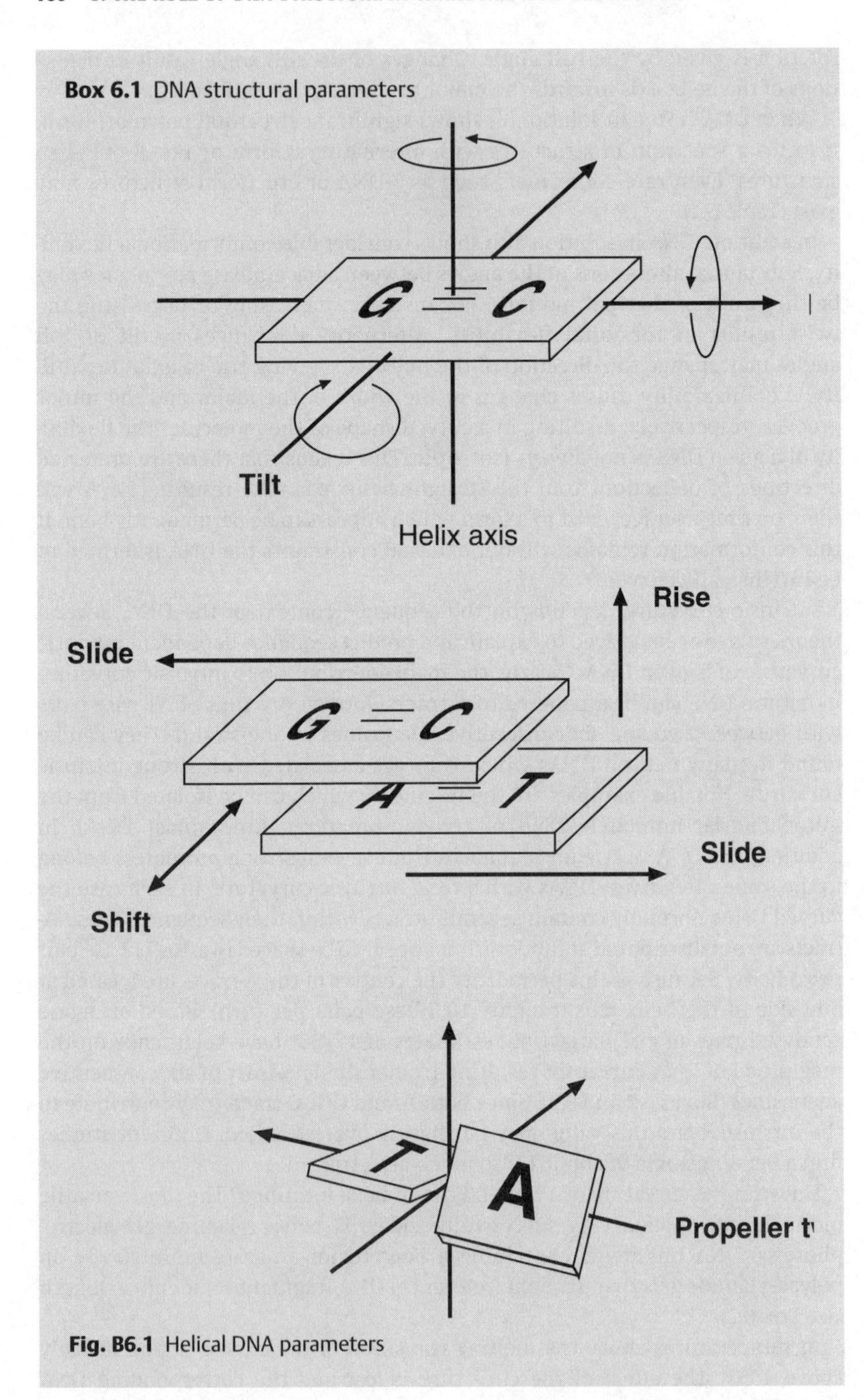

Fig. B6.1 Helical DNA parameters

DNA typically has a double-stranded structure of two antiparallel strands wound **plectonemically** around each other in right-handed orientation. The structure is held together by hydrogen bonds formed between bases. Base pairs are formed between each purine on the one strand and a complementary pyrimidine on the other strand (see Fig. 6.2). The purine A is complementary to the pyrimidine T (forms two hydrogen bonds) and the purine G is complementary to the pyrimidine C (forms three hydrogen bonds). The structure is additionally stabilized by **stacking interactions** between the parallel base pairs which are perpendicular to the helix axis in the centre of the double strand. The negatively charged phosphate residues (sugar–phosphate backbone) point to the outside of the helix. The helical structure is characterized by a **minor** and a **major groove**, which separate the two phosphodiester chains on the outside of the cylindrical molecule (see Fig. 6.2). DNA can exist in different helical structures (e.g. A-, B-, Z- or D-forms). With the exception of Z-DNA, all helical structures are right-handed, however, with different helical parameters (see Table 6.1). Deviations from the standard structure are described by a set of *rotational* (**twist, tilt, roll**) and *translational* (**rise, shift, slide**) parameters which are defined in Fig. B6.1. The same parameters are important to describe the conformation of curved (sequence-dependent) or bent (consequence of protein or ligand binding) DNA.

fragments migrate exactly according to their length. The anomalous migration of curved DNA can therefore be detected at low temperature gel electrophoresis. In this way the kinetoplastid DNA fragments, but also many DNA fragments obtained from regulatory regions of the genome, have been characterized as being curved. Sequence inspection of such curved DNA fragments normally reveals clusters of A-tracts, in line with the conclusion that A-tracts deform the path of the DNA helix.

It should be noted, however, that not only runs of As and Ts but other nucleotide steps may also cause intrinsic curvature. GGCC sequences, for instance, when repeated in helical phase, have also been shown to deform DNA and to produce stable curvature. The direction of curvature induced by A-tracts or GGCC repeats differs, however. It has been shown that the centre of A-tracts is found in an orientation such that the minor groove is facing the *inward* side of the curvature. In contrast, the minor groove of the centre of GGCC sequences points to the *outward* direction of the curvature. A different model, which has significant experimental support, predicts that A-tracts are straight and the intervening sequences are gently bent with positive roll angles.

All together quite a number of models have been developed to explain and predict regions of curvature within a given DNA sequence. Two of these models will be introduced in a simplified way. One model has been termed the **ApA wedge model**. This model relies on the assumption that every dinucleotide ApA step within a DNA sequence opens the base stack to form a wedge by a combination of changes in the tilt and roll angles. Such a deflection may yield

Table 6.1 Helical parameters for different DNA structures

	DNA A	DNA B	DNA Z	DNA D
Handedness	Right	Right	Left	Right
Diameter (Å)	23	19	18	21
Grooves	Major: deep Minor: shallow	Major: deep Minor: deep	Only one deep groove	Major: shallow Minor: deep and narrow
Periodicity (base pairs per turn)	11	10	12	8
Rise (Å) per base pair	2.56	3.38	3.71	3.03
Roll (°)	5.9 ± 4.7	−1.0 ± 5.5	3.4 ± 2.1	
Twist (°) per base pair	32.7	36.0	−30	45
Base tilt (°)	13–19	0	8	−16
Propeller twist (°)	15.4 ± 6.2	11 (GC), 17 (AT)	4.4 ± 2.8	
Sugar pucker	3′ endo-2′ exo	3′ exo-2′ endo 3′ endo-2′ exo	Pu:3′ endo-2′ exo Py:3′ exo-2′ endo	3′ exo-2′ endo
Examples	RNA douplex, RNA:DNA heteroduplex	Native DNA (Watson–Cricktype)	Alternating Pu:Py stretches, Poly (GC)	Modified (5-hydroxymethyl cytosine) and glycosylated phage T2 DNA, alternating AT-rich regions

Pu, purine: Py, pyrimidine; endo, exo, see Box 4.1, Fig. B4.1b p. 95

between 5 and 15° deviation from the straight helix axis. Several such elements of a bend must be repeated with the helical periodicity of the DNA (10–11 base pairs) to obtain a planar curvature. Otherwise, a randomly kinked trajectory may be the result. Any sequence region with periodically repeated runs of As longer than three will cause notable curvature.

An alternative model has been put forward which is called the **B junction model**. It starts from the observation that relaxed B-DNA is straight. However, polydA:polydT duplex molecules are structurally abnormal and exist in a non-B conformation. At the junction between the B and the non-B sections a marked change in the direction of the helix axis occurs, resulting in a kink. Several such

Box 6.2 Electrophoretic gel mobility shift analysis

Gel electrophoresis has proven to be a very versatile method of studying conformational transition of nucleic acids or the binding of proteins to nucleic acids. The method depends on the different mobilities of the naked DNA or RNA and the protein–nucleic acid complex. This change in mobility is determined by several factors: the ratio of mass of the protein to that of the nucleic acid, the alteration of the charge and changes in the conformation of the complex with respect to the free components. In almost all cases the mobility of a protein–nucleic acid complex is reduced compared to the free nucleic acid. Interestingly, the contribution of the conformational effect can exceed the increase in molecular weight and often results in highly aberrant mobilities (see Box 6.3).

Gel retardation depends also on the polyacrylamide concentration and the pore size of the gel (see also Box 6.3). Polyacrylamide gels between 4% and 10% have average pore sizes of between 50 and 200 Å, depending on the degree of cross-linking. It follows that there will be significant frictional drag on complexes with comparative dimensions. In comparison, the average pore sizes of agarose gels are in the range of 700–7000 Å and allow complexes of very large dimensions (DNA fragments over 1000 base pairs) to resolve; the conformational effect is lost in most cases, however.

It can generally be observed that protein–DNA complexes are more stable during electrophoresis than would be expected from their kinetic stability in solution. This apparent increase in stability is considered to be brought about by impeded diffusion of dissociated components within the gel matrix. The effect is vividly described as the **'caging effect'**.

Because of easily accessible methods for the quantification of bands in acrylamide gels (densitometry after staining or autoradiography), gel retardation can most successfully be applied to determine association, dissociation or equilibrium binding constants. It is also possible to determine thermodynamic parameters of nucleic acid–protein complexes if gels are run at different temperatures or if a special technique, e.g. **temperature-gradient gel electrophoresis** (TGGE), is employed (Hall and Kranz, 1995, Fried and Daugherty, 1998, and Rosenbaum and Riesner, 1987.

kinks spaced in helical phase will yield a molecule with a more or less planar curvature (Fig. 6.3).

DNA curvature can be influenced by the ionic environment. The presence of Mg^{2+} and Ba^{2+} ions, for instance, can have a profound effect on curved DNA. The DNA curvature can also be changed by drugs. One of the most widely used agents to change the conformation of DNA is probably **distamycin**. The drug distamycin binds to the minor groove of AT tracts. Binding extends the width of the minor groove and straightens out existing curvature. Distamycin has therefore

Box 6.3 Biochemical methods for determination of DNA curvature

Gel electrophoresis represents a powerful method for the conformational analysis of nucleic acids. In a first approximation the migration of a DNA molecule in a polyacrylamide gel can be described by the following equation:

$$v \cdot h^2 \cdot Q \cdot \frac{E}{L^2 \cdot f}$$

where v is the migration velocity, Q the effective charge, E the electric field, L the contour length, f the frictional coefficient, and h the end-to-end distance of the molecule. For straight molecules the empirical observation holds true that the mobility is a monotonic function of the molecule's length. Curved molecules, however, deviate from this rule. Their migration velocity is reduced compared to straight molecules. The magnitude in mobility change normally correlates with the altered end-to-end distance of a curved *versus* a straight molecule. The reduced distance of the ends apparently impedes with the worm-like fashion that a linear molecule is considered to migrate through a polyacrylamide or agarose gel (impeded reptation). The effect of curvature on gel electrophoretic mobility furthermore depends on the percentage of the acrylamide in the gel (higher concentrations decrease the pore size; see also Box 6.2) and also on the temperature (above 50°C static curvature is almost lost). Divalent metal ions, such as Mg^{2+} or Ba^{2+}, which are known to enhance curvature, or drugs like distamycin which reduces DNA curvature, exhibit a corresponding effect on the gel electrophoretic mobility.

The aberrant mobility of curved DNA fragments can be used to characterize the degree of curvature. Usually the *expected mobility* (μ_{act}), determined by the migration distance of a DNA fragment of identical length without curvature, divided by the *observed mobility* (μ_{obs}), the experimentally measured migration distance of the curved fragment, is expressed as the ***k*-factor** ($\mu_{act}/\mu_{obs} = k$). *k*-factors or *k*-values greater than one ($k > 1$) are indicative of a curved DNA structure. The maqnitude of a *k*-factor critically depends on the end-to-end distance of the fragment and thus on the degree of the bend. It also depends on the position of the bend with respect to the fragment ends. Gel electrophoretic migration is slowest when the bend is in the middle of the molecule compared to the same bend angle close to the ends of the fragment. At the middle position the end-to-end distance of a DNA fragment is minimal. From this observation an empirical equation has been derived from which bending angles might be obtained. The end-to-end distance of a rigid DNA fragment of the contour length L with the bend in the middle of the fragment is $L \cos \alpha/2$, with α defined as the angle at which the DNA is bent. The end-to-end distance of the same fragment with the bend at the end will be almost identical to the contour length L, however. When the relative electrophoretic mobilities (μ) of the same fragment in acrylamide gels are determined with the bend in the middle (μ_M) or the bend at the end (μ_E) the ratio of the mobilities can be expressed as:

$$\mu_M/\mu_E = \cos \alpha/2$$

Reasonable bending angles α can thus be obtained for fragments between 300 and 500 base pairs if acrylamide gels between 8% and 10% are employed. The procedure is suitable to determine protein-induced bending angles. Special vectors have been designed which allow a bent DNA or a protein binding site to be placed at different positions within a series of DNA fragments of identical length. In this way the centre of bending can be determined be **circular permutation assays**. The DNA fragments are obtained from restriction digests of a direct repeat with circular permutated restriction sites and a cloning site in the central region of the direct repeat (Figure B6.3).

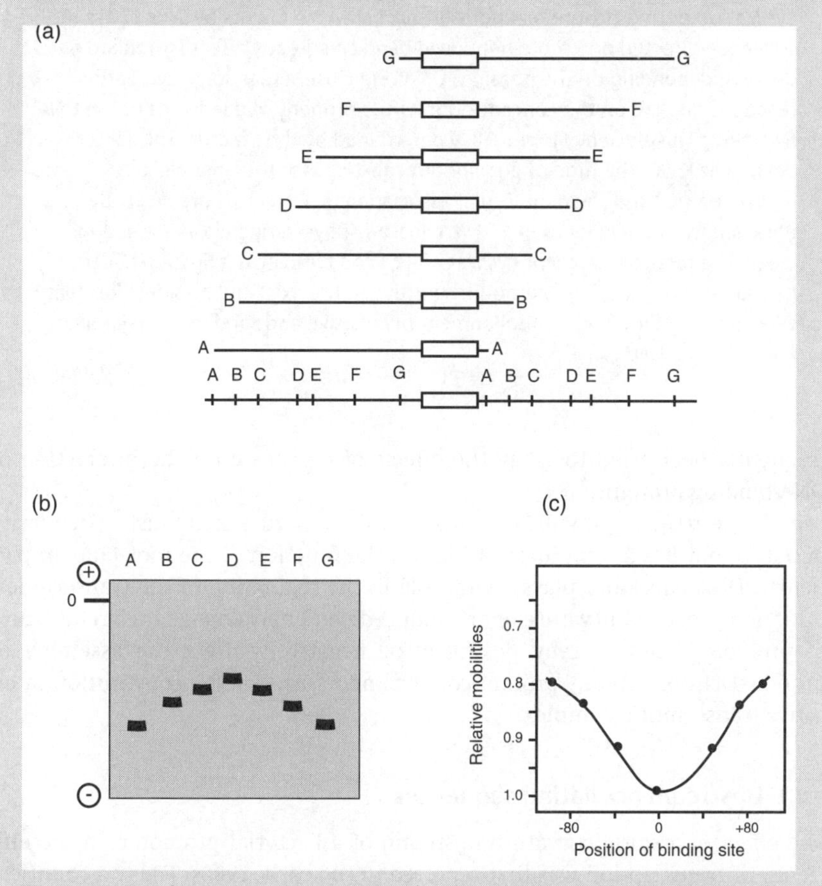

Fig. B6.3 Circular permutation assay to determine DNA-bending centres

The figure shows in (a) the schematic arrangement of DNA fragments of identical length. The fragments are obtained by restriction digestion with different enzymes (A to G). The same restriction sites are located as direct repeats flanking a cloning site (open box) which can be used for the insertion of a protein binding site or a piece of DNA with possible curvature. The use of each enzyme results in a DNA fragment of identical length but with the target site at a different position with respect to the fragment ends. Part (b) represents the gel electrophoretic separation of a family of DNA fragments obtained as shown in (a) where the centre of curvature has been permutated. The relative mobilities from this experiment are presented as a function of the position of the cloning or target site in (c). The minimal mobility apparent in lane D indicates that the region of DNA curvature must be most central in this fragment. The ratio of mobilities can then be used to estimate the bending angle by the equation given above.

DNA bending can also be studied in solution because it affects the efficiency of cyclization or ring closure reactions of short DNA fragments by DNA ligase. A sequence-directed or a protein-induced bend can greatly affect (stimulate or depress, depending on the phasing) DNA ring closure reactions. Cyclization reactions depend on the concentration of the properly aligned ends of the DNA fragment. The efficiency is generally determined by the ***J* factor**. This factor is defined as K_c/K_a, the ratio of equilibrium constants for the unimolecular cyclization (K_c) and the bimolecular association (K_a). *J* factors are normally obtained from the kinetics of T4 DNA ligase-catalysed ring closure reactions where the rate constants for cyclization (k_c) and bimolecular ligation (k_l) are measured ($J = k_c/k_a$). The method is sensitive to the axial and torsional flexibility, allowing the study of the misalignment of cohesive ends and the **persistence length** of the DNA.

frequently been used to study the effects of curvature on the interaction of DNA-binding proteins.

In the next section it will be seen that DNA curvature is not just a peculiarity of nature but has strong implications on the function of the molecule. In particular, DNA curvature plays a vital role in the regulation of the transcription initiation process. Curvature or any induced bend may change the architecture of promoters and thereby positively or negatively affect the assembly of protein–DNA or protein–protein contacts necessary for the constitution of an active transcription complex.

6.2.1 Upstream activating sequences

The effect of a static curvature upstream of a bacterial promoter on the efficiency of transcription was first reported in the early 1980s. Today a countless number of examples is known, indicating the importance of curved DNA sequences on different steps of transcription. DNA conformation is almost universally recognized as a parameter in gene regulation.

As outlined in Section 3.3, the binding of RNA polymerase to a promoter involves wrapping of the DNA around the surface of the polymerase. Thus, structural deviations from a straight linear DNA molecule appear to be obligatory for the formation of a transcription complex. To make this clear, a few facts from Section 2.4, where information on the three-dimensional structure

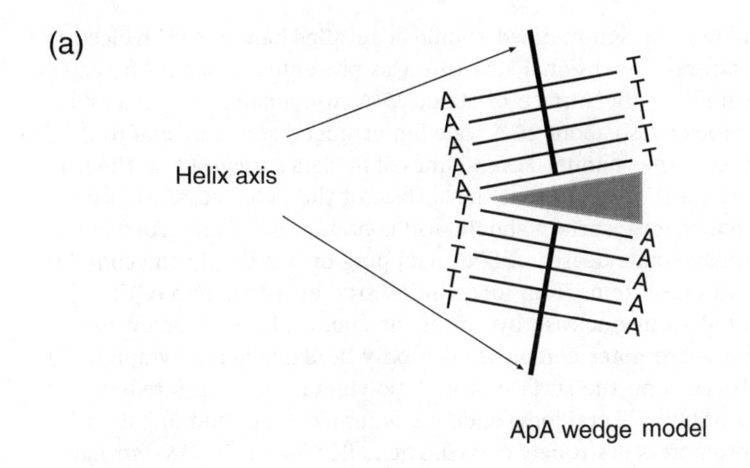

Figure 6.3 DNA bending according to the ApA wedge and the B-junction models. (a) Two clusters of AT base pairs are schematically depicted. The structure represents the B-form with the base pairs perpendicular to the helix axis (thick line). The two adjacent AT/TA base pairs are not parallel, thus forming a wedge that changes the direction of the DNA axis. If such wedges are inserted at regular intervals (e.g. after each full helical turn or multiples of the helical pitch) the DNA contour will result in a smooth curvature in one plane of the helix. (b) At the junction of a B-DNA section (with base pairs at right angles with respect to the helix axis) and a section of non-B-DNA, where base pairs are at tilt, the path of the helix is kinked. Continuous base stacking is observed between the two DNA segments. As in (a) periodic kinks after each full helical turn will result in a smooth DNA curvature.

of RNA polymerase was summarized, should be recalled here. Based on electron micrographs a three-dimensional structure was presented which is characterized by a channel on the surface of about 25 Å in diameter. The channel is flanked by a groove, also about 25 Å wide but arranged at an angle of roughly 60° with respect to the channel. Several lines of evidence suggest that the path of DNA follows these structures on the surface of the polymerase. Hence the DNA on the polymerase surface should also contain a 60° angle. Further evidence for a considerable degree of DNA wrapping or bending in the complex with RNA polymerase stems from footprinting studies. The length of the DNA region protected from nuclease hydrolysis or chemical modification within RNA polymerase–promoter complexes can only be explained if wrapping or bending of the DNA on the surface of RNA polymerase is considered. In line with this assumption, several independent studies have reported that the DNA structure of promoters is strongly curved. The deflection of the DNA structure may not always be the result of intrinsic curvature. Binding of RNA polymerase itself is able to reorientate the helix axis of DNA (this is a property of many DNA-binding proteins). The change of the DNA helical axis resulting from protein binding or other external forces is generally termed a **bend**, as opposed to the term **intrinsic curvature**, which only depends on the sequence. It is clear that the degree of correspondence between the magnitude and direction of intrinsic curvature and protein-induced bend has a great implication on the energetics of transcription initiation complex formation.

Regions of curved DNA or bends induced by protein binding are not only restricted to the core promoter. A large number of bacterial promoters have been shown to contain intrinsically curved regions upstream of the promoter core sequences. These curved sequences are found at a region between positions –40 to about –150 with respect to the transcription start site. Often the centre of curvature is located between –50 and –100. Sequence algorithms have been developed in the past that are able to predict the location and direction of intrinsic DNA curvature. When a survey employing such algorithms was performed looking for curved sequences within bacterial DNA, a remarkable result was obtained. About 50% of the strongly curved sequence regions of the screened DNA were found upstream of promoters, with the centre of curvature located approximately at nucleotide position –50. Numerous subsequent studies have shown that the curved sequences upstream of bacterial promoters are almost always related to the activity of the particular promoter (Fig. 6.4).

Several *in vitro* studies with isolated templates and purified RNA polymerase have led to the conclusion that bent DNA alone might be responsible for promoter activation. In some cases the curved upstream sequences have been exchanged between different promoters without loss of their activating function. It follows, therefore, that the activating capacity directly resides within the curved DNA sequence elements. This does not rule out that additional protein factors, in combination with the curved DNA, contribute to the activation *in vivo*. The family of curved sequences able to activate promoters independ-

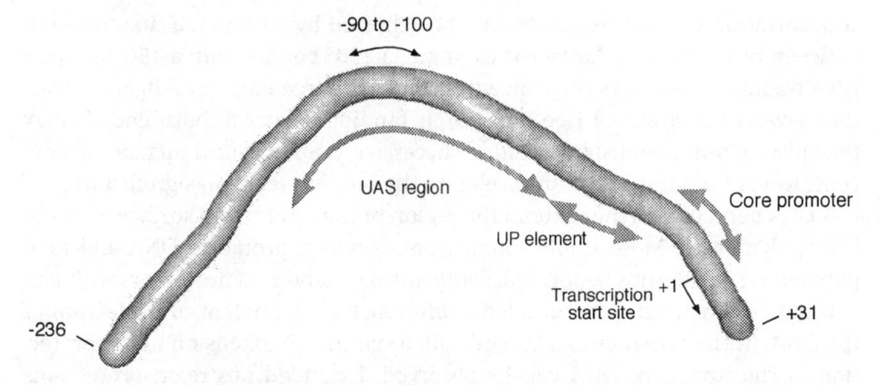

Figure 6.4 Curved UAS region of the *rrnB* operon. The curvature of the upstream sequence of the *rrnB* operon ranging from position −236 to +31 has been calculated using the program DIAMOD (Dlacik and Harrington, 1998). The DNA is presented as a worm-like structure with two centres of curvature, one between positions −90 and −100 and a second curvature in a different plane at the core promoter region. The AT-rich UP element upstream of the core promoter does not show a significant bend. The data is completely consistent with experimental findings (see Section 8.7.3).

ently has been termed **upstream activating sequences** (UAS). The effects of distance and phasing variations of UAS have been studied for several promoters in detail. The phasing of the curved upstream sequences with respect to the core promoter elements poses a stringent prerequisite of the activating function. This result has been obtained from studies where the distance between the centre of curvature and the promoter core elements was varied systematically by deletions or insertions of non-curved DNA stretches. Deletions or insertions of a random sequence of 10 or 21 base pairs (integral number of helical turns) did not significantly affect activation. In contrast, changes in the distance of 5 or 15 base pairs (non-integral number of helical turns) abolished activation. These studies indicate that in order to activate promoters the curved upstream sequences have to be in a fixed spatial orientation, while the distance of the centre of curvature relative to the transcription start seems to be of minor importance.

What is the mechanism by which curved DNA sequences upstream of the core promoter activate transcription? To answer this question some of the essential conclusions of Chapter 3 are reviewed. In principle, the activity of a promoter can be influenced by changing the binding affinity between RNA polymerase and the promoter, by changing the rate of isomerization from a closed to an open promoter complex or by changing the rate of promoter clearance. What then is the effect of curved upstream sequences on these possible steps of transcription initiation? From a physicochemical point of view, bending may cause an increase in the local concentration of interacting domains (this could be protein as well as DNA domains). An increase in the local concentration of reacting groups can be achieved by an optimal juxtaposition between interacting surfaces. For example, a 120° bend within a 150-base pair DNA fragment increases the concentration of the free ends for a ligation reaction several hundredfold (see Box 6.3). If binding occurs to both ends it may take place simultaneously, instead of successively. Such a local increase in concentration of binding sites can accelerate the binding reaction significantly.

A DNA bend can further extend the region of the interacting surface with the RNA polymerase. More stable interactions between promoter DNA and RNA polymerase might thus be formed. Footprinting analyses of promoters with and without UAS regions have revealed a difference in the extent of protection of the DNA. In the presence of a UAS region, a significant extension of the protection of the upstream DNA can be observed. Extended upstream protections have for instance been reported for the rRNA promoter *rrnB* P1 or the promoter for the tyrosyl tRNA gene (*tyrT*). In both cases the RNA polymerase–DNA contacts are clearly larger than those in promoters without UAS regions. The extended DNA contact probably results in a tighter binding between DNA and RNA polymerase. This has been verified, for instance, by measurements of the RNA polymerase concentration-dependent association constant K_B in the presence and absence of the corresponding UAS regions (see Section 3.6.1).

It should be noted that the analysis of the effects of UAS regions on promoter activity is often complicated by the high complexity of functional elements close to the promoter core region. For instance, for many strong bacterial promoters (the rRNA promoter *rrnB* P1 may serve here again as a prototype) AT-rich sequences of about 20 base pairs in length have been identified. These sequences are located just upstream of the –35 hexamer. These sequences are termed **UP elements**. Although UP elements are rich in AT, they do not display sufficient bending to be detected by gel electrophoresis. However, the presence of an UP element can increase promoter activity between two and 20-fold. The activating effect of UP elements has been largely solved. DNA bases within the UP element provide a binding target for the C-terminal domain of the RNA polymerase α subunit (αCTD) (see also Section 2.2.1). Since RNA polymerase recognizes and binds the –10 and the –35 promoter regions, the UP element is sometimes referred to as the third promoter recognition element. Hence, for promoters with an UP element an increase in K_B can be easily explained by the stabilizing interaction between the αCTD and the UP element. In the case of promoters that have an upstream curvature in addition to an UP element, such as the *rrnB* P1 promoter, it is questionable or even impossible to dissect how far the individual structural components contribute to the observed effects. The situation becomes even more complex when the participation of proteins which specifically interact with UAS regions are considered. Discussions of this type will be addressed in more detail in later sections below.

Curved UAS elements have been demonstrated to affect the function of many promoters. The observed activating effect is not always restricted to an increase in the primary binding of RNA polymerase (K_B), however. For some promoters an increase in the isomerization rate from the closed to the open RNA polymerase–promoter complex (k_2) has been determined as a result of upstream curved sequences. It has also been noted that both parameters, the affinity for RNA polymerase (K_B) *and* the isomerization rate (k_2), can be affected simultaneously by the presence of a UAS region. An increase in the rate of the isomerization reaction can be envisaged when promoter melting is facilitated. Unwinding of the DNA may be supported if the DNA which is wrapped around the polymerase in the initiation complex undergoes a conformational change. Such a conformational change of the enzyme, like a movement of the downstream edge, for instance, will cause torsional stress to the DNA, which is fixed to the polymerase at the far ends. The strain from such an intermediate structure can be released by unwinding the DNA in the –10 region. Additional contact sites at the upstream edge of the polymerase may increase the stress of the intermediate complex and thus facilitate melting. In this way, extended RNA polymerase–DNA contacts with the upstream curved sequences can also explain an increase in k_2.

Curved upstream sequences may also affect the promoter clearance. It is known that an extreme increase in the stability of the initiation complex may slow down or inhibit the subsequent steps of transcription, namely promoter clearance and the formation of an elongating complex (see Section 3.5). In fact, the strength of promoters with curved upstream sequences is often limited at the step of promoter clearance. On the other hand, promoter clearance or promoter escape may also be facilitated by UAS regions. The respective function may depend greatly on the position and the extent of the upstream contacts. Additional contacts between RNA polymerase and the upstream DNA could act as a spring, helping to release the transcribing complex from the promoter.

It is important to note that the analysis of the effects of UAS regions *in vitro* using purified RNA polymerase and isolated DNA templates can generally reveal only part of the events that affect the transcription reaction. In the majority of cases studied, UAS regions do not simply exert their effects directly because of their curved structures. Often they provide binding sites for regulatory proteins. A number of such regulatory proteins bind preferentially to curved DNA or to DNA with anisotropic flexibility, where the proteins can induce a bend upon binding. For a more complete picture of the effects of UAS regions on transcription potential protein binding must usually be taken into account. This means that effects determined *in vitro* with purified components must be verified *in vivo* before a complete picture of the activating effect of UAS regions can be drawn. In the next section indirect effects of DNA conformation on transcription activation because of protein binding are summarized.

6.2.2 Transcription factor-induced bends in DNA

The previous section described how the DNA in the vicinity of a promoter is often characterized by intrinsic curvature. Sometimes, however, structural deviations of the DNA are brought about through the binding of specific proteins. When binding of such proteins changes the efficiency of transcription in a specific way the proteins are considered to be transcription factors, even when they do not directly interact with RNA polymerase. Figure 6.5 summarizes a collection of several of the more important transcription factors which are known to induce DNA bends and some of their typical DNA binding targets.

A great number of examples are now known where the recognition and binding of proteins to their target DNAs is not only mediated by sequence specificity of the amino acid residues of the protein and the DNA bases. Rather, intrinsic properties of the DNA, such as the tendency to form a curved structure, often play an important role in DNA–protein interacting systems. Moreover, the interaction between DNA and proteins does not always involve static structures. Increasing evidence supports the notion that there is a considerable dynamic aspect in DNA–protein binding, and conformational deformations of both interacting molecules, DNA as well as protein, often affect the interaction. Based on such observations two modes of recognition have been defined. If the recognition between DNA and protein involves a precise steric match between exposed functional groups of the static structure of the interacting molecules the interaction is generally referred to as **direct read-out**. Such interactions may involve any functional groups of the amino acid side chains or the peptide backbone that are accessible at the surface of the protein, and any of the corresponding reactive positions in the minor or major groove of the DNA. In contrast, **indirect read-out** describes the necessary structural adaptation or fit induced by conformational changes, which makes additional contributions to the binding free energy. Hence, indirect read-out contains a dynamic component based on the conformational deformation of the interacting DNA and/or protein components.

Several structures of complexes between DNA binding proteins and their DNA target sites have been resolved at high resolution. In a number of cases it could be inferred that binding of the respective protein changes the trajectory of the target DNA. Notable examples include the nucleoid structuring protein HU, the catabolite regulator protein CRP, and several bacterial or phage repressor proteins (see Chapter 7). Changing the trajectory of the DNA to which the proteins are bound can generally not be explained by a single mechanism. The specificity of binding is not unique to DNA bending proteins. Some recognize palindromic sequences or conserved primary structural motifs (e.g. the *lac* repressor, CRP or IHF). Others require no sequence specificity at all and seem to recognize specifically curved (or flexible) DNA regions (e.g. HU or H-NS). Moreover, within DNA binding proteins different amino acid secondary structural motifs have been identified which are responsible for DNA interaction (see Box 2.2). The most common element has been described as the helix-turn-helix motif (two short α helices separated by a glycine residue), which is found in the majority of prokaryotic DNA-binding proteins (e.g. FIS and CRP). However, other structural motifs occur frequently, such as the looped β-sheet structure,

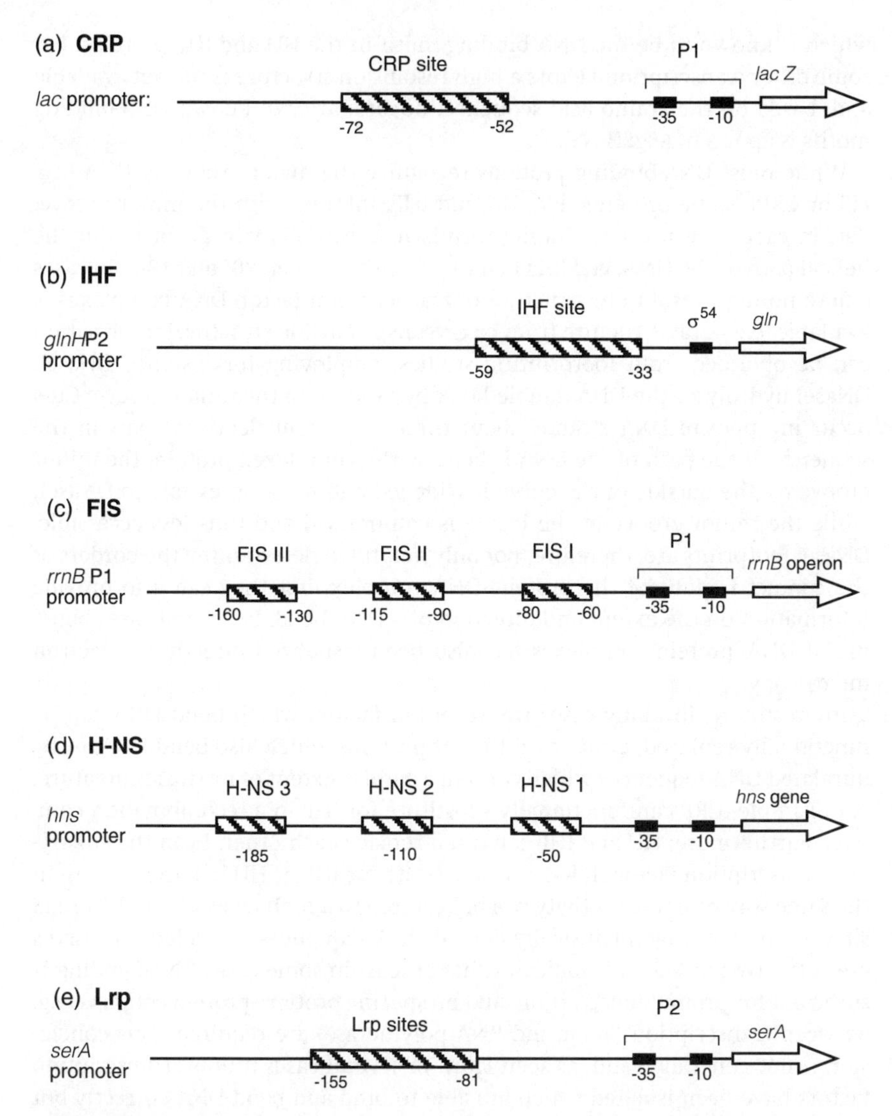

Figure 6.5 Transcription factors affecting DNA curvature. The five transcription factors CRP, IHF, FIS, H-NS and Lrp are presented as examples of DNA-bending proteins. For each protein an example of a regulated gene and the respective position of the binding sites on the DNA (shaded bars) are indicated. With the exception of IHF, binding of the proteins to a single site results in an average DNA bending of about 90°. Bending induced by IHF is stronger, and 160–180° are observed for a single IHF dimer bound to DNA (see Figs. 7.12 and 7.14).

which is known to be the DNA binding motif in the HU and IHF proteins. For some other transcription factors a high resolution structure is not yet available and, based on the amino acid sequence, no similarity to known DNA-binding motifs is apparent (e.g. H-NS).

While most DNA-binding proteins recognize the major groove of DNA (e.g. FIS or CRP), some proteins, like HU and IHF, interact with the minor groove. Yet, in each case protein binding results in a considerable distortion of the helical path of the DNA, yielding bending angles between 90° and 180°. In cases where no high resolution structure of transcription factor–DNA complexes is available (e.g. X-ray structure from co-crystals), valuable structural information can be obtained from footprinting studies, employing for instance DNaseI. DNaseI hydrolyses the DNA double helix by binding to the minor groove. Cuts occur in opposite DNA strands about three to four nucleotides apart in the sequence. If the path of the DNA is bent by the complexed protein, the minor groove on the outside of the curve is widened and more accessible to DNaseI, while the minor groove at the inside is compressed and thus less accessible. DNaseI footprints are, therefore, not only helpful in determining the borders of the contact regions of the protein–DNA complex, but they can also provide information on the extent and direction of a DNA bend. In several cases bending of DNA–protein complexes has also been visualized directly by electron microscopy.

Interestingly, in many cases transcription factors which bend DNA can be functionally replaced, either by different proteins which also bend DNA, or by unrelated DNA sequences which contain a similar extent of intrinsic curvature. For example, CRP can functionally substitute for IHF in a recombination complex. Furthermore, HU and IHF can often replace each other. Even the eukaryotic transcription factor HMG1 can substitute for IHF or HU in special cases. In the same way promoter activity can be restored to a high level when CRP or FIS sites are replaced by intrinsically curved DNA sequences, provided the curves are correctly phased. This indicates that at least in some cases DNA-bending is sufficient for promoter activation, and no specific protein–protein contacts (e.g. between transcription factor and RNA polymerase) are required. This conclusion is not generally valid, as seen later. In several cases mutant transcription factors have been isolated which are able to bind and bend DNA correctly but which fail to activate or repress the respective promoters. Examples include mutant CRP or FIS proteins. This indicates that these transcription factors must interact with RNA polymerase to exert their function. In fact, direct interactions between transcription factors and RNA polymerase have been mapped in several cases. The CRP protein at Class I promoters interacts for instance with the αCTD of RNA polymerase (see Section 2.2.1). An exposed loop at the surface of the protein (amino acids 156–162) appears to be responsible for the contact with the αCTD (see section 7.3.2). On the other hand, mutations in the αCTD which abolish binding of CRP or other transcription factors of the same group to the polymerase have no effect on FIS-mediated activation. Therefore, interaction between FIS and RNA polymerase either involves different portions of the α subunit or it does not interact with the α subunit at all. It should be noted in this respect that there is indirect evidence for FIS binding to the σ^{70} subunit (see Section 7.5.2).

It is not always clear whether DNA-bending supports contacts between transcription factors and RNA polymerase or if bent DNA has a regulatory function *per se*. For many systems studied the available evidence suggests that both RNA polymerase transcription factor contacts and DNA curvature are required for full promoter activity.

6.3 DNA supercoiling and transcription

Covalently closed circular DNA molecules or DNA that is fixed with both strands at the ends give rise to a new structural dimension. Such molecules can exist as different **topological isomers** which may form different superhelical structures. The principal definitions of supercoiled DNA are presented in Box 6.4.

Some fundamental aspects of DNA structure are now recalled. In a covalently closed circular DNA the two strands cannot be separated without breaking the phosphodiester chain. The molecule can be characterized by three parameters:

1. The **linking number** (L_k). This number indicates how many times one DNA strand is linked with the other. It represents a topological constant which can only be altered upon strand scission. This number is necessarily an integer.
2. The **twist** (T_w). The twist gives the rotations of the two strands around the helix axis. In B-form DNA the twist is one turn per 10.5 base pairs, for instance.
3. The **writhe** (W_r). The writhe gives the path of the helix axis in space and thus describes the number of superhelical turns.

A very simple relationship exists between the above three parameters:

$$L_k = T_w + W_r \qquad \text{(eqn 6.1)}$$

In a relaxed circular molecule with no torsional forces applied L_k always equals T_w. Under this condition $W_r = 0$ and no supercoiling exists. Imagine now that the two DNA strands are overwound or underwound before they are covalently closed, such that overwinding or underwinding becomes trapped in the molecule by the circularization step. The linking number L_k has therefore been changed. It is either increased (overwinding) or it is decreased (underwinding). Because of the tendency of DNA to adopt a B-form-like structure the twist will remain constant within a narrow range. It follows that the writhe W_r will

Box 6.4 Basic parameters of DNA supercoiling

Covalently closed circular DNA molecules have new topological properties compared to the same molecules with linear structure. The resulting topological isomers can be described by three parameters: the linking number L_k, the twist T_w, and the writhing number W_R.

The *linking number* defines how often in a closed DNA molecule two strands are crossed. It is necessarily an integer. The linking number is positive ($L_K > 0$) for right-handed helical turns and negative ($L_K < 0$) for left-handed helical turns. Note that the linking number is a topological constant and changes in the linking number can only be obtained after strand scission.

Twist indicates the total number of turns in a DNA helix. It is positive ($T_W > 0$) for right handed helical DNA. In a relaxed circular molecule the twist is given by the ratio of the total number of base pairs per molecule to the number of base pairs per turn, which is about 10 for B-DNA.

The *writhing number* defines the crossovers within a superhelical DNA molecule. For a relaxed circular DNA molecule the writhing number is zero ($W_R = 0$). Under this condition the twist is identical with the linking number ($T_W = L_K$). For right-handed superhelical DNA the writhing number is negative ($W_R < 0$), while it is positive ($W_R > 0$) for left-handed supercoiled structures.

There is a simple relationship between these three parameters and superhelical structures given by the equation:

$$L_K = T_W + W_R$$

When normal B-form DNA is closed in a circle it is relaxed, not supercoiled. How does a relaxed circular DNA adopt a superhelical structure? Assume that a double-stranded helical B-form DNA is underwound, for example by fixing both strands of the left end and rotating the free right ends in a clockwise direction. This operation will reduce the twist (T_W) and in the same way the linking number (L_K). If the ends are now joined, forming a circle, L_K and T_W are reduced by the number of unwindings and $W_R = 0$. The DNA has the tendency, however, to reform a B-DNA structure. L_K, which is a topological constant, cannot be changed without strand scission. The existing deficit in L_K can either be compensated by base pair disruption (the same number of base pairs that correspond to the unwound helical turns would no longer be in a helical form) or, to satisfy the equation $L_K = T_W + W_R$, the writhing number will be reduced by the same number of turns that had been unwound. Through this operation W_R becomes negative ($W_R < 0$) and a right-handed supercoiled structure is formed with a number of superhelical crossovers that corresponds to the number of turns that had been unwound before with the linear molecule (Fig. B6.4).

Another operation with interesting implications on DNA topology is pertinent. Consider that a double-stranded DNA is wound in *left-handed* turns around a core of proteins, as in a nucleosome structure where DNA is wound around histone octamers. When the free ends of the DNA are now joined and the protein core is

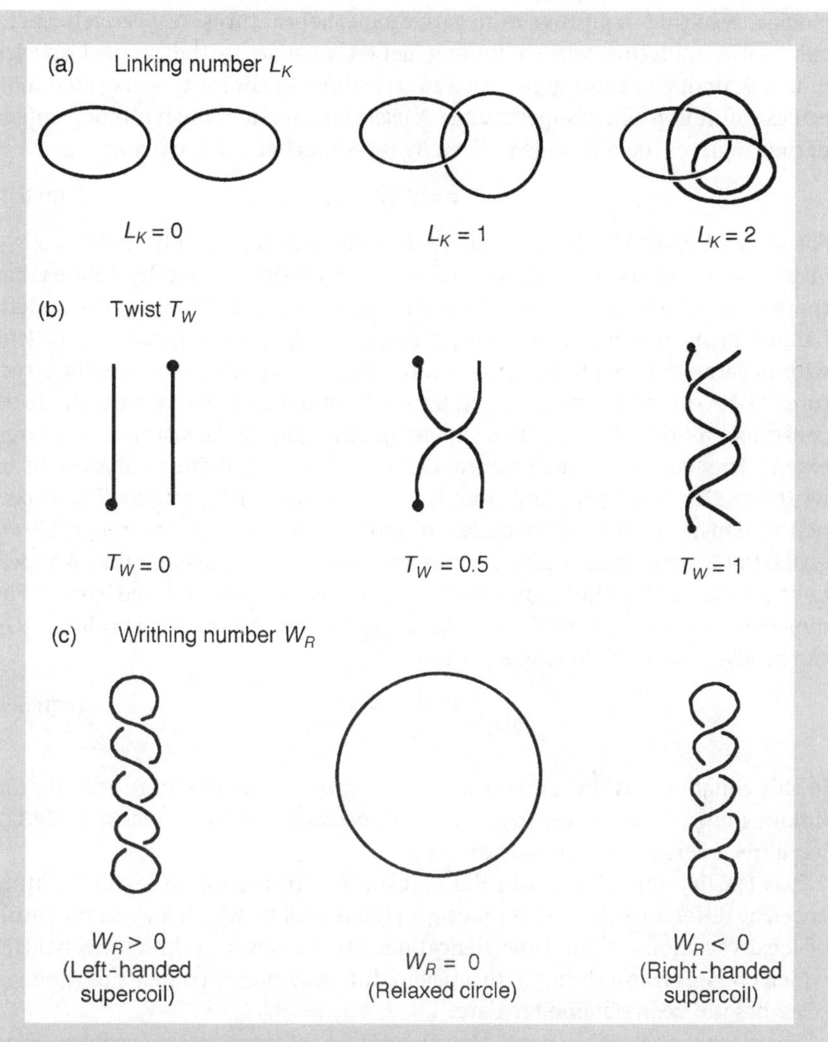

Fig. B6.4 Parameters defining DNA supercoiling

removed the DNA snaps into a supercoiled structure. The number of windings correspond to the number of supercoiled crossovers. Most surprisingly the superhelical windings describe a *right-handed* helix; this means that the writhe is negative ($W_R < 0$). This can easily be demonstrated with a rubber tube wound in left-handed turns around a cylinder, which is subsequently removed after the ends have been joined. Note that left-handed turns around a core of protein are topologically equivalent to negatively supercoiled DNA.

change, resulting in positive or negative superhelical turns, respectively. A circular DNA molecule with a linking deficit relative to the relaxed species (underwound) gives rise to negative supercoiling. In contrast, overwound molecules result in positive supercoiling. A useful parameter which is independent of the length of the molecule is given by the **superhelical density σ**:

$$\sigma = W_r/T_w \qquad \text{(eqn 6.2)}$$

For isolated plasmid DNA the superhelical density is generally similar to $\sigma \approx -0.05$. Assuming a twist T_w of one turn per 10 base pairs this number tells us that there is about one superhelical turn per 200 base pairs. Winding DNA in left-handed turns around a protein (e.g. histone core) is topologically equivalent with negative supercoiling. Thus, supercoiling may reside in the specific structure of DNA–protein complexes. Under such conditions supercoils are said to be constraint. Within the cell DNA supercoiling is normally constraint to a large extent. This means the superhelical density σ is changed due to alterations in writhe and/or twist, resulting from the interactions with proteins. The superhelical density in the cell is therefore more likely to be in the range of $\sigma = -0.03$. Deformations in writhe or twist are energetically unfavourable. A superhelical molecule has therefore a higher free energy than the relaxed isomer. For molecules larger than 2000 base pairs the free energy of supercoiling (ΔG_s) can be given by the following equation:

$$\Delta G_s = \frac{1050 \times RT}{N} \times \Delta L_k^2 \qquad \text{(eqn 6.3)}$$

In this equation R is the gas constant, T the absolute temperature and ΔL_K the linking number difference. To give an example: at $\sigma = -0.05$ the plasmid pBR322 has a free energy of about 100 kcal/mol.

It is feasible that changes in the efficiency of transcription might be influenced by differences in the free energy of the template, which may be the result of local variations of the superhelical density. However, a direct mechanism which couples transcription efficiency to different energy contents of the template has not been obtained to date.

It is clear, on the other hand, that the process of transcription requires local strand separation of the template. Each transcribing RNA polymerase unwinds the template by about 1.5 helical turns. Certainly, this local change in the helical geometry is coupled to the topology of the template. It is not surprising, therefore, that each step of the transcription cycle, like initiation, elongation, pausing and termination is more or less affected by the superhelical density of the template (some of the effects have already been reported in Section 4.3.3). The question of how transcription initiation is affected by supercoiling is first addressed before the problem of how the transcription elongation process itself influences the template topology.

It is important to know how supercoiling in the cell is maintained and how it can be influenced to understand effects on transcription. The topology of supercoiled DNA in the cell is generated or released by enzymatic activities. The enzymes responsible for creating and changing DNA topology are termed **topoisomerases**. In bacteria, two different types of topoisomerase act with opposite activities to maintain a 'normal' degree of supercoiling (Box 6.5).

While DNA gyrase (topoisomerase II) introduces negative superhelical turns (reduces the linking number), topoisomerase I is responsible for relaxation of negative supercoils (increases the linking number). Gyrase requires energy in the form of ATP hydrolysis for its activity. The mutually antagonistic action of the two types of topoisomerases dynamically controls a balanced state of superhelicity in the cell. This antagonistic system has been described as **homeostatic regulation**. To study the effects of supercoiling on transcription *in vivo* this homeostatic regulation can be disturbed. This can be achieved by drugs, such as **novobiocin** or **coumermycin**, two inhibitors of gyrase which cause relaxation of the DNA. Alternatively, mutations in the genes for topoisomerase I (*topA*) or gyrase (*gyrA* and *gyrB*) disturb the net level of supercoiling. Already, from early studies employing gyrase inhibitors or topoisomerase mutants it has become clear that the expression of many genes is in fact dependent on the degree of cellular supercoiling. It has also been shown that the observed regulation is because of a change in transcription. Among the genes affected by supercoiling are the topoisomerase genes. Their expression is autoregulated by a feedback mechanism involving supercoil-dependent promoters. Other examples that require a negatively supercoiled template for optimal activity are the rRNA promoters (see Section 8.7.2). Comparison of the different promoters dependent on supercoiling reveals that there is quite some variability as to the direction and degree of regulation. For instance, negative supercoiling can activate one group of promoters while at the same time other promoters are inhibited, and still others are apparently left unaffected. Moreover, the activities of supercoil-dependent promoters are not changed linearly with increasing or decreasing superhelical density but their activity is often maximal at a specific intermediate value (see below).

Two questions immediately result from the above observation. First what are the mechanisms underlying the different effects of supercoiling? Second, how is supercoiling affected in the cell?

The general *ad hoc* explanation for the effect of supercoiling on activating promoters takes into account that negatively supercoiled DNA, which is underwound, is more easy to melt. The deficit in linking number can thus be released by base pair opening. In other words, high negative superhelical densities facilitate strand separation. Consequently, negative supercoils should increase the rate of promoter melting (isomerization; k_2) and thus increase the efficiency of transcription (see Section 3.6.1). This assumption may be valid for many promoters that are stimulated by negative supercoils. The actual situation is more complex, however. Detailed studies of several promoters *in vitro* employing templates with variable superhelical densities have revealed that the activity of promoters does not simply correlate with the degree of supercoiling.

Box 6.5 DNA topoisomerases

DNA topoisomerases are essential enzymes that solve the helical winding and tangling problems of DNA that occur during replication, recombination and transcription. During their reaction topoisomerases must cleave and seal DNA strands. This is a prerequisite to interconvert topological isomers. All topoisomerases perform two consecutive transesterification reactions with a protein–DNA intermediate. This intermediate is formed by a covalent bond

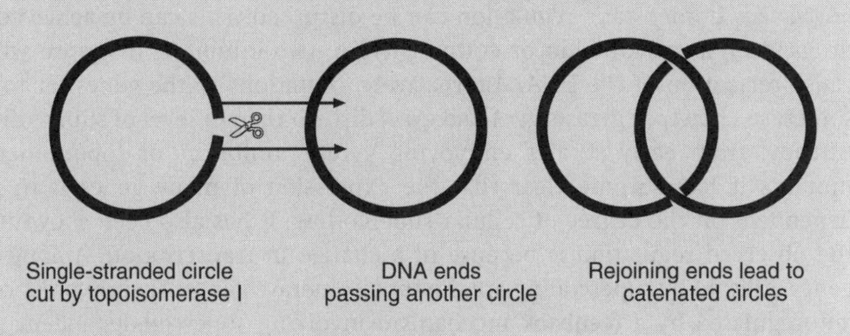

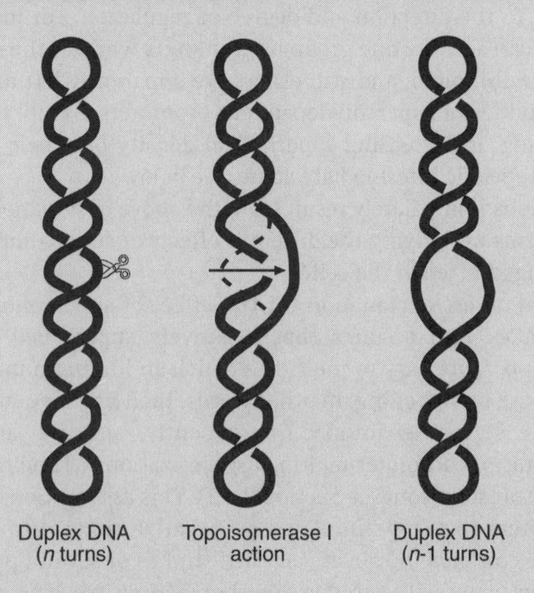

Fig. B6.5a Topoisomerase I action

between a tyrosine hydroxyl group of the topoisomerase and the phosphate group at the DNA cleavage site. The energy for the religation is stored in this intermediate and no extra energy is required for the DNA rejoining step.

DNA topoisomerases have been grouped into three independent families: type I-5′, type I-3′, and type II topoisomerases. The prototype of the type I-5′ topoisomerases is bacterial **topoisomerase I.** This 97-kDa protein relaxes negatively coiled DNA in an ATP-independent reaction by creating *single strand breaks* in DNA. The linking number of DNA is increased by increments of one through the consecutive nicking–closing steps. As an intermediate a tyrosyl group is covalently linked to the 5′ phosphate at the break site. The enzyme is also able to knot single-stranded rings or to link two single-stranded complementary rings into one double-stranded ring. Other members of the type I-5′ family are the *E. coli* DNA topoisomerase III or the reverse gyrase found in thermophilic archea. This enzyme *overwinds* duplex DNA in an ATP-dependent way. Topoisomerases of the type I-3′ family form covalent intermediates between a tyrosyl OH-group and the 3′ phosphate site of the DNA break. This type of enzyme can relax both overwound and underwound DNA duplexes. Members of the type I-3′ family seem to occur exclusively in eukaryotes and not in bacteria. The eukaryotic DNA topoisomerase I is a prototype of this family.

In contrast to the type I topoisomerase families, DNA **topoisomerases II** function as dimeric proteins and require ATP hydrolysis for catalysis. Type II topoisomerases cleave *both DNA strands*, forming a pair of covalent 5′ phosphoryl tyrosine intermediates. After ligation the linking number of the DNA is *decreased by two*. The bacterial enzyme is also termed **gyrase** and consists of two subunits, of molecular weights 105 kDa (GyrA) and 95 kDa (GyrB). Gyrase serves to introduce *negative* supercoils into DNA. The enzyme is specifically inhibited by antibiotics such as **novobiocin** or **oxolinic acid**. Several other bacterial topoisomerases have been discovered recently. Among them, **topoisomerase IV** bears a large resemblance to the eukaryotic type II topoisomerases and is homologous to gyrase. It contains two subunits encoded by the *parC* and *parE*

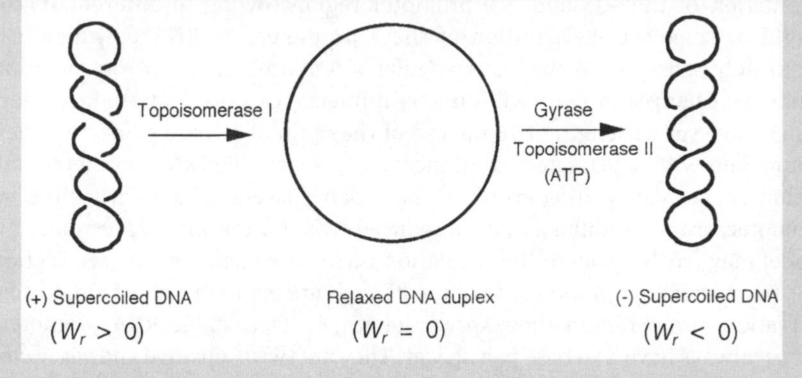

Fig. B6.5b Differential activities of topoisomerase I and topoisomerase II (gyrase)

genes. Like gyrase, topoisomerase IV has an important function in relaxing negatively supercoiled DNA in *E. coli* and plays a major role in decatenation (Figs B6.5a and b).

In most cases a maximal activity is found at medium superhelical densities $\sigma \approx -0.05 \pm 0.01$, similar to the condition expected *in vivo*. Moreover, not all promoters respond to an increase in k_2 with increasing negative superhelical density (e.g. the *lac* UV5 promoter). Some promoters show a decrease in k_2 and also an increase in primary RNA polymerase binding (K_B). An example is the *ada* promoter controlling the genes for the adaptive response. The latter two results demonstrate that the effects of supercoiling require a more complex explanation. Subtle structural changes of the promoter conformation can be caused when the DNA is distorted by changes in the superhelicity. Torsional strain can also affect bending and thereby influence the contact between RNA polymerase and the promoter. Extended DNA contacts may be the reason that for some promoters the affinity (K_B) to the RNA polymerase is changed.

It is known from several carefully studied examples that the binding of regulatory proteins in the vicinity of the promoter is affected by changes in supercoiling. In these cases activation or inhibition of the promoter is a secondary effect of the template topology. In rare cases a high superhelical density may also give rise to abnormal DNA structures. For instance, palindromic sites may fold into cruciform structures at very high superhelical torsions. These structures could cause a kind of roadblock to RNA polymerase and thus may probably inhibit transcription elongation.

Recently a model has been proposed that environmental changes on the topology of DNA directly affect twist. This means the superhelical density remains constant because of homeostatic regulation and changes in the linking number directly correspond to local changes in twist. The model further proposes that there are classes of promoters which differ in their relative angular orientation of the –35 and –10 promoter regions owing to different spacer length or sequence. Recognition of these promoters by RNA polymerase is exquisitely sensitive on the exact angular orientation of the two recognition elements. Changes in twist will directly influence this orientation. Promoters which have spacer length of 16 instead of the optimal 17 base pairs are underwound and will be activated by an increase in twist. The rRNA promoters fall within this category. In contrast, 18 base pair spacers, as known for the *his* promoter, are overwound and need reduced twist for optimal expression. This model may partly apply to the regulation under stringent control (see Section 8.5), because many promoters that are affected under conditions of amino acid starvation also differ in their spacer length. In fact, stable RNA promoters almost always have a 16 base pair spacer. They are twist-sensitive and *inactivated* when the template is relaxed. In contrast, promoters which are under positive stringent regulation, like those directing the amino acid biosynthesis operons, appear to have 18 base pair spacers. These promoters are *activated* under the same conditions.

Another exiting example for regulation by twist-sensitive promoters occurs at the divergent pair of ***mer* promoters** and involves the regulatory protein **MerR**. The MerR protein acts as a dual regulator. First it controls its own transcription from the *merR* promoter (P_R), and secondly it controls the *merT* promoter (P_T), which regulates transcription of a set of genes needed for the detoxification of mercury. Mercury is highly toxic (not only) for bacteria. The two promoters are directly adjacent with divergent orientation. MerR binds in between, such that its binding site overlaps both promoters. Activation of transcription from promoter P_T occurs in the presence of mercury which binds to the regulator MerR. In the absence of mercury the P_R promoter is activated and transcription from P_T is repressed. The switch between activation and repression does not involve dissociation of the MerR protein. Instead, the dual function of the regulator is allosterically triggered by mercury binding. It is known that Hg–MerR is able to alter the twist of the DNA to which it is bound. The *mer* P_T promoter contains an unusual 19 base pair spacer, which makes it a rather inefficient promoter. The dihedral angle of the centres of the –10 and –35 hexamer sequences are about 140° compared to the usual 68–76° for consensus promoters with a 17 base pair spacer. Binding of mercury to MerR causes an allosteric *underwinding* of the spacer sequence by about 0.1 helical turn. This brings the phases of the –10 and –35 elements into a location comparable with the consensus promoter structure with normal spacer length (Fig. 6.6). Thus, binding of mercury to the regulator MerR induces underwinding of the DNA-binding site, which in turn causes a change in the activity of the promoter responsible for the detoxification genes. This example illustrates that changes in twist which influence promoter activity can either be brought about by changes in local supercoiling or by transcription factors, as in the case of the MerR-dependent promoter.

Regulation of transcription through the topology of the template requires the existence of a mechanism in the cell which is able to alter the linking number of the DNA in a global or local manner. What is known about the effectors or mediators that change the superhelical character of DNA? As discussed above, net supercoiling in the cell is dependent on the relative activity of the enzymatic systems (topoisomerase I and II) responsible for the maintenance of DNA topology. Negative supercoils are introduced by the action of gyrase at the expense of ATP hydrolysis. The degree of supercoiling is therefore sensitive to the energy state of the cell which is often expressed as the ratio of the concentrations of ATP:ADP. Alterations in this ratio affect the activity of gyrase, which consequently affects the degree of supercoiling in the cell. The ATP:ADP concentration ratio drops, for instance, when *E. coli* cells are shifted from aerobic to anaerobic growth conditions. Concomitantly, the degree of supercoiling will be reduced. Every environmental effect which influences the energy state of the cell will be reflected by changes in the superhelicity of the DNA. The

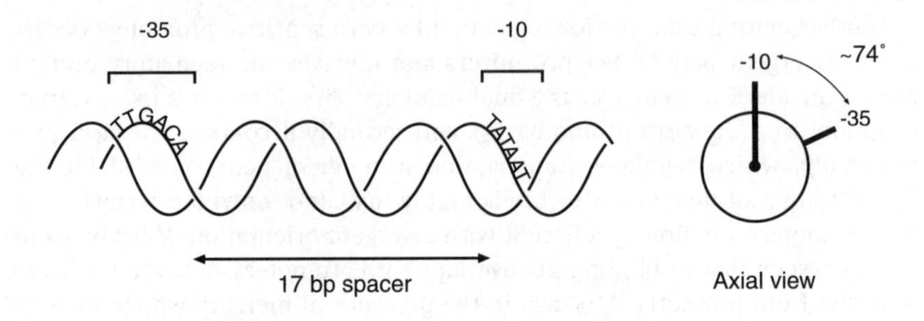

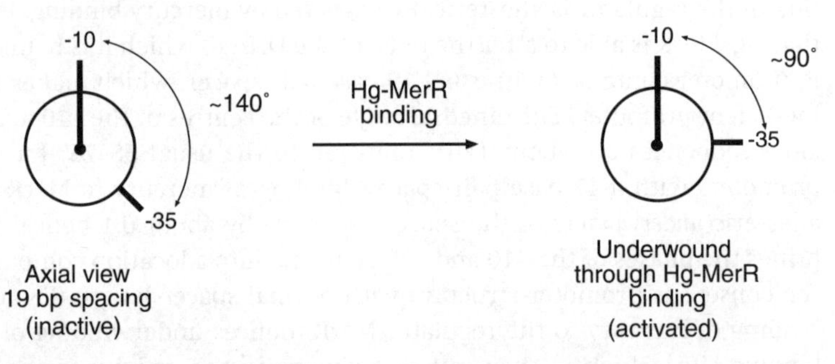

Figure 6.6 Regulation at the twist-sensitive *mer* promoters. (a) The positions of the centres of the −10 and −35 recognition elements are illustrated for the ideal consensus *E. coli* promoter with 17 base pair spacer. The centres of the two hexamer sequences are thus 23 base pairs apart. Given a helical pitch of 10.4 base pairs the central positions of the recognition elements are separated by a dihedral angle of roughly 74°, as illustrated by an axial view. (b) In case of the *mer* P_T promoter which has a 19 base pair spacing, the dihedral angle is enlarged to roughly 140°. This makes *mer* P_T a very inefficient promoter. Binding of the transcription factor MerR in the presence of mercury causes underwinding of the spacer DNA, resulting in a dihedral angle of approximately 90°, much closer to the consensus promoter. As a consequence of this underwinding the *mer* P_T promoter is activated. The figure is adapted from Ansari and O'Haloran (1994).

degree of supercoiling also, however, responds to other environmental changes. One important parameter is the osmolarity of the growth medium. Since DNA supercoils are strongly dependent on the ionic environment such changes can be sensed directly. Increased osmolarity of the growth medium causes an increase in negative supercoils. A situation where such a regulation is of great importance is given at a rapid osmotic increase. The cell responds to such an osmotic shock by activating the supercoil-dependent *proU* promoter, which directs the synthesis of the transport system for the osmoprotective substance glycine betaine. Additional effects on supercoiling are mediated by temperature shifts or nutritional deprivation. These latter parameters cause relaxation of the supercoils in the cell.

Wrapping DNA in left-handed turns around a protein core, as it occurs in eukaryotic histones, is topologically equivalent with a negative superhelical DNA structure (see Box 6.4). The superhelical density of such protein–DNA complexes is smaller than the free DNA. The superhelical character is said to be **constraint** (see above). A great deal of the supercoils in a prokaryotic cell are actually constrained by protein interactions due to wrapping. Small DNA-binding proteins which are often termed histone-like or nucleoid-structuring proteins are responsible for a great deal of the constraint supercoils (see also Section 7.4). These proteins are generally rather abundant and one of these proteins, HU, exists at about 20 000–30 000 dimers in the cell. Assuming that the binding of about 10 dimers is required to create one superhelical turn the number of HU molecules in the cell could account for 2000– 3000 superhelical turns of the bacterial chromosome. The total *E. coli* chromosome which is 4×10^6 base pairs in length is considered to have a superhelical density of –0.05. This corresponds to about 20 000 superhelical turns. The HU molecules alone would thus account for maximally 15% of the actual superhelical density.

DNA supercoils can also be constrained by strand separation. Transcribing RNA polymerase causes DNA strand separation. There are also about 3000 polymerase molecules in an *E. coli* cell, most of which are engaged in transcription. Only a very small proportion of these are free and not bound to DNA (see Section 2.5). Each elongating RNA polymerase is considered to unwind the DNA in the transcription bubble by about 1.7 turns, corresponding to about 5000 supercoils. Thus 25% of the total supercoils may be constrained through the binding of RNA polymerase. It is thus clear that the local superhelical structure of DNA can be drastically affected due to the binding and dissociation of RNA polymerase. A similar conclusion is valid for nucleoid structuring proteins.

At present it is not known how the different DNA binding proteins are involved in changing local supercoils, and if and how they interact in concert with topoisomerases to regulate DNA topology. However, recent observations have clearly shown that transcription itself has a direct influence on the local supercoiling of the cellular DNA. In the next section it will be seen how transcription and DNA topology are mutually coupled.

6.3.1 The twin-supercoiled domain model

The coupling between DNA topology and transcription is not a one-way route. In the same way that supercoiling affects transcription, the transcription process has a significant effect on local supercoiling. A model has been developed to describe this intricate coupling between transcription and the DNA topology, termed the **twin-supercoiled domain** model. The model is based on the premise that transcribing RNA polymerase has to rotate around the DNA helix axis. Given an average transcription elongation rate of 50 nucleotides per second would mean that the RNA polymerase and the template DNA must rotate at roughly five turns/s around each other. Because of the complexity of the process and the fact that transcription is coupled to translation, a free rotation of the transcription machinery and the DNA would be almost impossible, however. The growing RNA chain still attached to the polymerase is a target for translating ribosomes or polyribosomes which are engaged in the synthesis of proteins. Sometimes the proteins may even be associated with the cellular membrane. Such an arrangement surely cannot rotate. Similarly, a giant molecule like the bacterial chromosome cannot rotate either (the same might be true for a large plasmid). As a consequence, the translocating RNA polymerase causes overwinding of the DNA in front of the transcription complex. At the same time the DNA behind the moving transcription complex is left underwound. The result is a local region of positive supercoiled DNA in front of the elongating RNA polymerase and a comparable region of negative supercoiling in the wake of it. Thus, divergent waves of transcription-induced supercoiling move with the transcription complex along the template DNA. This situation exactly describes the twin-supercoiled domain model. The local domains of divergent supercoils on both sites of the translocating RNA polymerase will not always be distributed fast enough by diffusion to be levelled out. Nevertheless, transcription does not cease due to the constantly increasing torsional strain. This is mainly because of the action of the different topoisomerases, which equalize the two domains of supercoils with opposing signs. Topoisomerase I relaxes the negative supercoiled domain behind the moving RNA polymerase. At the same time, gyrase (topoisomerase II) relaxes the positive supercoiled domain upstream of it (Fig 6.7).

The fact that transcription and supercoiling are intimately coupled poses the interesting question, what happens when two adjacent transcription units are transcribed simultaneously? This can occur in tandem in a convergent or a divergent arrangement. How does transcription-induced supercoiling affect the different ways of coupled transcription? Several systems have been analysed where two promoters are located on the same template within a distance that is covered by the twin-supercoiled domains. In such a situation the two promoters are characterized as being **topologically coupled**. One example that has been studied in some detail is represented by the mutant *leu-500* promoter. The *leu-500* promoter has an A to G transition in the –10 region of the leucine promoter. This mutant promoter is inhibited in its isomerization reaction. This means that the formation of open transcription complexes is energetically unfavoured (see Section 3.3). It was observed that this inhibition was overcome in strains which are deficient in the *topA* gene, encoding topoisomerase I. Cells with mutations in the *topA* gene contain DNA with generally higher negative superhelical density. Negative supercoils facilitate strand separation. The explanation for the lack of inhibition of the mutant *leu-500* promoter in *topA*$^-$ cells

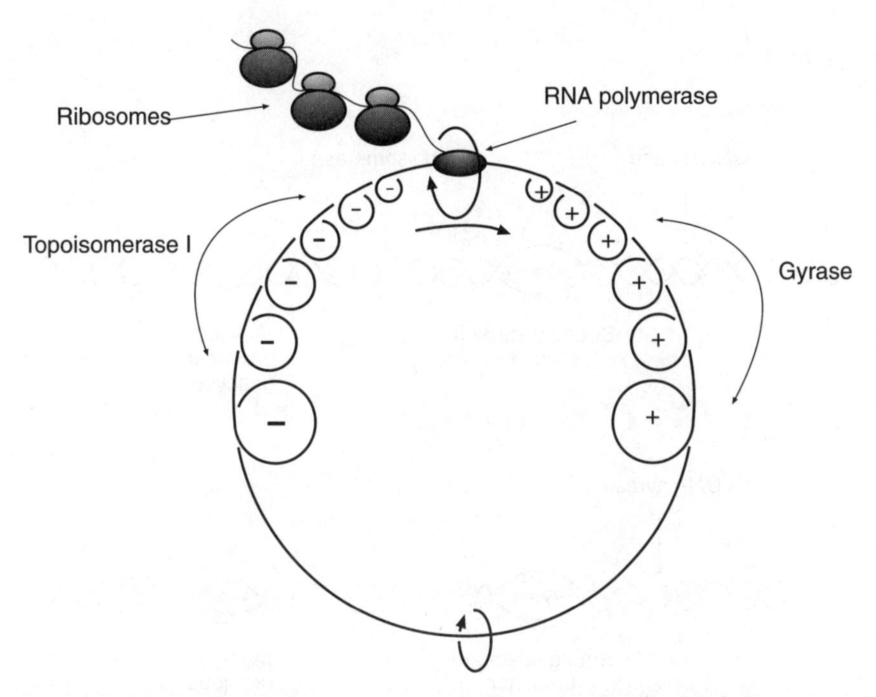

Figure 6.7 The twin-supercoiled domain model. A circular DNA (bacterial chromosome or plasmid DNA) is shown with a single transcription complex transcribing a mRNA. Binding of several translating ribosomes to the growing transcript imposes a strong rotational resistance. Because of the restricted rotation of the moving complex a domain of positive superhelical DNA in front of the transcription complex and a similar domain of negative superhelical DNA in the wake of the RNA polymerase is created. By homeostatic regulation these different supercoiled domains are released through the action of topoisomerase I and gyrase. Generally the different domains are too far apart to release the difference in superhelicity through diffusion. The figure is adapted from Wu *et al.* (1988).

appeared to be straightforward, therefore. Surprisingly, the lack of inhibition is not observed when the *leu-500* promoter was displaced from its natural chromosomal location. The mutation in the *topA* gene could therefore not be the direct cause of the altered promoter activity. A second *divergent* promoter upstream of the *leu-500* promoter was found to be responsible for the accumulation of local negative supercoils. Initiation at the *leu-500* promoter was thus facilitated by the negative supercoiling resulting from divergent transcription from the adjacent promoter (Fig. 6.8). In the natural chromosomal context the *ilvH* promoter, which is surprisingly far apart (1.9 kilobase pairs), has been identified as the likely divergent promoter. This example illustrates how transcription can influence the activity of distant promoters on the same DNA molecule through local supercoiling. It also underlines the importance of the activity of topoisomerases which balance local DNA supercoiling. Modulation

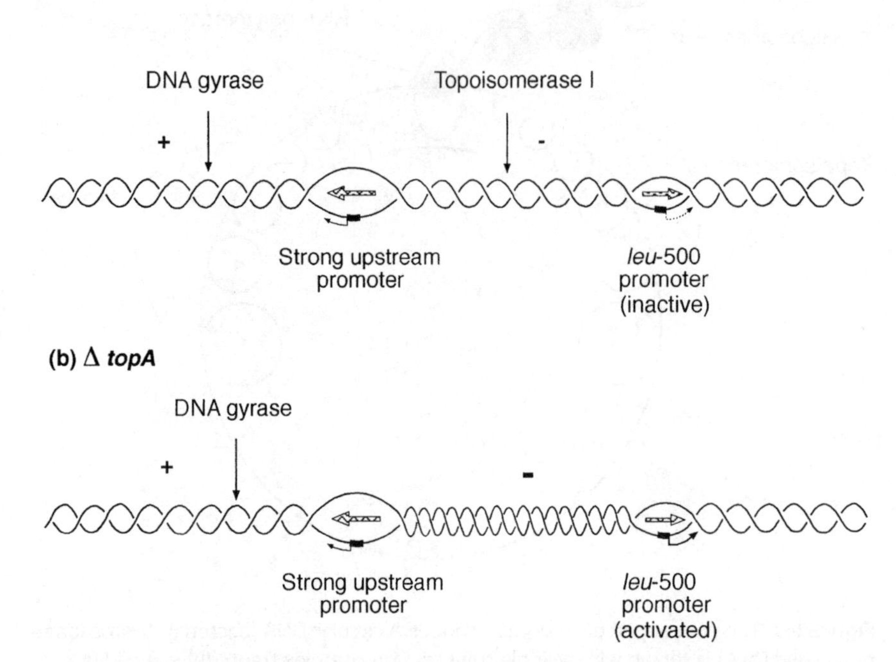

Figure 6.8 Topological coupling at the *leu-500* promoter. Activation of the *leu-500* promoter is believed to result from a divergent upstream promoter that causes a negatively supercoiled DNA domain necessary for the efficient isomerization of the *leu-500* promoter. In strains with an active topoisomerase I this negative supercoiled domain is rapidly relaxed, resulting in an average local superhelical density and an inactive *leu-500* promoter. In strains devoid of topoisomerase I high negative superhelical density resulting from the strong upstream transcription process accumulates and activates promoter melting at *leu-500*.

of local supercoiling is thus an important regulatory process for transcription, especially for coupled promoters.

Summary

The informational character of DNA does not only reside in the primary nucleotide sequence. Signals determining the efficiency of transcription initiation and elongation are additionally encoded in the three-dimensional structure and conformation of the DNA molecule. Conformational changes are based on deviations from the standard DNA B-form, which is normally represented as a straight rod-like molecule. Structural alterations may occur because of changes in the helical parameters, such as the *twist*, *tilt* or *roll* angles. Moreover, the DNA structure may contain *bends* or *supercoils*. The information stored in DNA can further be modified by *base methylation*. The bacterial *dam* methylase system, for instance, is known to convert adenosines in GATC sequences to 6-methyl adenosine. A number of promoters which contain GATC sites within their core region can be regulated by the state of methylation. Because *dam* methylation occurs after DNA replication, transcription initiation at methylation-sensitive promoters is linked to the cell cycle.

Because of axial and torsional deflections from a straight helical molecule DNA can exist in a temporary or permanent curved structure. Such DNA molecules are designated as *intrinsically curved*. Intrinsic DNA curvature normally results from specific sequence steps, such as AT clusters or GGCC sequences, which are repeated in helical phase. Curved DNA sequences are often found within promoters and between 150 and 50 base pairs upstream of strong promoters, where they either enhance RNA polymerase–DNA interactions because of extended contacts or facilitate DNA unwinding during open complex formation. These curved regions, which are commonly called *upstream activating sequences* (UAS), often contain binding sites for regulatory proteins. Binding of proteins can induce DNA curvature in straight DNA molecules or increase already existing curvature. Protein induced deflections from the straight DNA structure are designated as *bends*. Depending on the phasing or the direction in three-dimensional space, DNA bends can either enhance or inhibit the formation of transcription initiation complexes at many promoters. The angular orientation of a DNA bend relative to the promoter sites that bind to RNA polymerase is of great importance for the efficiency of transcription initiation, and characterizes conformation-sensitive promoters.

The binding of a number of transcription factors, such as HU, IHF or H-NS, depends predominantly on a defined DNA conformation rather than on a primary sequence element. These factors often act as *architectural components* supporting a favourable structural arrangement of the transcription initiation complex. Several of these factors can act in concert or functionally replace each other at sensitive promoters.

DNA can exist as different *topological isomers* characterized by different superhelical structures. *DNA supercoils* result from overwinding or underwinding of DNA molecules, which are either circular or whose ends cannot rotate freely. Natural DNA in the cell is normally underwound, resulting in negatively supercoiled structures. The introduction of supercoils into relaxed DNA molecules requires energy. This energy can be used to facilitate DNA melting during the conversion of closed to open promoter complexes. The mechanism of facilitated melting probably involves the conversion of supercoils to *changes in twist*. Promoters whose activity depends on supercoiling are therefore considered to be twist-sensitive. The degree of supercoiling in bacteria is regulated in response to the cell growth and the environmental conditions through the enzymatic activities of *DNA topoisiomerases*. Topoisomerases I are responsible for the relaxation of negative supercoils and do not require energy. In contrast, topoisomerases II introduce negative supercoils at the expense of ATP hydrolysis. The degree of DNA supercoiling can be constrained by protein binding or by strand opening, as it occurs during transcription. Transcription and supercoiling are intricately linked processes. Supercoiling affects many promoters and also influences the elongation rate. On the other hand, transcribing RNA polymerases influence the superhelicity of the template DNA. Elongating transcription complexes and the template DNA cannot rotate freely around each other. As a consequence, transcription elongation causes the formation of separate regions where the DNA is either overwound or underwound. This situation is best characterized by the *twin supercoiled domain model*, which explains a high degree of positive supercoils in front and a negative supercoiled domain behind a moving transcription complex. Local supercoiling resulting from divergent or convergent transcription processes often controls the activity of neighbouring promoters.

References

Ansari, A. Z. and O'Haloran, T. V. (1994) An emerging role for allosteric modulation of DNA structure in transcription, In: Conaway, R. C. and Conaway, J. W. (eds) *Transcription: Mechanisms and Regulation*. New York: Raven Press, pp. 369–86.

Dlacik, M. and Harrington R. E. (1998) DIAMOD: display and modeling of DNA bending. *Bioinformatics* **14**: 326–31.

Fried, M. G. and Daugherty, M. A. (1998) Electrophoretic analysis of multiple protein–DNA interactions. *Electrophoresis* **19**: 1247–53.

Hall, K. B. and Kranz, K. K. (1995) Thermodynamics and mutations in RNA–protein interactions. *Methods in Enzymology* **259**: 261–81.

Plumbridge, J. (1987) The role of *dam* methylation in controlling gene expression. *Biochimie* **69**: 439–43.

Rosenbaum, V. and Riesner, D. (1987) Temperature-gradient gel electrophoresis: thermodynamic analysis of nucleic acids and proteins in purified form and in cellular extracts. *Biophysical Chemistry* **26**: 235–46.

Wu, H.-Y., Shyy, S., Wang, J. C. and Liu, L. F. (1988) Transcription generates positively and negatively supercoiled domains in the template. *Cell* **53**: 433–40.

Further reading

Ansari, A. Z. and O'Halloran, T. V. (1994) An emerging role for allosteric modulation of DNA structure in transcription. In: Conaway, R. C. and Conaway, J. W. (eds) *Transcription: Mechanisms and Regulation*. New York: Raven Press, pp. 369–386.

Ansari, A. Z., Bradner, J. E. and O'Haloran, T. V. (1995) DNA-bend modulation in a repressor to activator switching mechanism. *Nature* **374**: 371–5.

Bracco, L., Kotlarz, D., Kolb, A., Diekmann, S. and Buc, H. (1989) Synthetic curved DNA sequences can act as transcriptional activators in *Escherichia coli. EMBO Journal* **8**: 4289–96.

Crothers, D. M., Gartenberg, M. R. and Shrader, T. E. (1991) DNA bending in protein-DNA complexes. *Methods in Enzymology* **208**: 118–46.

Dlacik, M. and Harrington, R. E. (1998) Diamod: display and modeling of DNA bending. *Bioinformatics* **14**: 326–31.

Drlica, K. (1990) Bacterial topoisomerases and the control of DNA supercoiling. *Trends in Genetics* **6**: 433–7.

Drlica, K. (1992) Control of bacterial DNA supercoiling. *Molecular Microbiology* **6**: 425–33.

Hagerman, P. J. (1990) Sequence-directed curvature of DNA. *Annual Reviews of Biochemistry* **59**: 755–81.

Harrington, R. E. (1992) DNA curving and bending in protein–DNA recognition. *Molecular Microbiology* **6**: 2549–55.

Harrington, R. E. and Winicov, I. (1994) New concepts in protein–DNA recognition: sequence-directed DNA bending and flexibility. *Progress in Nucleic Acid Research and Molecular Biology* **47**: 195–270.

Hilchey, S., Xu, J. and Koudelka, G. B. (1997) Indirect effects of DNA sequence on transcriptional activation by prokaryotic DNA binding proteins. In: Eckstein, F. and Lilley, D. M. J. (eds) *Nucleic Acids and Molecular Biology*, Vol. 11. Berlin-Heidelberg: Springer Verlag, pp. 115–34.

Lane, D., Prentki, P. and Chandler, M. (1992) Use of gelretardation to analyze protein–nucleic acid interactions. *Microbiological Reviews* **56**: 509–28.

Lilley, D. M. J. (1997) Transcription and DNA topology in eubacteria, Vol. 11. In: Eckstein, F. and Lilley, D. M. J. (eds) *Nucleic Acids and Molecular Biology*, Berlin-Heidelberg: Springer Verlag, pp. 191–217.

Lilley, D. M. J. and Higgins, C. F. (1991) Local DNA topology and gene expression: the case of the *leu-500* promoter. *Molecular Microbiology* **5**: 779–83.

Matthews, K. S. (1992) DNA looping. *Microbiological Reviews* **56**: 123–36.

Palecek, E. (1991) Local supercoil-stabilized DNA structures. *Critical Reviews in Biochemistry and Molecular Biology* **26**: 151–226.

Pérez-Martin, J. and de Lorenzo, V. (1997) Clues and consequences of DNA bending in transcription. *Annual Reviews of Microbiology* **51**: 593–628.

Pérez-Martin, J. and Espinosa, M. (1993) Protein-induced bewnding as a transcriptional switch. *Science* **260**: 805–7.

Pérez-Martin, J., Rojo, F. and de Lorenzo, V. (1994) Promoters responsive to DNA bending: a common theme in prokaryotic gene expression. *Microbiological Reviews* **58**: 268–90.

Plumbridge, J. (1987) The role of *dam* methylation in controlling gene expression. *Biochimie* **69**: 439–43.

Rees, W. A., Keller, R. W., Vesenka, J. P., Yang, G. and Bustamante, C. (1993) Evidence of DNA bending in transcription complexes imaged by scanning force microscopy. *Science* **260**: 1646–9.

Schmid, M. B. and Sawitzke, J. A. (1993) Multiple bacterial topoisomerases: specialization or redundancy? *BioEssays* **15**: 445–9.

Travers, A. A. (1989) DNA conformation and protein binding. *Annual Reviews of Biochemistry* **58**: 427–52.

Travers, A. A. (1990) Why bend DNA? *Cell* **60**: 177–80.

Travers, A. A. (1991) To bend or . . .? *Current Biology* **1**: 171–3.

Trifonov, E. N. (1991) DNA in profile. *Trends in Biochemical Sciences* **16**: 467–70.

Tsao, Y.-P., Wu, H.-Y. and Liu, L. F. (1989) Transcription-dependent supercoiling of DNA: direct biochemical evidence from *in vitro* studies. *Cell* **56**: 111–18.

Wang, J.-Y. and Syvanen, M. (1992) DNA twist as a transcriptional sensor for environmental changes. *Molecular Microbiology* **6**: 1861–6.

Wu, H.-Y. and Liu, L. F. (1991) DNA conformational changes during RNA transcription. In: Eckstein, F. and Lilley, D. M. J. (eds) *Nucleic Acids and Molecular Biology*, Vol. 5. Berlin–Heidelberg: Springer Verlag, pp. 187–93.

Wu, H.-Y., Shyy, S. and Wang, J. C. (1988) Transcription generates positively and negatively supercoiled domains in the template. *Cell* **53**: 433–40.

Yang, Y., Westcott, T. P., Pedersen, S. C., Tobias, I. and Olson, W. K. (1995) Effects of localized bending on DNA supercoiling. *Trends in Biochemical Sciences* **20**: 313–19.

Zuber, F., Kotlarz, D., Rimsky, S. and Buc, H. (1994) Modulated expression of promoters containing upstream curved DNA sequences by the *Escherichia coli* nucleoid protein H-NS. *Molecular Microbiology* **12**: 231–40.

7 Regulation by transcription factors

Transcription factors have already been mentioned frequently throughout the preceding chapters of this book. Here, in a more detailed presentation, their special importance in regulating transcription initiation is described. The chapter begins with the classical concept of repressors and operators defined by the operon model. The general properties of transcription factors, which often exhibit dual functions, that of activators and repressors are then described. Moreover, the modular structure of transcription factors, which consist of different functional domains, like DNA binding, oligomerization or protein recognition domains is explained, and the modification of transcription factors by inducers or corepressors is exemplified. There follows a brief section in which different families of transcription factors, based on structural and functional similarities, are presented. Examples of the function of transcription factors in dynamic aspects of regulation, involving DNA structural distortions and the formation of DNA loops, are presented. There is a summary of a family of transcription factors also known as DNA-structuring or histone-like proteins. The chapter also includes information on two-component regulatory systems where the activity of transcription factors is modulated through phosphorylation. Finally, the special situation of the control of nitrogen-fixation genes will be described as an example of σ^{54} promoter-specific transcription factors.

7.1 Repressors and operators—the classical model

Understanding of transcription regulation started from the early paradigm of the **operon model** which was ingeniously proposed by Jacob and Monod, first in 1961. Their operon model was exemplified for the regulation of the expression of the *lac* genes and the genes of bacteriophage λ. At that time regulation was almost entirely explained by direct repression, which today is known to be only one facet of the many regulatory concepts that have since been discovered. The operon paradigm can be considered as an idea that has initiated and stimulated valuable research leading to the present understanding of transcription regulation. The chapter starts with a brief description of the concept of the operon model. As the chapter continues it will be seen how studies over the past 30 years have changed ideas of the initial operon model.

The *lac* operon model has served as an archetype for explaining the mechanism of transcription regulation and is still the central example in many textbooks. The *lac* operon represents a regulatory entity (transcription unit), which consists of the structural genes *lacZ*, *lacY* and *lacA*, encoding the β-galactosidase, lactose permease and a transacetylase, respectively (Fig. 7.1a).

Transcription of the operon is initiated from the *lac* promoter which is controlled by a sequence termed the **operator**. The operator is a palindromic DNA sequence immediately downstream from the –10 hexamer of the *lac* promoter. The operator serves as a target site for a DNA-binding protein, which is termed the ***lac* repressor** (**LacI** or **LacR**). Many operator sequences for different transcription units have been characterized today and they are generally palindromic sequences between 20 and 30 base pairs in length. Often the palindromes are imperfect and interrupted by short sequence elements (Fig. 7.1b).

Operators either overlap with the promoter sequence or are located in close proximity to the promoter. In some cases, however, operators are also found more distant to the promoter, within the structural genes (see below).

The *lac* repressor (LacI or LacR) is the product of the *lacI* gene which is transcribed from a separate site located at a short distance upstream of the other *lac* genes. LacI is a 38-kDa protein which binds to the operator as a tetramer. The monomeric protein is composed of three structural domains. Two subdomains can be obtained by mild proteolytic digestion of the repressor. The N-terminal domain (NTD), comprising amino acids 1–59, is termed the **headpiece**. It contains the structural motif responsible for DNA-binding (a helix-turn-helix motif). The headpiece is linked by a flexible **hinge** region (amino acids 60–80) to the **core** domain which consists of the residual amino acids 81–360. The core domain contains a binding site for low molecular weight effector molecules called **inducers**. Binding of an inducer to LacI alters the affinity of the repressor for the operator sequence. In the case of the *lac* repressor the natural inducer is **allolactose**. In the laboratory, however, allolactose is usually replaced by the analogous compound **isopropyl thiogalactose** (**IPTG**) (see Fig 7.2), which is not metabolized.

Inducers are generally metabolic substrates or end products, like sugars or amino acids, of the enzymatic systems that are encoded in the respective operon.

The classical concept of repression resides on the assumption that the repressor binds with high affinity to the operator sequence which is close to, or even overlaps, the transcription initiation site. When the affinity of the repressor for the operator is higher than that of RNA polymerase for the promoter, repressor binding will sterically block RNA polymerase binding, and thereby efficiently repress transcription. The affinity of the *lac* repressor for its operator sequence

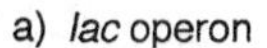

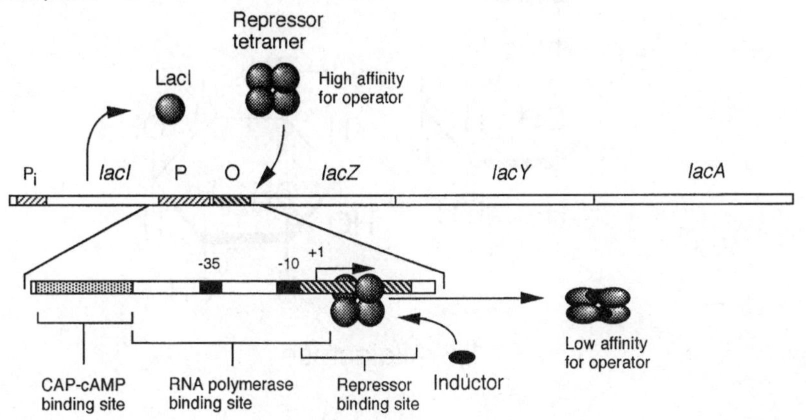

b) Operator sequence

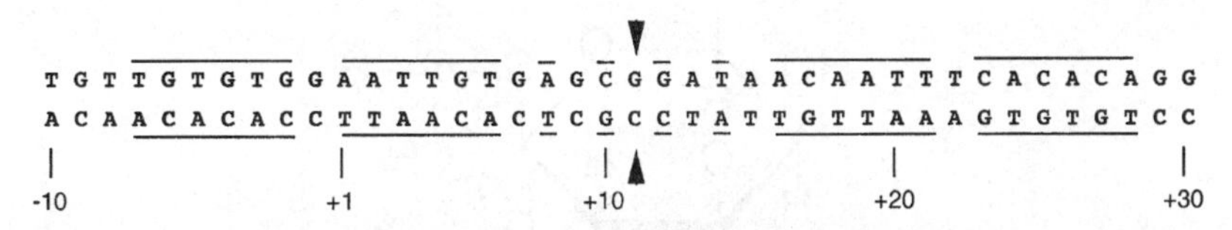

Figure 7.1 The *lac* operon. (a) The physical map of the *lac* operon with the *lacZ* (β-galactosidase), *lacY* (lactose permease) and *lacA* (transacetylase) genes is schematically depicted. The *lac* repressor LacI is encoded by the upstream *lacI* gene. LacI forms a tetramer which binds with high affinity to the operator region. (O). P denotes the *lac* promoter region. The *lacI* gene is controlled by the P_i promoter. The promoter-operator region is shown enlarged below. It contains a binding site for the catabolite regulator protein. The −10, −35 and +1 sequence elements of the *lac* promoter are marked. The repressor binding site (operator) overlaps with the transcription start region. Binding of the inductor isopropyl thiogalactose or allolactose (shown as dark oval) reduces the affinity of the tetrameric repressor which dissociates from the operator and enables transcription of the *lacZ*, *lacY* and *lacA* genes. (b) The sequence of the *lac* operator is shown. The interrupted palindromic sequence regions are underlined or overlined, respectively. Numbers indicate sequence positions relative to the transcription start site. The centre of diad symmetry is indicated by two arrowheads.

is very high ($K_{ass} \sim 1 \times 10^{13}$ M^{-1}). Hence only 10–20 tetramers of the repressor per cell are sufficient to cause efficient operator occupancy.

The important point for regulation is that repression must be *reversible*. Consequently, the repressor must be released from the DNA when the products encoded in the respective operon are required by the cell. Bacteria have evolved two principal mechanisms to release the inhibitory effect of repressors. One way to terminate repression is to alter the affinity of the repressor for its

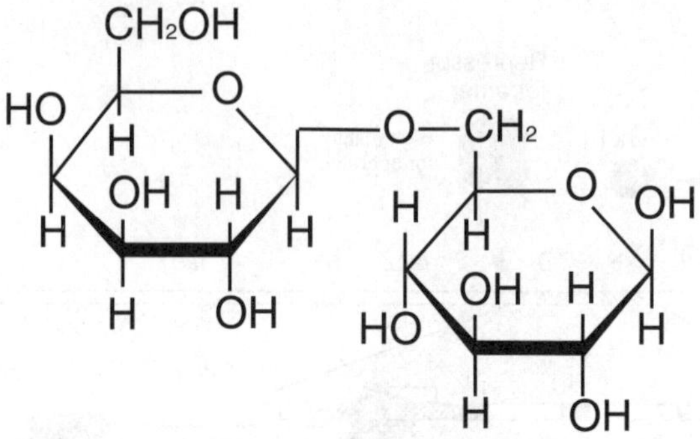

1,6-Allolactose

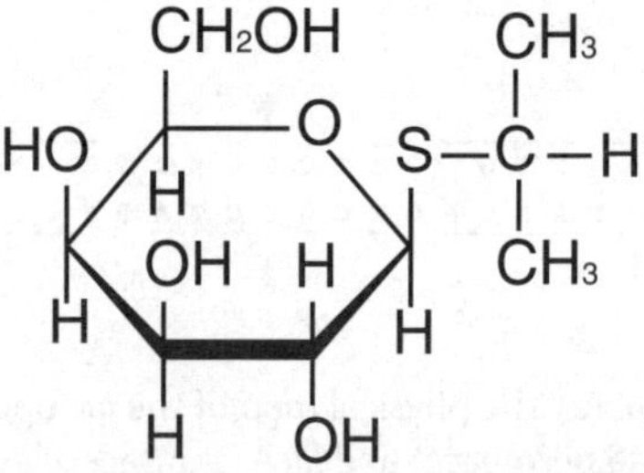

Isopropylthiogalactose (IPTG)

Figure 7.2 Structure of the *lac* inducers. The chemical structures of the natural inducer substances of the *lac* operon 1,6-allolactose and the laboratory compound isopropyl thiogalactose (IPTG), which is not metabolized, are shown.

operator. This is the way most repressors are inactivated. For many repressors a reduction in the affinity for the operator is achieved by binding of low molecular weight inducer substances, for example IPTG, which has been mentioned above. The natural inducer substances are often intermediates of the synthesis cycle which ought to be repressed. The situation corresponds to a typical feedback loop. At high concentrations the inducer will bind to the repressor and modulate through allosteric conformational changes the affinity of the repressor for its operator. The affinity of the repressor can in principle be reduced or enhanced over a significant range. If the inducer is a substrate for the biosynthesis enzymes which should be controlled, its presence will signal the induction of the operon by *reducing* the repressor affinity for the operator. Hence, repression is released (**substrate induction**). The inducer can also be an end product of the synthesis reaction. In this case, repression is induced by *increasing* the affinity for the operator. This situation corresponds to a typical feedback inhibition (**end product inhibition**). The effector molecules for such systems are often designated as **co-repressors** rather than inducer. In the case of the *lac* operon the inducer allolactose (or IPTG) reduces the affinity of the *lac* repressor for the *lac* operator more than 1000-fold. As a consequence, inducer-bound *lac* repressor is inactivated, transcription of the operon can take place and the enzymes for the metabolic conversion of lactose are synthesized. In contrast, the *trp* repressor (TrpR), which regulates the biosynthesis of tryptophane, is activated by the co-repressor tryptophane which increases the affinity for the operator and thus causes repression of transcription.

A second way to release repression is to reduce the concentration of the repressor itself. This is most efficiently achieved by proteolytic digestion of repressor molecules. Note that this type of repressor inactivation is irreversible. The regulated expression of the repressor itself is thus an important feature of the control system. The **LexA repressor** may serve as a typical example for this type of inactivation. The LexA repressor coordinately regulates the genes that belong to the **SOS response** (see Section 8.1). Another example is the bacteriophage λ **repressor cI**. In both cases degradation of the repressors is caused by the inducible proteolytic activity mediated by a protein termed **RecA**. This protein normally has important functions in recombination. Under detrimental environmental conditions (for instance, DNA damage), RecA can be converted to a form which stimulates proteolytic cleavage of LexA (or λ cI). The reduction in the concentration of intact repressor molecules removes the bound repressor species from the operator, and thereby causes de-repression of the operon in question.

If regulation depends on the affinity of the repressor molecules for their cognate operator sequences, the question arises as to what the actual affinities are, and to what degree can they be modulated through effector molecules. To answer these questions consider that, like any DNA-binding protein, repressors may interact with the DNA in a specific or an unspecific way. Of course, both modes of binding have different affinities. Although the *affinity* of the repressor for its operator is very high, the considerations presented in Box 7.1 show that the preference for binding of the *lac* repressor to the cognate operator in competition with the total *E. coli* DNA is only 20.

It follows that a 20-fold reduction in the affinity of the repressor will cause a redistribution of the repressor between the excess random DNA and the operator. Repressor mutants with about 20-fold lower affinity can thus cause constitutive expression of the *lac* genes. Binding of the inducer IPTG changes the affinity of the *lac* repressor about 1000-fold. The affinity for unspecific DNA-binding is not affected, however. Therefore, the presence of inducer does not lead to free repressor molecules in the cell. Instead, a re-distribution between specific sites (operator DNA) and unspecific sites (non-operator DNA) occurs. The concentration of free repressors in the cell at any time can thus be neglected. It

Box 7.1 The preference of repressor-operator binding

As with any DNA-binding protein, repressors may interact with the DNA in a specific or unspecific way. Of course, both modes of binding have different affinities. Unspecific binding is almost entirely the result of electrostatic interactions between positively charged amino acid side chains of the DNA binding protein and the negatively charged phosphate backbone of DNA. Unspecific binding occurs thus without particular sequence specificity. In contrast, specific interactions are sequence- or structure-specific and involve additional hydrophobic elements and hydrogen bonding of precisely orientated functional groups. Usually repressors have rather high affinities for specific binding to their operator sequences. It was shown in the example in Section 7.1 that the association constant K_{ass} for the *lac* repressor–operator interaction is roughly 10^{13} M^{-1}. The ratio of association constants for specific binding (to the operator) and non-specific binding (to non-operator DNA) is in the region of 4×10^6. It should be borne in mind, however, that there is often only one or at most a few operator sites in the genome. For the entire *E. coli* chromosome (4.2×10^6 base pairs) and an average operator length of 26 base pairs, as in the case of the *lac* operator, this would mean a large excess of unspecific over specific repressor binding sites. In the example of a single *lac* operator there are 2×10^5 unspecific over one specific sites. In other words, there is one high affinity site (operator) for every 2×10^5 low affinity sites (non-operator DNA). Consider now the **preference for operator binding** which is given by the ratio of association constants for specific *versus* unspecific binding divided by the number of low affinity sites. In the example above, the following is derived:

$$\frac{4 \times 10^6}{2 \times 10^5} = 20.$$

The preference for specific binding to a single operator within the complete *E. coli* genome is thus similar to 20.

Although the *affinity* of the repressor for its operator is very high the above calculation shows that the *preference* for operator binding is not expressed by a particularly large value. The preference for binding of the *lac* repressor to the cognate operator in competition with the total *E. coli* DNA is only 20. It follows that a 20-fold reduction in the affinity of the repressor will cause a redistribution of the repressor between the excess random DNA and the operator. The repressor is thus released from the operator and no longer inactivates the promoter. As a result constitutive expression of the *lac* genes is observed.

should be emphasized that the above considerations are not restricted to the *lac* repressor but are valid for many DNA binding proteins.

Different repressors can vary considerably in their affinities for operator sites. The spectrum of affinities ranges from very strong binding ($K_{ass} \sim 10^{11}$ M^{-1} to 10^{13} M^{-1}), which causes almost stoichiometric binding under physiological conditions (e.g. *lac* repressor), to moderate binding ($K_{ass} \sim 10^8$ M^{-1} to 10^9 M^{-1}; e.g. λ cI repressor) or even rather weak binding ($K_{ass} \sim 10^7$ M^{-1}), for which the phage P22 Arc repressor may be given as an example. The range covered by repressor affinities therefore comprises regulation from a complete shut-off to a moderate downregulation.

7.2 The modular structure of transcription factors

Normally, prokaryotic transcription factors are organized as symmetrical molecules composed of identical (or homologous) subunits. This means that transcription factors are active as multimers and, most remarkably, these multimers are assembled from an even number of monomers; e.g. CRP (dimer), LacI (tetramer), DeoR (octamer). This symmetrical arrangement is probably important for the recognition of the operator DNA which itself is characterized by a symmetrical architecture (e.g. palindrome).

To summarize, transcription factors have at least three independent functions. They are able to bind specific DNA sequences, their activity can usually be modulated either by effector molecules or protein modifications, and they are able to multimerize in a symmetrical fashion. As seen below in more detail, transcription factors can have an additional function. In many cases their activity depends on direct contacts with the RNA polymerase. The structural analysis of many transcription factors has revealed that these different functions are normally reflected in a domain structure of the protein molecules. The different functions can often be assigned to conserved amino acid motifs which are separated into independent structural units. Transcription factors, therefore, frequently show a modular organization. The modular elements consist of DNA-binding motifs, like the helix-turn-helix motif, or other structures capable of recognizing DNA (see Box 2.2). Zn-fingers, often found in eukaryotic transcription factors, are rare in prokaryotes, however. Another domain is usually responsible for dimerization or oligomerization of the transcription factor monomers. The oligomerization domains are characterized by protein–protein recognition elements. Typical structures for protein–protein recognition are **amphipathic helices** or **leucine zipper** elements. For prokaryotic transcription factors leucine zipper elements are the exception (e.g. MetR). They are very frequent, however, for eukaryotic transcription factors. The structures of the sites of interaction with cofactors (inducers, co-repressors or coactivators) vary depending on the chemical nature of the compounds in question. Nonhomologous amino acid regions are presumed to provide the necessary variation to confer cofactor specificity. Specific amino acid residues are arranged in a sterical conformation to accommodate interactions with either charged, hydrophobic or aromatic compounds or molecules capable of forming specific hydrogen bonds. In many cases these sites are characterized by an amino acid arrangement which forms a kind of binding pocket similar to the active centres of metabolic enzymes. For instance, in case of the transcription factor AraC the active modifying ligand (inducer) is the sugar arabinose (see Section 7.4.2). The crystal structure of AraC dimers in the presence and absence of arabinose has been elucidated. In the AraC dimer arabinose is bound at the N-terminal part of the protein and completely buried within the structure. In contrast, in the absence of arabinose, the N-terminal part of the protein is disordered and the amino acids that form the arabinose binding pocket are free to serve as an oligomerization interface. Often, cofactor binding sites are located at the interacting surfaces of the multimers. Binding at the interface may be especially suitable to cause an allosteric change of the three-dimensional structure of the multimer. Domains known to interact directly with RNA polymerase (often referred to as **activation domains**) are again similar to known protein–protein binding motifs (see above). A common motif found in the activation domain of prokaryotic transcription factors is a cluster of acidic amino acids (e.g. λ cI or the phage 434 repressor). This so-called **acidic patch** is known to make contacts with RNA polymerase.

The individual structural motifs appear to be conserved and are functionally interchangeable (see Box 2.2 for DNA-binding and protein–protein binding motifs). They are not always located at similar positions with respect to the N- or C-terminal ends of the transcription factors, however. As shown in Fig. 7.3, for instance, some transcription factors have their DNA-binding domain at the N-terminus (e.g. λ cI repressor), while in other cases it is localized at the C-terminal part (e.g. CRP).

In a number of cases the activity of special transcription factors has been changed successfully by elegant 'domain shuffling' experiments. In these cases different functional modules, for instance the DNA-binding or oligomerization domains, were exchanged (shuffled) between different proteins. In the examples studied the activity and specificity could be transferred with the respective protein domain. The specificity of prokaryotic transcription factors has also been transferred to eukaryotic activators by fusion of DNA-binding domains. For instance, a peptide of the yeast Gal4 activator or the Gal11 protein has been fused to the DNA-binding domain of the bacterial LexA repressor. By using these constructs eukaryotic transcription can be activated by the insertion of the LexA operator sequence near to a eukaryotic promoter (Box 7.2).

7.2.1 To activate or to repress: what makes the difference?

The activity and specificity of prokaryotic transcription factors can be completely altered by changing the amino acid motifs responsible for the specific regulator functions. However, many transcription factors *per se* have dual functions depending on the sequence context of their operator sites. That is, a given transcription factor can act as an activator for one transcription unit and as a

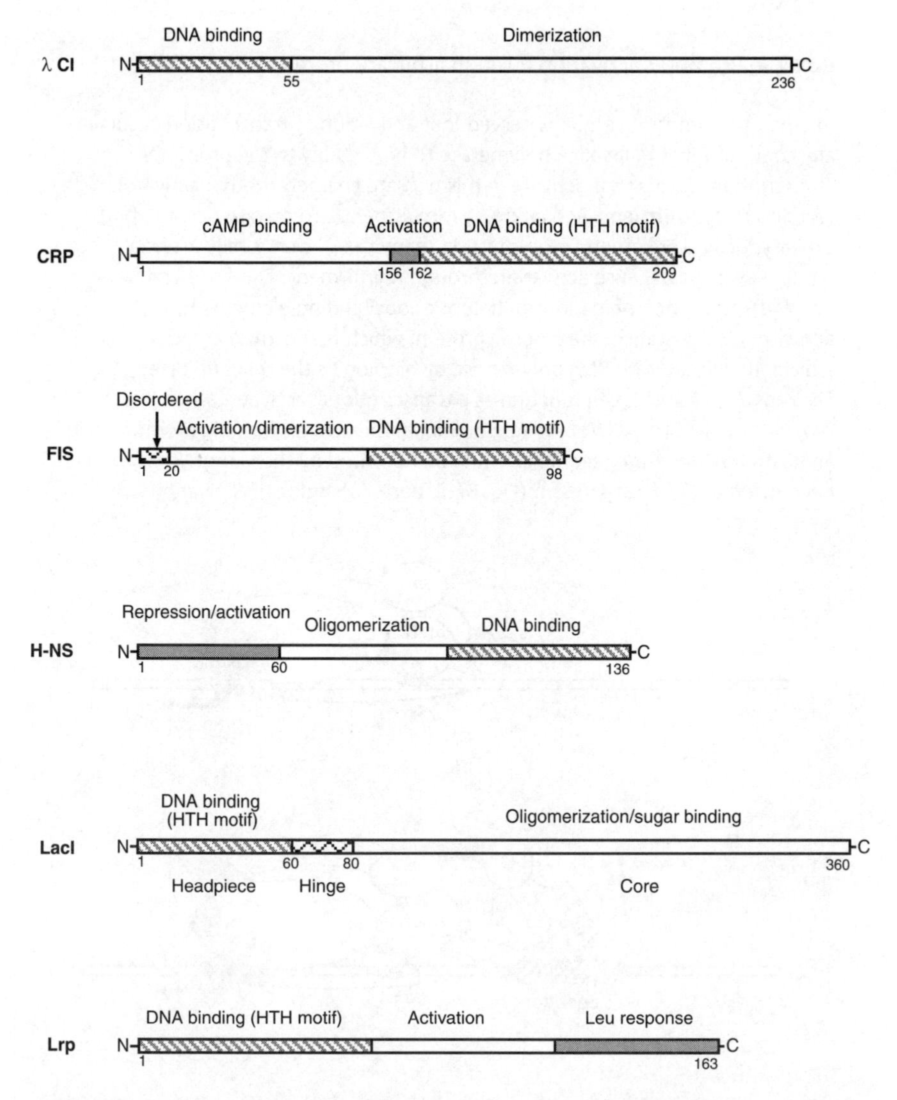

Figure 7.3 Modular organisation of transcription factors. A collection of representative prokaryotic transcription factors with their functional domains highlighted is shown. The N- and C-terminal ends are marked. Numbers indicate amino acid positions. HTH, helix-turn-helix DNA-binding motif.

repressor for another. The transcription factor CRP may be taken as a representative example (see also Section 7.3.2). CRP functions as an activator in the case of the *lac* P1 promoter and simultaneously as a repressor in the case of the *gal* P2 promoter. Such dual activity of transcription factors is by no means an exception. It is rather a very common phenomenon, and many transcription factors

Box 7.2 Promoter activation through arbitrary protein–protein contact

At some promoters it can be observed that activation of transcription occurs in any condition that helps RNA polymerase to bind stably to the promoter. Transcription factors that activate in this way are characterized as activators working by **recruitment**. Activation by recruitment is known to occur in both prokaryotic and eukaryotic promoters. In many cases even totally artificial binding events can cause activation through recruitment. This has been demonstrated by combinations of heterologous binding elements fused to activators. For instance, the *E. coli* ω protein, which is known to associate stoichiometrically with RNA polymerase by binding to the β′ subunit (see Sections 2.2 and 2.5), can function as a transcriptional activator when it is covalently linked to a DNA-binding domain. In this way a synthetic promoter with an upstream λ operator sequence could be activated by the ω protein which had been fused to the λ cI repressor (Fig. B7.2, parts (a) and (b)). Activation by

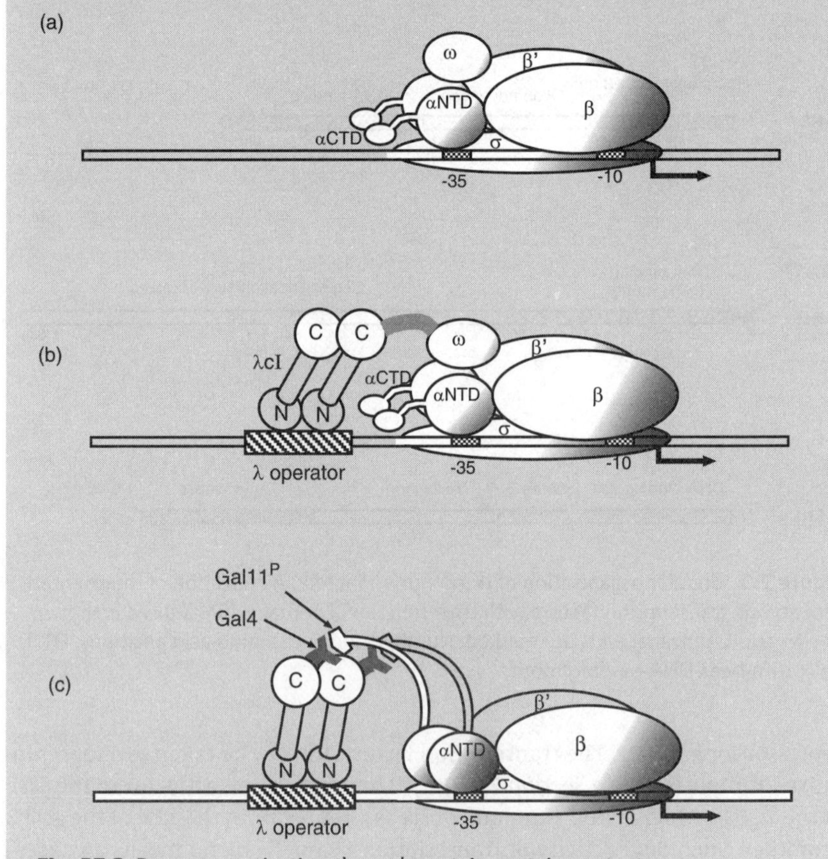

Fig. B7.2 Promoter activation through protein-protein contacts

recruitment has also been observed when two chimerical proteins were used to create an arbitrary protein–protein contact. For instance, the yeast Gal4 protein forms dimers with a mutant component of the yeast RNA polymerase II holoenzyme (Gal 11^P). Fusion of Gal 11^P to either the N-terminal domain of RNA polymerase α subunit (αNTD) or to the ω protein results in activation of a promoter which contains an upstream λ operator when the corresponding dimerization domain of Gal4 is fused to the λ repressor. In other words, DNA-binding is mediated through λ repressor-operator interaction and RNA polymerase is tethered by Gal 11^P–Gal4 dimerization (Fig. B7.2, part (c)). These, and similar, results obtained in yeast have led to the conclusion that some promoters can be activated by any arbitrary protein–protein contact between a DNA bound factor and an available target element associated or coupled with the RNA polymerase holoenzyme. It should be noted, however, that promoters which are not limited by stable RNA polymerase binding (e.g. promoters limited in their isomerization rate to open complexes) may not be activated by this mechanism.

characterized today are known to be bifunctional. Table 7.1 lists several examples of regulatory proteins for σ^{70} promoters which have been characterized as dual regulators.

Transcription factors are known that act purely as repressors or activators. Pure activators are clearly the rarest species, and only five activators have been described to date for which no natural repressor function has been demonstrated.

Are there general rules that determine whether a transcription factor functions as an activator or a repressor? The answer can in part be given when the activity of transcription factors and their sites of interaction with the DNA (the location of the operator) are compared. From such a systematic comparison it was found that activation or repression mediated by regulatory proteins is linked to exclusive distances of the regulator binding sites (operators) relative to the promoters. The preferred sites for activators are located between nucleotide positions –80 and –30. For a number of transcription units operator sites further upstream exist. Such remote sites are only found, however, when an additional proximal operator site is also present. Repressor binding sites are generally located downstream from nucleotide position –30. This zone seems to be exclusive for repressors, and *no* activators are known which bind downstream from –30. If, for instance, an activator binding site is artificially moved to a site downstream from template position –30, the activating function is lost. Moreover, it has been observed that an activator binding site, when displaced to a site downstream from –30, causes the original activator now to act as a repressor. Hence, by moving an activator binding site to a typical repressor zone a switch from activation to repression may occur.

In many cases regulators are involved in the control of their own synthesis (see, for example, the family of LysR type regulators below). Activators can

Table 7.1 Dual functions of transcription factors confined to σ^{70} promoters

Pure activators	Pure repressors	Dual regulators
Ada	ArcA	AraC
KpdE	ArgR	CRP
MalT	BioB/BirA	CysB
RhaR	CytR	DynA
PhoB	DeoR	FadR
	Fur	FIS
	GalR	FNR
	GalS	IlvY
	GlpR	Lrp
	IclR	MetR
	LacI	NarL
	LexA	OmpR
	MetJ	OxyR
	NR_1/NtrC *	PabB
	PigC	TyrR
	PurR	
	PutR	
	PutA	
	RafA	
	TetR	
	TrpR	

* Repression of NtrC is confined to σ^{70} promoters, not to σ^{54} promoters, where it acts as an activator (see Section 7.6.2). The data is taken from Gralla and Collado-Vides (1996).

therefore often repress their own synthesis. Hence, in these cases, the synthesis of the regulatory proteins is balanced by autogeneous regulation to maintain the desired concentration within fixed limits. This type of autoregulation involving the dual activities of regulatory proteins is very common. For instance, FIS, the activator protein for stable RNA synthesis, acts as a repressor of its own synthesis. Other examples include the global regulator CRP or the regulators CysB, FNR, MetR, OxyR or TyrR (see Table 7.1).

For a given repressor the exact locations of the operator sequences with respect to the different promoters of a regulon are not identical. This means that when a single repressor controls a family of promoters the distance of the repressor binding sites to the core promoters can vary within certain limits. This variation in operator distances may be the basis for a differential capacity of repression at the individual promoters. Thus individual genes of the same regulon can be appropriately adapted to cell physiology by simply varying the promoter–operator distance within a given limit.

What are the possible mechanisms by which transcription factors change the efficiencies of promoters? For repression the most obvious and self-explanatory mechanism is steric hindrance. Binding of a regulatory protein that overlaps with the RNA polymerase binding site at the promoter will inhibit transcription if the affinity of the regulatory protein for its operator is high compared to the affinity of the RNA polymerase for the promoter. Steric hindrance reduces the primary association of RNA polymerase with the promoter. Hence, less closed complex can be formed and the binding constant K_B is reduced (see Chapter 3). Several examples have been found where repression can be exclusively explained by a decrease in K_B (LexA at the *uvrA* promoter, LacI at the *lacUV5* promoter or the λ cI repressor at the λ P_R promoter). The early step in RNA polymerase–promoter binding, which affects K_B, is not only the target for *repressor* action. This step can also be influenced by *activators*. In such cases activators increase the stability of the closed complex. This is a common way in which activators stimulate transcription. What is known about the mechanism by which an activator can increase K_B? The small number of pure activators suggests that the mechanism of activation may require specific features not usually found in repressor molecules. In fact, many activators have sequence elements distinctly different from pure repressor molecules. These are the activation domains which recognize specific sites within the RNA polymerase (e.g. acidic patches). Some activators are therefore able to bind directly to RNA polymerase. In such cases the RNA polymerase–promoter complex is stabilized by additional protein–protein contacts between sites within RNA polymerase and the transcription factor, which is tethered to the DNA through operator binding. This additional contact enhances formation of the closed RNA polymerase–promoter complex (increase in K_B). Both the α and the σ subunits of RNA polymerase have been identified as sites to which activators can bind (see Sections 2.2.1 and 2.2.4). Binding to the α subunit occurs to the C-terminal 85 amino acids (αCTD), which form an independent structural domain linked to the N-terminal part of the α subunit by a flexible linker of about 10 amino acids. For several activators the sites of interaction with the αCTD have been identified. Not all activators that interact with the αCTD were affected to the same extent when amino acids within the critical region (amino acids 249–329) of the αCTD were substituted. Different transcription factors probably make slightly different contacts with the αCTD or other RNA polymerase domains. Interacting surfaces have also been mapped within the primary sequence of some transcription factors. In the case of the regulator CRP at Class I CRP-dependent promoters (see Section 2.2.1) it is assumed that direct protein–protein contacts are maintained through binding of amino acids 156–162 of the promoter-proximal CRP subunit to amino acids 261 and 265 of αCTD. DNA-binding sites of the RNA polymerase and CRP are orientated such that both proteins are located on the same face of the DNA helix close to the –35 promoter region (Fig. 7.4). The flexible linker region which connects the αCTD to the rest of the molecule is considered to provide the necessary flexibility to accommodate direct contacts between RNA polymerase and other activators which bind at slightly different DNA positions relative to the promoter. At some promoters (e.g. at Class II CRP-dependent promoters, see below) CRP has additional contacts to the N-terminal domain of the α subunit (αNTD).

It should be noted that direct contacts between transcription activators and RNA polymerase are not restricted to the α subunit. Several transcription factors are known to interact with subregion 4.2 of the σ subunit. These proteins include AraC, MalT, λ cI PhoB and possibly also FIS (see Section 2.2.4). There are also some interesting reports that the replication initiator protein DnaA functions as a transcription activator in a number of cases. DnaA binds to a conserved DnaA box with nine base pairs (5′-TT$^{A}/_{T}$TNCACA-3′), and can either activate or repress transcription of several genes. In case of activation of the λ P_R promoter it has been shown that the β subunit of RNA polymerase serves as a transcription factor contact site. From a series of experiments summarized in Box 7.2 it has been concluded that any specific interaction between RNA polymerase and a favourably orientated DNA binding protein may be sufficient to increase promoter activity. This phenomenon has been termed **activation by recruitment**.

Regulation by transcription factors is by no means restricted to a change in the RNA polymerase affinity (K_B). In principle, any step during the initiation pathway can be affected (see Chapter 3). Alterations in the rate of isomerization from the closed to the open RNA polymerase–promoter complex or changes in the promoter escape rate may result from allosteric changes in the initiation complex geometry. Such a structural change in the initiation complex can result from direct contacts between the transcription factor and RNA polymerase. However, DNA conformation may also be involved. Transcription factor-induced DNA-bending can clearly influence the architecture of the initiation complex. Bending can, for instance, promote DNA wrapping and support additional RNA polymerase DNA contacts (increase in K_B). Bending or changes in twist may further directly influence promoter melting and thus give rise to changes in the isomerization rate k_2 (see Section 6.2.2). In some cases transcription factors can bind to several operator sites, some of which are located at quite some distance upstream. In such cases the formation of DNA loops, stabilized by protein–protein interaction, may occur. A more detailed discussion of this type of regulation involving dynamic structures will be given in Section 7.4.

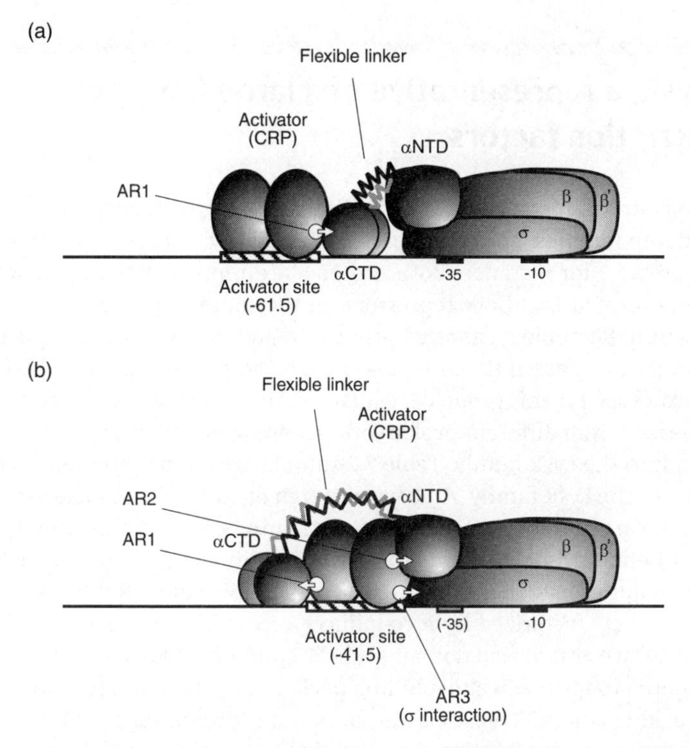

Figure 7.4 Transcription activator contacts to RNA polymerase. Two transcription intiation complexes involving the activator catabolite regulator protein (CRP) are illustrated as cartoons. (a) A CRP dimer, presented as grey spheres, is bound to the activator site (shaded box) upstream of the promoter with the binding centre at position –61.5 relative to the transcription start. RNA polymerase bound to the –10 and –35 promoter elements is schematically drawn with the β, β′ and σ subunits marked. The two α subunits are presented by separated spheres, representing the N-terminal (αNTD) and the C-terminal (αCTD) domains, respectively. The αNTDs make multiple RNA polymerase contacts while the αCTDs bind to DNA between the activator binding site site and the –35 region. Each of the two domains are connected by flexible linkers indicated by zigzag lines. The downstream subunit of CRP makes contacts to the αCTD domain by activator region 1 (AR1) which is shown as a white circle and marked by an open arrow. (b) Schematic arrangement of a transcription initiation complex with the centre of activator proteins (CRP) bound to position –41.5. Such promoters do not normally have a convincing –35 consensus sequence. The –35 element is therefore shown in brackets. With the activator site closer to the core promoter a set of different contacts between RNA polymerase and the activator protein is made. The upstream subunit of the activator (CRP) binds to the αCTD, which contacts the DNA upstream of the activator. This interaction occurs via activating region 1 (AR1) of CRP, indicated by a white circle and an open arrow. The downstream subunit of CRP makes different contacts. Activating region 2 (AR2) is bound to the αNTD domain, while activating region 3 (Ar3) of CRP may interact with the σ subunit of RNA polymerase. The figure is adapted from Busby and Ebright (1997).

7.3 LysR, a representative of a large family of transcription factors

Based on structural and functional similarity, regulatory proteins can be grouped into families. The LacI repressor described above is very similar to the GalR repressor, for instance. Both proteins are representatives of a family of regulators termed LacI/GalR repressors. In the same way, based on structural and functional homology, another group of regulators has been assigned as the **LysR family** of transcriptional regulators. The transcription factor LysR serves as the prototype for this group of regulators. More than 50 members have been characterized from different prokaryotic genera which fit by size, structure and function into the LysR family. Table 7.2 summarizes some representative *E. coli* members of the LysR family. All have a stretch of about 60 characteristic amino acids which are highly conserved at the N-terminus. Within this stretch there is a typical helix-turn-helix motif (roughly 20 amino acid residues) for binding into the major groove of a target DNA. The structure of LysR-type regulators is furthermore characterized by a co-inducer recognition or **response domain**, comprising two amino acid domains (95–173 and 196–206). The latter sequence is considered to form a ligand-binding pocket. A conserved C-terminal domain (amino acid residues 227–253) is also important for DNA interaction and probably functions as a multimerization domain. Both activators and repressors can be found within the family of LysR regulators.

Table 7.2 *E. coli* LysR regulators

Regulator	Function	Co-inducer
CynR	Cyanate detoxification	Cyanate
IciA	Inhibition of DNA replication	?
IlvY	Isoleucine/valine biosynthesis	Acetolactate or acetohydroxybutyrate
LeuO	?	?
LysR	Lysine biosynthesis	Diaminopimelate
MetR	Methionine biosynthesis	Homocysteine
NhaR	Na^+/H^+ antiporter	Na^+ or Li^+
OxyR	Oxidative stress response	Oxidation of OxyR, H_2O_2?
TdcO	Theonine degradation	Anaerobic metabolite?
XapR	Xanthosine phosphorylase	?

?, not known. Data are taken from Schell (1993).

Generally, members of the LysR family function by recognizing an operator sequence (roughly 15 base pairs) of interrupted dyad symmetry. The centre of this operator or **recognition site** is usually located at position –65 relative to the transcription start site of the regulated transcription unit. In addition to this recognition site, there is often a second downstream binding site with less similarity to the operator sequence. This site, termed the **activation site**, is located on the same face of the DNA helix as the recognition site. It is thus very close to the –35 promoter hexamer sequence. For some but not all members of the LysR family binding to the activation site requires prior binding of the co-inducer for transcription activation. In the presence of the co-inducer, interaction with both operator sites occurs cooperatively, and probably induces a structural change of the DNA complex. This conformational change apparently causes significant DNA bending. Thus, DNA conformation and/or RNA polymerase contacts are certainly important aspects in the mechanism of LysR regulator-dependent activation. Interestingly, the genes for the LysR-type regulators, and the target genes controlled by the same regulators, are often arranged in *divergent orientation*. The distances between the promoters for the regulators and the genes to be controlled are in the range of 25–70 base pairs. The divergent pairs of promoters are therefore not functionally independent. In each case the recognition site for the target gene directly overlaps with the promoter of the regulator gene. Binding of the regulator will, therefore, autogeneously repress the synthesis of the regulator while activating the target gene (Fig. 7.5).

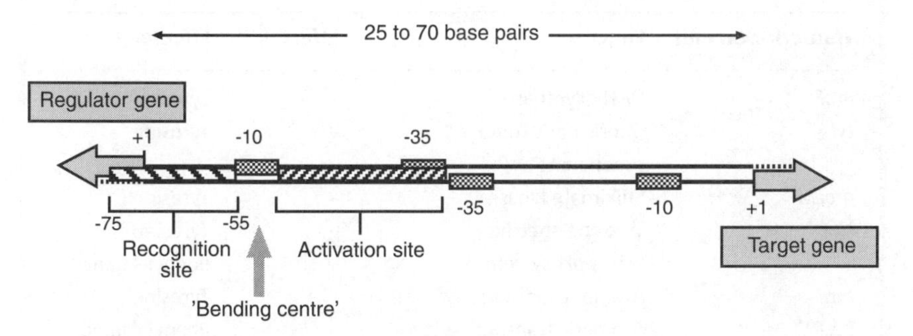

Figure 7.5 Divergent LysR-type autoregulatory region. The divergent arrangement of promoters and control elements within the LysR-type regulatory region is schematically depicted. Generally the autoregulatory region extends between 25 and 70 base pairs and separates the divergently orientated genes for the regulator and the target. Two DNA-binding regions for the interaction with the regulator are indicated (recognition site and activation site), and their sequence positions relative to the transcription start of the target gene are given. The regulator binding sites overlap with the core promoter elements of the regulator gene such that activation of the target gene automatically inhibits transcription of the regulator gene. Generally a region of DNA-bending is observed between the recognition site and the activation site.

7.3.1 The TyrR regulon

As a typical example for a dual regulatory system, the **TyrR regulon** should be introduced here. The transcription factor **TyrR** regulates the genes for the biosynthesis and transport of aromatic amino acids encoded in several different transcription units. The regulon comprises eight operons (Table 7.3).

TyrR acts as a pure repressor for six of these eight transcription units. It acts for one operon as a dual regulator, either activating or repressing transcription, and for one operon it functions exclusively as an activator. The TyrR regulator is the product of the *tyrR* gene. This gene codes for a protein which is 513 amino acids in length. The protein has a helix-turn-helix DNA-binding motif at the C-terminus, and the amino acid sequence for a typical ATP binding site in its centre. It thus shows considerable homology with other regulators, such as NtrC (see Section 7.6.2). However, unlike NtrC, which activates σ^{54} promoters, TyrR acts at σ^{70} promoters and is clearly functionally different from NtrC. In addition to the DNA-binding motif and the ATP binding site, TyrR contains binding sites for the aromatic amino acids **tyrosine** and **phenylalanine**. Each transcription unit regulated by TyrR contains one or more interrupted palindromic operator sequences termed **TyrR boxes**. The TyrR boxes have a TGTAAAN$_6$TTTACA consensus sequence. Binding to closely spaced TyrR boxes occurs cooperatively. In line with this notion, the TyrR boxes need to be located

Table 7.3 Members of the TyrR regulon

Transcription unit	Function	Effect	Effector
aroF	DAHP synthase *	–	Tyrosine
tyrA	Chorismatic mutase, prephenate	–	Tyrosine
aroLM	Shikimate kinase	–	Tyrosine
tyrP	Tyrosine-specific	–	Tyrosine
	transport system	+	Phenylalanine
tyrB	Aromatic aminotransferase	–	Tyrosine
aroP	Aromatic transport system	–	Phenylalanine, tyrosine or tryptophan
tyrR	TyrR regulator	–	None
mtr	Tryptophan-specific transport system	+	Phenylalanine or tyrosine
aroG	DAHP synthase *	–	Phenylalanine

+, Activation; –; Repression. * DAHP, 3-deoxy-D-arabino-heptulosonate 7-phosphate synthase. The data is taken from Pittard and Davidson (1991).

on the same face of the DNA helix. Changing the helical phasing of the boxes or inserting a long stretch of DNA between the boxes abolishes repression. Operons which are repressed by TyrR have at least one TyrR box that overlaps with the core promoter sequence or is located downstream of the transcription start site. One member of the TyrR regulon can be both activated and repressed by TyrR. This is the *tyrP* operon which encodes the tyrosine-specific transport gene. This transcription unit contains two TyrR boxes. One is located upstream of the core promoter sequence, while a second box overlaps the –35 promoter region. Binding of TyrR to the two boxes does not occur with the same affinity. The interaction with the box that overlaps the promoter region requires much higher TyrR concentrations. A switch between activation and repression can thus be mediated by different concentrations of TyrR protein.

The *mtr* transcription unit, which encodes a tryptophan-specific transport gene, is the only member of the TyrR regulon for which TyrR acts as a pure activator. In this case two TyrR boxes are found upstream of the promoter consensus hexamer sequences. However, only a single box located between 18 and 28 nucleotides upstream of the –35 region is necessary for activation. The binding of TyrR must occur at the same face of the DNA helix as RNA polymerase. Activation thus almost certainly involves direct protein–protein interaction between TyrR and the RNA polymerase. The ability of TyrR to act as an activator or as a repressor depends additionally on the binding of the cofactors: *repression* requires the binding of *tyrosine and ATP; activation* is obtained upon binding of *phenylalanine in the absence of ATP*.

7.3.2 The catabolite regulator protein CRP—many tasks for a small protein

The fact that **CRP**, the **catabolite regulatory protein**, has already been mentioned in many chapters of this book underlines the central importance of this protein in the process of transcription regulation. Countless studies from many different laboratories have elucidated more information about CRP than about any other transcription factor. CRP, sometimes also called **CAP** (for **catabolite activator protein**), is involved in the regulation of more than 100 different promoters. It acts both as a repressor and an activator of transcription. The protein has a molecular weight of 22.5 kDa (209 amino acids) and functions as a dimer. CRP requires the cofactor **cyclic adenosine monophosphate (cAMP)** for activity. cAMP is known as a second messenger in eukaryotic gene expression and activates cAMP-dependent protein kinases. In *E. coli* this nucleotide appears to be specifically required to turn on the activity of CRP. The complex cAMP–CRP bound to the *lac* operator has been crystallized and the X-ray structure deduced. From this structure it can be inferred that the cAMP binding site is located in a distinct domain separated from the DNA-binding region at the C-terminus. The DNA-binding region consists of a typical helix-turn-helix structure. cAMP is not involved in DNA-binding, and structural changes within CRP and the DNA-binding domain observed upon cAMP interaction must be of an allosteric nature. CRP binding occurs at two consecutive major grooves of the target DNA. From the many CRP binding sites known a 22-base pair consensus binding sequence with the interrupted palindromic sequence **AA-TGTGA--TA--TCACA-TT** has been derived. Each CRP monomer binds to one half-site of the dyad sequence. Binding of CRP to its operator induces a static bend into the DNA of approximately 90°. This is apparent both from the crystal structure of the cAMP–CRP complex with the *lac* operator and also from biochemical experiments employing cyclization kinetics (see Box 6.3). The bound DNA is curved towards the protein by two kinks of approximately 45° which are spaced in helical repeat. The kinks appear to be caused by changes in the roll angle between TG bases of the operator (see Box 6.1). Amino acids Arg180, Glu181 and Arg185 appear to make contacts with the GTG bases of the conserved TGTGA operator sequence. Bending of the DNA by CRP is not the only factor that mediates activation. Rather protein–protein interaction between CRP and RNA polymerase has proven to be an important factor. Presumably DNA-bending and protein–protein interaction synergistically contribute to promoter activation.

The CRP binding sites for activation can be grouped into three distinct classes according to their distances relative to the promoter. The two major groups contain operons where the CRP site is either centred around positions –41.5 or –61.5 relative to the transcription start site. Promoters with the CRP binding site at –61.5 or –41.5 are grouped into **Class I** or **Class II CRP-dependent promoters**, respectively. Direct interaction between CRP and RNA polymerase occurs at promoters where the CRP sites are located at either –61.5 or –41.5. At some promoters CRP sites are found further upstream, around position –100. In the latter situation activation usually involves a second regulatory protein (e.g. AraC in the *araBAD* operon or MalT at the *malE/malK* promoters). These promoters are sometimes termed **Class III CRP-dependent promoters**. The binding situations of CRP to the different promoter classes are depicted schematically in Fig. 7.6.

Activation at Class I CRP-dependent promoters requires only cAMP–CRP which is bound on the same face of the DNA helix as RNA polymerase. Direct interaction occurs between the αCTD and the CRP protein (see Section 2.2.1). This interaction involves amino acids 156–164 of CRP. These amino acids are located in a surface-exposed β-turn of the protein which has been termed the **activating region 1**. Binding of the downstream subunit of the CRP dimer to the αCTD seems to be sufficient for activation. Activation is explained by an increase in RNA polymerase binding affinity (enhanced formation of closed complexes: increase in K_B). The *lac* promoter can be taken as a prototype of Class I CRP-dependent promoters.

At Class II CRP-dependent promoters the CRP site is closer to the promoter. In fact, it overlaps or replaces the –35 region. It has been shown that RNA polymerase interacts with DNA both upstream and downstream of the bound CRP dimer. The αCTD appears to bind DNA upstream of CRP and to interact with the *upstream* subunit of the CRP dimer *via* activating region 1. It is not known exactly whether the same amino acids of the αCTD are involved in protein–protein interactions as in the case of the Class I CRP-dependent promoters, but recent evidence suggests that several distinct amino acids closer to the C-terminus may be specifically responsible for the interaction at Class II-dependent promoters. Additional protein–protein contacts of the RNA polymerase are made to amino acids 19, 20, 96 and 101 of the *downstream* subunit of the CRP dimer. These amino acids form a cluster on the surface of CRP which has been termed **activating region 2**. The activating region 2 appears to interact with the αNTD of RNA polymerase. It has been shown that this type of interaction increases the rate of isomerization from the closed to the open RNA polymerase–promoter complex (increase in k_2). A typical example for a Class II CRP-dependent promoter is the *gal* P1 promoter.

Promoters where the CRP binding site is located further upstream (over 90 base pairs) are sometimes classified as Class III CRP-dependent promoters. CRP binding is too far apart for direct interaction with RNA polymerase. Activation at these promoters requires the simultaneous interaction of additional regulators. For the *malT* promoter, which belongs to the Class III CRP-dependent promoters, it has been shown that steps after isomerization are affected. In this case CRP acts by enhancing promoter clearance. The activating mechanism, which involves dynamic changes of the DNA structure, e.g. the formation of DNA loops, is the subject of the next section. The *araBAD* promoter might be taken as a typical example of a Class III CRP-dependent promoter.

Repression by CRP occurs when the CRP–cAMP complex is bound between positions –6 and +20 but outside the phasing from where it activates. Moving the CRP binding site from an activating position by roughly five nucleotides, corresponding to half a helical turn, will change the effect of CRP from an activator to a repressor. Obviously, the obligatory RNA polymerase–CRP interactions for activation cannot take place. Moreover, favourable interactions of RNA polymerase to the promoter upstream DNA are blocked by CRP binding.

CRP is also implicated in the repression of the enzyme adenylate cyclase, which is responsible for the synthesis of cAMP. A CRP binding site is present at the core promoter sequence of *cya* P2, the major promoter for the adenylate cyclase gene. Binding of cAMP–CRP to this site is responsible for repression of the *cya* gene. Other examples of CRP-dependent repression involve competition between overlapping promoters and include the promoter for the *crp* gene itself. The *crp* promoter overlaps a divergent promoter to which CRP–cAMP can bind and block the synthesis from the *crp* promoter while activating the divergent promoter. The transcript of the divergent promoter might also be involved in the regulation of CRP expression by forming an antisense RNA which inhibits CRP transcription. Inhibition is proposed to result from the formation of a Rho-independent terminator through complementary base pairing between the antisense transcript and the *crp* mRNA. Furthermore, the

(a) Class I CRP-dependent promoters

e.g. *lac* P1 promoter

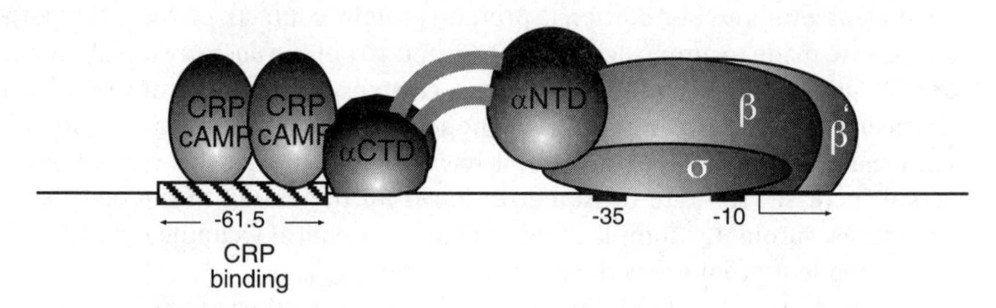

(b) Class II CRP-dependent promoters

e.g. *gal* P1 promoter

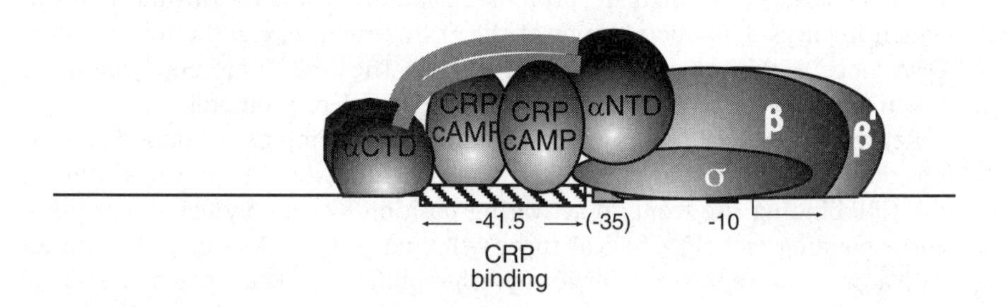

(c) Class III CRP-dependent promoters

e.g. *araBAD* promoter

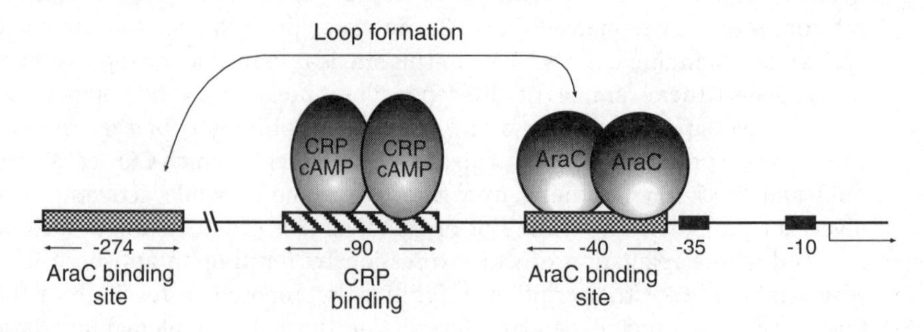

activity of the transcription factor FIS has been shown to be essential for repression of *crp* transcription (Fig. 7.7).

There are many more examples where the distribution of RNA polymerase between overlapping promoters is directed by the binding of cAMP–CRP to one of the promoter sites. The histidine utilization operon (*hutUH*), encoding the genes for the degradation of the amino acid histidine to glutamate, ammonia and formamide, is one such example. The transcription start region of the *hutUH* operon contains two divergently orientated overlapping promoters, P_C and P_{hutU}. A CRP binding site is located close to the P_C start (Fig. 7.8).

In the absence of cAMP–CRP transcription proceeds primarily from promoter P_C. In the presence of cAMP–CRP transcription is largely favoured from P_{hutU}. Binding of RNA polymerase at P_{hutU} and P_C appears to be mutually exclusive. cAMP–CRP binding may either directly activate P_{hutU} by facilitating RNA polymerase binding to the P_{hutU} promoter or indirectly by blocking transcription from P_C.

An interesting case of regulation by virtue of the cooperation between CRP and a second regulator is represented by the **CyrR regulon**. The genes for the uptake and catabolism of ribonucleotides and deoxyribonucleotides are encoded in a regulon which is negatively controlled by the regulators **CytR** and **DeoR**. CytR belongs to the LacI/GalR family of repressors, contains a helix-turn-

Figure 7.6 Different classes of CRP-dependent promoters. Depending on the site of CRP interaction with the upstream DNA several classes of CRP-dependent promoters can be distinguished. (a) The *lac* P1 promoter is presented as an example of Class I CRP-dependent promoters. The centre of the CRP binding site (shaded box) is located at position –61.5 relative to the transcription start (indicated by an arrow). The –35 and –10 promoter regions are shown as black boxes and marked. RNA polymerase, with the β, β′ and σ subunits labelled, is bound to the promoter. The α subunits are shown with their two separate N- and C-terminal domains (αNTD and αCTD) separated by flexible linkers (shown as thick grey lines). The downstream subunit of the CRP dimer makes contacts with the αCTD which is bound upstream of the –35 sequence (see Fig. 7.5). (b) At Class II CRP-dependent promoters, for which the *gal* P1 promoter is shown as an example, the centre of the CRP binding (–41.5) site is located very close to the promoter, which usually lacks a good –35 consensus. At Class II CRP-dependent promoters the αCTD binds upstream of the CRP binding site. The upstream and downstream subunits of the CRP dimer make different contacts to the RNA polymerase. The downstream subunit binds to the αNTD or the σ subunit while the upstream CRP subunit makes contact to the αCTD (see Fig. 7.5). (c) At some promoters CRP binding sites are found further upstream (roughly –100). Usually at such promoters a second regulator protein is required for transcription. Promoters of this type are sometimes designated as Class III CRP-dependent promoters. The *araBAD* promoter is shown as an example. The CRP binding site, located at position –90, is flanked by two binding sites for the regulator protein AraC. One operator sequence for AraC is centred at position –40, the other far upstream at position –274. Regulation at Class III CRP-dependent promoters usually involves DNA loop formation. Part of the figure is adapted from Busby and Ebright (1997).

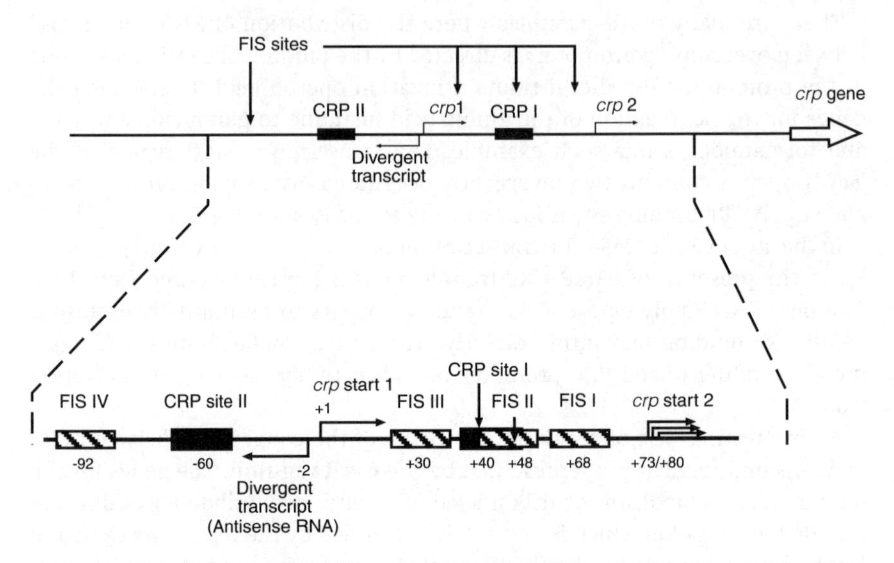

Figure 7.7 Regulatory region of the *crp* gene. The complex regulatory region for the transcription of the *crp* gene is schematically illustrated. The sequence contains four binding sites for the transcription factor FIS and two CRP binding sites, involved in autoregulation. Transcription is initiated at two tandem promoters, *crp*1 and *crp*2, in the rightward direction. A third promoter produces a divergent transcript. An enlarged illustration is presented in the lower part of the figure. The positions of the transcription factor binding site centres are indicated. Transcription initiation at *crp*2 occurs from several closely spaced positions between +73 and +80. The divergent transcript starts at position –2. Numbers correspond to the start site of the major promoter *crp*1. The first 11 nucleotides of the divergent transcript and the *crp* mRNA show strong complementarity, which has led to the suggestion that the divergent RNA may act as antisense RNA. The data is taken from Hanamura and Aiba (1992) and Gonzáles-Gil *et al.* (1998).

helix DNA-binding motif and regulates at least eight genes. Repression by CytR is released in the presence of the inducer cytidine. The mechanism of repression by CytR is better described as *anti-activation* because it inhibits the activation of cAMP-CRP. The CytR operator sites are flanked by two close-by CRP binding sites (one in case of the *cytR* promoter). One CRP site is always located at around –40, resembling the Class II CRP-dependent promoters. Interestingly, CytR and cAMP–CRP bind cooperatively to the DNA. Binding of CytR is stimulated more than 1000-fold in the presence of cAMP-CRP. CytR is thus recruited to its binding site by CRP. There is good evidence for protein–protein interactions between CytR and CRP when bound to their operator sites. This interaction probably masks amino acid side chains that constitute the activation patch within CRP normally involved in binding to RNA polymerase. This kind

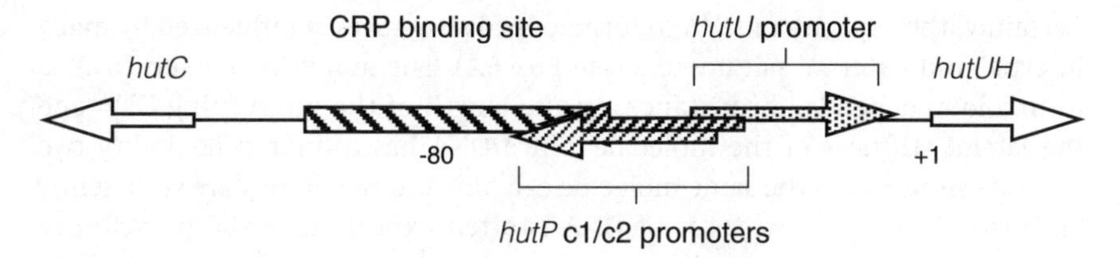

Figure 7.8 The *hutUH* operon structure. The regulatory region of the *hutUH* operon with the divergent transcription units *hutUH* and *hutC* is shown. Transcription of the *hutUH* genes starts from the *hutU* promoter. Two overlapping promoters *hutP* c1 and *hutP* c2 are arranged in divergent orientation and partially overlap the *hutU* promoter. A CRP binding site with its centre at position –80 with respect to the *hutU* start site overlaps the *hutP* c1 and c2 promoters. Binding of CRP simultaneously blocks transcription from the P_c promoters and activates the *hutU* promoter.

of regulation, where an activator and a repressor act in concert, has many parallels with eukaryotic systems in which activators are often inhibited through the interaction with a repressor (compare also the interaction of σ factors and anti-sigma factors; see Section 2.3.7).

7.4 DNA loops—regulation through dynamic structures

Several mechanistic concepts by which transcription factors activate or repress transcription initiation have been discussed. They either block or recruit RNA polymerase binding to the promoter, thereby inhibiting or activating transcription, respectively. Transcription factor binding to RNA polymerase may furthermore change the geometry of the initiation complex and thus influence later steps in the initiation cycle, such as isomerization or promoter clearance. These mechanisms fail, however, to explain the presence of multipartite DNA control elements frequently found in several regulatory regions. Often more than one operator structure is found at several distant sites upstream of a promoter. Usually, one site is close-by while the other operator is located more than 100 base pairs apart. The existence of such multipartite operator sites at great distance and their necessity for regulation has led to the concept of **DNA loops**. DNA loops are considered to bring regulatory proteins which are bound to distant sites into close proximity. The distant proteins can thus interact with each other and through such protein-protein interactions a new topological structure is formed with the intervening DNA looped out. The concept of DNA loops has therefore led to the expression **action-at-a-distance**.

What are the physical requirements to describe DNA loop formation?

Certainly, the capacity of DNA to form looped structures is influenced by many internal and external parameters (see Box 6.1). The ability of a given DNA to form a loop depends, for instance, on the length of the intervening DNA and the lateral stiffness of the molecule. Since DNA has a limited flexibility two separate sites within the same molecule can only interact if they are sufficiently far apart. The lateral stiffness of DNA is often expressed as the **persistence length**. The persistence length is a measure of the tendency of a stiff polymer to continue in the same direction without bending. The persistence length of random B-DNA is about 150 base pairs. Loop formation of DNA at about the persistence length is very difficult because of the stiffness of the molecule. On the other hand, loop formation is also improbable for kinetic reasons if the two interacting sites of the molecule are too far apart. The optimal distance for two points within a DNA molecule to form a loop is about 3.6 times the persistence length which is approximately 500 base pairs for B-DNA.

There are two important aspects to explain how loop formation influences gene expression. First, DNA loops may significantly increase the local concentration of distantly bound regulators at their site of action (see also Box 6.3). Second, there is a clear topological effect of the looped DNA itself. Loops may render the DNA to conform in a way which precludes stable binding of RNA polymerase or later steps of the initiation cycle. Thus, DNA looping plays an *active* role in transcription regulation. It should be noted here that DNA loop formation is not restricted to transcription. Rather it is a general strategy and plays a vital role for many reactions involving DNA rearrangements, like replication and recombination.

How can loop formation be demonstrated experimentally? In fact, evidence for loop formation has been presented by many different techniques. These techniques include chemical modification and limited nuclease digestion studies, which indicate enhanced accessibility on the outside and protection on the inside surfaces of the looped DNA. DNA loops have also been characterized by cyclization experiments and gel retardation studies. Moreover, for several examples loops have been directly demonstrated by electron microscopy.

In recent years many examples have been studied where transcription regulation involves DNA loop formation and protein–protein interaction-at-a-distance. Loop formation has been shown to be important, for instance, for the expression of the *lac* genes. Another intensely studied example comprises the *deoCABD* operon which encodes the genes for ribonucleotide and deoxyribonucleotide catabolism (see CytR in the above Section). Here the regulation of the *galETK* and the *araBAD* operons are chosen as representative examples which are discussed in more detail in the next section.

7.4.1 Negative regulation of the *galETK* operon

The *galETK* operon encodes the three enzymes necessary for the degradation of galactose, namely galactokinase (*galK*), galactose-1-phosphate uridyl transferase (*galT*) and uridine diphosphogalactose-4-epimerase (*galE*). The operon is transcribed from two overlapping tandem promoters P1 and P2, which are separated by only five base pairs (Fig. 7.9). The two promoters P1 and P2 are differentially affected by cAMP–CRP for which a binding site is located with its centre at position –41.5 relative to the transcription start of promoter P1. As already seen in Section 7.3.2 binding of cAMP–CRP to this site activates transcription from P1. At the same time, transcription from P2 is repressed owing to a decrease in K_B. In addition to the regulation by cAMP–CRP, transcription of both promoters of the *gal* operon is also subject to negative control mediated by the repressor protein **GalR**. This repressor is encoded at some distance by the *galR* transcription unit. GalR is a 36.9-kDa protein which exists as a dimer in solution. Based on the similarity of GalR with LacI, CytR and several other related repressors it has been assigned as a representative member of the LacI/GalR family of repressors (see Section 7.3). GalR binds to operator sequences with the hyphenated dyad symmetry GTG-AA-CG-TT-CAC. The activity of GalR is modulated by binding the sugars galactose or fucose. Binding of these sugars releases GalR-mediated repression.

Two GalR operator sites, O_E and O_I, are present in the *gal* operon which are 114 base pairs apart. O_E is located 60 base pairs upstream of the promoter P1, while O_I is located downstream at position +54 (see Fig. 7.9). GalR binds to both operator sequences. This organization of regulatory elements suggests a mechanism different from classical repression by sterical hindrance. A dynamic

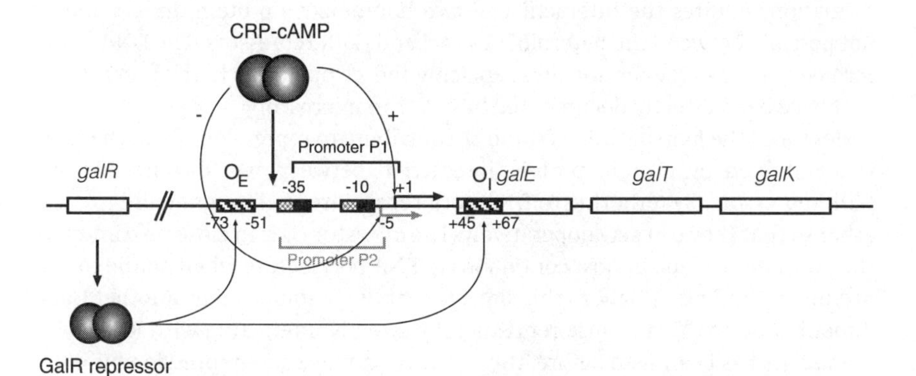

Figure 7.9 The structure of the *galETK* operon. The *gal* operon encoding the three genes *galE*, *galT* and *galK*, together with the regulatory region, is schematically presented. Transcription of the *gal* genes is controlled by two overlapping tandem promoters, designated P1 and P2, which are five base pairs apart. The *gal* repressor GalR is encoded at some distance upstream. The dimeric *gal* repressor binds to two operator regions O_E and O_I flanking the *gal* promoters at the indicated positions. A CRP binding site is located immediately upstream of promoter P1 (Class II CRP-dependent promoter; see Fig. 7.6). Binding of CRP to this site activates transcription from P1 (indicated by +) and represses transcription from P2 (indicated by –).

mechanism involving a DNA loop which contains both promoters has been demonstrated.

What has been the experimental evidence for loop formation at the *gal* operon? First indications of the involvement of dynamic DNA structures in the repression of the *gal* operon were derived from binding competition and footprinting studies in the presence of RNA polymerase, cAMP–CRP and GalR. DNA-binding studies and DNaseI protection analyses with RNA polymerase, CRP and GalR alone or in combination demonstrated that there is no competition by sterical hindrance, although the binding sites of cAMP–CRP and GalR at O_E partially overlap. Supported by the footprinting results this finding could be explained by the fact that the proteins occupy different faces of the DNA helix. It was then demonstrated that the helical phasing of the operator sites O_E and O_I with respect to each other were critical for repression. If DNA segments of non-integer helical turns which alter the angular orientation of the operator sites are inserted between O_E and the promoter or O_I and the promoter, the operon is no longer repressed, although GalR still binds to the operators (Fig 7.10). The involvement of protein–protein interactions between GalR repressor molecules in maintaining a DNA loop has been discerned by operator conversion experiments. If one of the *gal* operator sites is replaced by a *lac* operator site no repression is observed even in the presence of high concentrations of both GalR and LacI repressors. If, however, both *gal* operator sequences are converted to the corresponding *lac* operators repression is observed. This repression is dependent on the presence of LacI, not GalR repressors. Obviously, loop formation requires the interaction of two homologous protein dimers and is not possible between the heterologous LacI and GalR repressors. The DNA loops formed with the *lac* operator sites replacing the *gal* operators in the presence of LacI have been directly documented by electron microscopy.

How does the loop between O_E and O_I contribute to repression? First, the loop is maintained by protein–protein interaction between two distantly bound GalR repressors. By tethering to DNA both repressors are brought close to each other so that they can act cooperatively. The operator O_E is in close proximity to the promoter P1, and almost contiguous to RNA polymerase when bound to the promoter. Binding of GalR to this site, cooperatively stabilized by another GalR dimer bound to O_I, can cause repression by directly interacting with RNA polymerase. As has been seen before, this interaction may cause repression by altering the stability of the RNA polymerase–promoter complex (K_B). Alternatively, repression may result from reduced rates of isomerization to the open complex (k_2) or inhibition of RNA polymerase promoter escape. A second explanation for the repression of the *gal* operon via DNA loop formation takes into account the structure of the DNA loop itself. The loop, stabilized by protein–protein interactions of operator-bound GalR repressors, forms an independent topological domain. Because of the small size of the loop the conformation of the promoter may be altered in such a way that it is inadequate for any one step of transcription initiation. RNA polymerase binding may be inhibited by the distorted

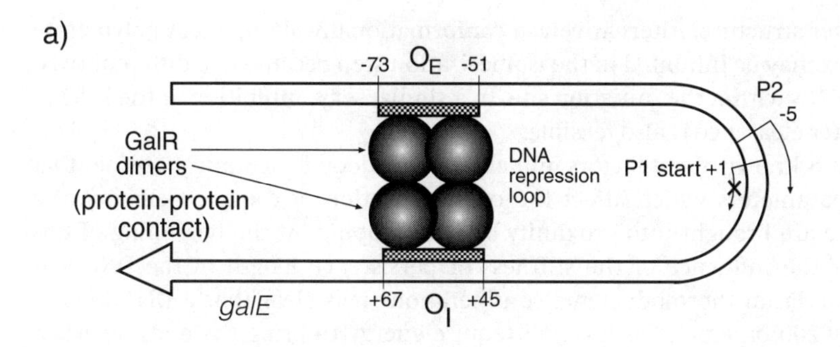

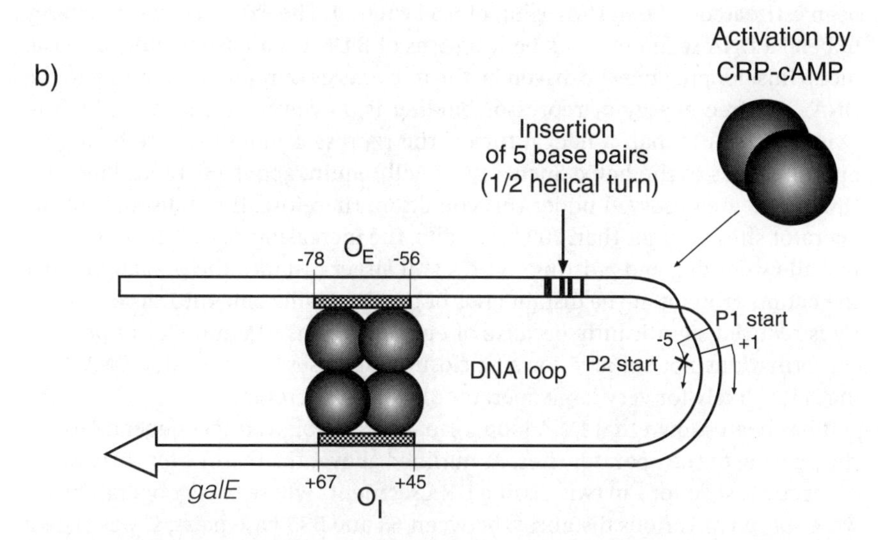

Figure 7.10 Loop formation at the *galETK* regulatory region. Transcriptional control of the *gal* genes involves DNA loop formation of the regulatory region. The promoter–operator region and the transcription start site of the *galE* gene are schematically presented as a DNA loop. The loop is fixed through protein–protein interaction between GalR repressor dimers (indicated as grey spheres) bound to the operators O_E and O_I (indicated as shaded boxes). (a) As a result of the loop formation promoter P1 is orientated on the inside of the topologically closed loop, while promoter P2 is located on the outside. Thus, only P2 is in a favourable conformation to interact with RNA polymerase, whereas transcription from P1 is inhibited (repression loop). (b) Changing the distance between O_E and O_I by insertion (or deletion) of five base pairs, equivalent to half a helical turn, for instance upstream of P1, will cause a change in twist after the loop is topologically closed through interaction of the GalR dimers. As a result the recognition elements of promoter P1 are now facing the outside of the loop while P2 is located at the inside. Transcription from P2 is now inhibited, while P1 is in a favourable conformation for RNA polymerase binding and transcription initiation. Moreover, the CRP binding site, necessary for activation, is also located on the outside of the loop.

promoter structure. Alternatively, a conformationally altered RNA polymerase complex may be inhibited in the isomerization step because of a different twist of the DNA within the initiation site. In a similar way, inhibition at the level of promoter clearance is also feasible.

What is known about factors influencing DNA loop formation? Consider that some parameters which affect the communication of control elements at a distance are brought into proximity by DNA looping. At the beginning of this chapter the influence of the stiffness or persistence length of the DNA was dicussed. From thermodynamic considerations it is clear that a DNA loop of roughly 200 base pairs or less will require energy to bring the ends together. The free energy generated by the binding of two repressor dimers to DNA has been estimated to be in the region of 8.5 kcal/mol. The expected cost in energy to bend a DNA segment of six helical turns of B-DNA is approximately 6.5 kcal/mol. Thus, looping may be driven by the free energy of repressor binding to the DNA. The free energy of repressor binding is, however, not large enough to twist the DNA by half a helical turn if the repressor molecules are bound on opposite faces of the helix. Such a twist will require about 11–12 kcal/mol. No DNA loops are observed under this condition, therefore. If the distance of the operator sites is larger than 200 base pairs, the increasing flexibility of the DNA will allow curving and twisting. With a still larger distance the probability that the bound proteins at the distant ends of the DNA will come into close proximity is reduced significantly because of entropic terms (the number of possible conformations not suitable for ring closure increases enormously). DNA looping is less likely for very large operator distances, therefore.

It has been shown that DNA loop formation exhibits a direct dependence on the phasing of the operator sites. As outlined above, the reasons for this are the energetic restrictions in twist. Using DNA segments where two *lac* operator sites were spaced at various distances, between 63 and 535 base pairs, it was shown that loop formation is a matter of periodicity. Stable loop structures can only be detected when the distance of the bound repressors corresponds to an integral number of B-DNA helical turns.

It is clear that loop formation is greatly affected by intrinsic curvature or protein-induced bends within the looped out DNA segment. On the one hand, the phasing and the spatial distance of the interacting binding proteins may be influenced by static DNA curvature. However, the binding of transcription factors within the intervening DNA has strong implications on the probability of loop formation. Interacting proteins might bend the intervening DNA in the same direction of the loop trajectory. Loop formation would be facilitated in such cases. DNA-binding proteins may also bend the DNA in an opposite direction than that required for loop formation. This would result in the inhibition of loop formation or disruption of loops. The same DNA-binding protein, by shifting its binding site by half a helical turn, may facilitate or inhibit DNA loops. It is not surprising that the intervening DNA of many known repression loops contain binding sites for proteins known to induce DNA curvature. These proteins may either be necessary to enable loop formation or they may be additional regulatory elements modulating loop stability.

For many of the regulatory systems controlled by DNA loops it has also been shown that the local superhelical density of the DNA is important for loop formation. It can be shown that increasing negative superhelical densities (see Section 6.3) facilitate loop formation with decreasing length of the intervening DNA. Many studies on loop formation *in vitro* cannot be performed with linearized DNA, therefore.

7.4.2 Regulation of the *araBAD* operon

One of the best studied examples, and the prototype of transcription regulation by loop formation, is represented by the *araBAD* operon. The *araBAD* operon encodes the arabinose-inducible enzymes which convert L-arabinose to D-xylulose-5-phosphate. The operon contains the genes for *araB*, *araA* and *araD*, encoding an isomerase, a kinase and an epimerase, respectively. The transcription unit is schematically presented in Fig. 7.11.

Transcription of the *araBAD* operon starts from the promoter P_B and is controlled by the repressor **AraC**. AraC is the gene product of the *araC* gene, which is transcribed from the promoter P_C in divergent orientation to the *araBAD* P_B promoter. The 292-amino acid residue AraC protein has a modular organization typical of many transcription factors (see Section 7.2). It contains two potential helix-turn-helix DNA binding motifs at the C-terminus and, like most transcription factors, the AraC protein contains a dimerization domain. This domain is located at the N-terminus. Several other transcription factors showing homologies to AraC, in particular to the helix-turn-helix region but also at the N-terminus, have been classified as AraC-like regulators (e.g. XylS). The activity of AraC is allosterically modulated by the sugar arabinose. Arabinose binds to the cofactor binding domain of AraC near the N-terminus and converts it to an activator of P_B transcription. AraC recognizes two 17 base pair DNA half-sites in *directly repeated* orientation. This is unusual for a DNA binding sequence because most regulators recognize *inverted repeat* structures or *palindromes*. As will be seen later, AraC will cause DNA loops by bridging distant operator sites. Unlike the GalR repressor no tetrameric proteins are involved in DNA loop formation. A single AraC dimer is sufficient to generate a loop.

The *araBAD* operon is organized such that the *araBAD* promoter (P_B) is divergently orientated with respect to the promoter P_C which controls transcription of the *araC* gene. Both promoters are regulated by AraC. Three operator sequences for AraC are located within the *araBAD* regulatory region: *araI*, *araO*$_1$ and *araO*$_2$. The distance between *araO*$_1$ and *araO*$_2$ is roughly 170 base pairs, while *araO*$_1$ and *araI* are roughly 50 base pairs apart. The two operators *araI* and *araO*$_1$ are located between the promoters P_B and P_C. The third operator *araO*$_2$ is separated by about 210 base pairs from *araI* and is located within the region of *araC* transcription. In addition to the three AraC sites the operon contains a CRP

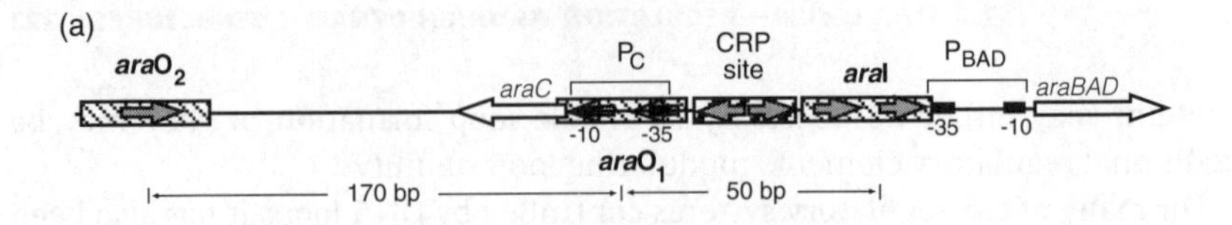

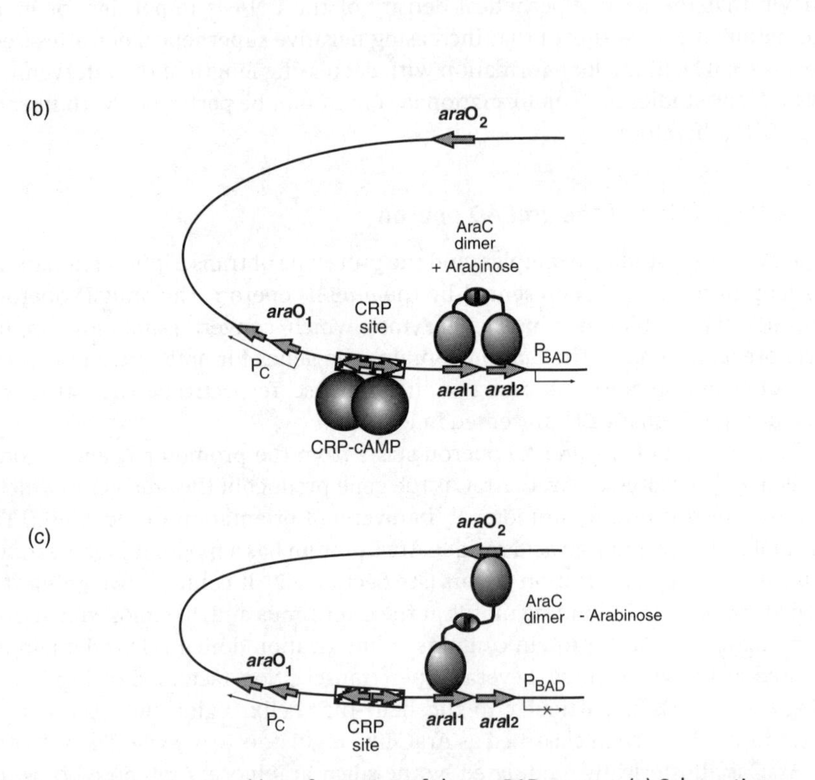

Figure 7.11 Structure and loop formation of the *araBAD* operon. (a) Schematic arrangement of the regulatory elements of the *araBAD* operon. Transcription of the *araBAD* and *araC* genes occurs in divergent orientation, as indicated by open arrows which are labelled correspondingly. Transcription in the rightward direction is started from promoter P_{BAD}, or in the leftward direction from promoter P_C. The –35 and –10 regions for the two promoters are shown. Three operator sites for the regulatory protein AraC are shown as hatched boxes and labelled *ara*I, *ara*O_1 and *ara*O_2. Grey arrows within the operator boxes pointing in the same direction correspond to operator half-sites arranged as tandem repeats rather than in the usual palindromic orientation. The distance of the three operators is given in base pairs below the drawing. A CRP binding site, indicated by a shaded box with two grey arrows pointing outwards, is located between the operator sites *ara*I and *ara*O_1. (b) A situation is depicted where transcription occurs in the presence of the inducer arabinose. Binding of arabinose to the dimeric AraC regulator causes AraC to occupy the two adjacent operator half-sites *ara*I_1 and *ara*I_2. Binding of CRP to its regulatory site activates both promoters P_C and P_{BAD}. (c) In the absence of arabinose an AraC dimer interacts cooperatively with the operator half-sites *ara*O_2 and *ara*I_1, forming a DNA loop. This repression loop inhibits transcription from the P_{BAD} promoter.

binding site (see Fig. 7.11). Binding of cAMP–CRP to this site activates both promoters P_B and P_C. In the absence of arabinose transcription from P_B is negatively regulated by AraC. This repression involves a DNA loop held together by binding of an AraC dimer to both operators *ara*O_2 and *ara*I. Apparently, the AraC dimer binds to a half-site of each operator. Formation of the loop between the *ara*O_2 and *ara*I sites is dependent on the correct angular orientation of the two AraC binding sites. AraC must occupy the same site of the DNA helix at both operators. Interaction of AraC at the two operator sites *ara*O_2 and *ara*I occurs cooperatively. Rotational deviations by inserting (or deleting) DNA segments corresponding to half a helical turn (or multiples thereof) abolish repression.

Interestingly, the base pair periodicity for one turn determined in the *araC* system *in vivo* is larger than 11, and not 10.5 base pairs per turn as found for B-DNA in many other systems. This notably larger number of the base pair periodicity might be explained by a linking number deficit resulting from supercoils of the DNA *in vivo*. An increase in linking number will cause a larger number of base pairs per turn. The measured superhelical density of the DNA does not fully account for the linking number deficit necessary to explain the increase in helical periodicity of more than 11 base pairs, however. It is assumed, therefore, that transcription of the DNA itself contributes to the increase in base pair periodicity due to transcription-induced supercoiling (see Section 6.3.1).

In the presence of the inducer arabinose the promoter P_B is activated (derepressed) more than 100-fold. Binding of arabinose to the AraC protein induces a change in the structure of the domains responsible for dimerization. The ability to contact both operators, *ara*O_2 and *ara*I, is lost. Instead of binding to two distant operator half-sites, AraC now prefers to bind to two adjacent half-sites of the *ara*I operator. The loop that was held together by AraC contacts between *ara*O_2 and *ara*I is disrupted and AraC remains firmly bound to *ara*I. Because both half-sites of *ara*I are now occupied by AraC the activation domain of the regulator is in a favourable position to activate transcription from P_B. The disruption of the DNA loop causes a transient activation of the P_C promoter until additional AraC molecules bind to *ara*O_1 and again repress transcription from this promoter.

It is assumed that one function of the cAMP–CRP complex bound between *ara*O_1 and *ara*I is to disrupt the repression loop formed between operator sites *ara*O_2 and *ara*I. This is probably accomplished by CRP-dependent DNA-bending in a direction unfavourable for loop formation.

Formation of a new loop by AraC which bridges operators *ara*O_1 and *ara*O_2 has also been proposed as an alternative explanation. Loop formation between these two operators is not observed *in vitro* with linearized DNA, and probably requires a superhelical structure of the template.

In summary, the change between repression and activation of the *araBAD* operon obviously entails two forms of AraC with different activities dependent on the presence of the inducer arabinose. These two forms of AraC undergo different binding combinations with the multipartite operator sites. As a result a switch in DNA loops occurs. Either a DNA loop formed by operator-bound AraC molecules in the absence of arabinose represses both promoters P_B and P_C (AraC-mediated loop between *ara*I and *ara*O_2) or a loop is formed by AraC between *ara*O_1 and *ara*O_2 in the presence of arabinose. This structure leads to the selective repression of the P_C promoter.

The examples presented so far have been selected to describe the involvement of DNA loops in *negative regulation.* DNA loops participate, however, also in the *activation* of transcription. Examples in which DNA loops and protein-protein interactions-at-distance are involved in positive regulation of transcription are presented in Section 7.6.2.

7.5 Regulation by DNA structuring proteins

Prokaryotes, unlike eukaryotes, do not contain **histones** which compact and structure the chromosomal DNA. Nevertheless, the DNA of prokaryotes is packed in a 1000-fold condensed form to fit within the dimensions of a bacterial cell. The size of a bacterial genome, like that of *E. coli*, for instance, is approximately 4×10^6 base pairs. This corresponds to a DNA contour length of more than 1 mm. The length of this molecule has to be reduced to fit within a cell with the approximate dimensions 0.75×2 μm. It is not yet completely clear how such a compaction of the bacterial nucleoid structure is achieved. However, a set of proteins has been characterized which, in analogy to the eukaryotic histone proteins, appears to be involved in structuring and condensing bacterial DNA. These proteins have been termed **histone-like** proteins. Generally these proteins are rather small and capable of binding to DNA. They do not share all the typical characteristics of histones, however. Their amino acid sequences are not homologous to eukaryotic histone proteins. Not all of the histone-like proteins have a basic isoelectric point (pI) as do their eukaryotic counterparts. Furthermore, they are not quite as abundant as histones. Altogether they do not suffice to cover the complete bacterial DNA. More detailed studies in the past have revealed a number of additional specific functions for some of the histone-like proteins. Many participate in site-specific recombination, in replication, and also in transcription. In fact, for many of the histone-like proteins important functions in gene expression have been uncovered, and some of the proteins function as versatile transcription regulators rather than stabilizing the structure of the nucleoid. Hence, the term **chromatin-associated** or **DNA structuring** proteins is often more appropriate than the term histone-like. Some even fulfil all the criteria of true transcription factors. Often these DNA-binding proteins have common or closely overlapping target sites and either act synergistically or as antagonists.

Moreover, because of considerable structural and functional homology several members of the DNA structuring proteins can functionally substitute each other.

Examination of the early literature reveals that some DNA structuring proteins have been described redundantly and named differently. This has caused some confusion in the past. For instance, one of the proteins (H) which had been erroneously assigned to the group of DNA-binding proteins was in fact the ribosomal protein S3. Another example is the protein H-NS, which is often termed H1 in the older literature. Today a rather unique nomenclature exists which will be used here. Table 7.4 lists the most important representatives of the group of DNA structuring proteins that will be discussed in more detail in the sections below.

7.5.1 The master-bender integration host factor (IHF) and its family of DNA-binding proteins

The protein IHF was first identified as an essential component for the site-specific integration of phage λ in the bacterial chromosome. It was called **integration host factor** (IHF), therefore. The protein consists of two heterologous subunits, IHF-α and IHF-β, which are encoded in different transcription units by the highly homologous *himA* and *hip* genes, respectively. Both genes appear to be negatively autoregulated by IHF. The heterodimeric protein binds to DNA covering about 35 base pairs. It recognizes a specific consensus sequence **WATCAANNNNTTR**, where W stands for either A or T and R for either of the purines A or G. Binding analyses employing two-dimensional gel electrophoresis revealed a minimum of 70 IHF binding sites within the total *E. coli* DNA. The distances of the IHF sites relative to the transcription start positions vary between −180 and +130. Many operons contain multiple IHF binding sites,

Table 7.4 Members of the family of DNA structuring proteins

Protein	Gene	Molecular weight (kDa)	Monomers/cell
H-NS	*hns*	15.6	20 000–60 000 *
HUα	*hupA*	9.5	60 000
HUβ	*hupB*	9.5	60 000
IHFα	*himA*	11.2	17 000–34 000
IHFβ	*hip*	10.6	17 000–34 000
FIS	*fis*	11.2	200–100 000 †

* Different numbers have been reported in the literature.
† The number of FIS molecules shows a strong growth phase-dependent variation (see Fig. 7.15).

and additional sites for other regulatory proteins, indicating complex mechanisms of regulation, probably quite variable between different operons.

Upon binding to DNA, IHF creates the sharpest bend of all prokaryotic DNA-binding proteins. The X-ray structure of IHF bound to a 35-base pair DNA fragment is shown in Fig. 7.12. According to this structure IHF bends DNA by more than 160°, probably by 180°, so that the DNA describes a complete **U-turn.**

The protein monomers consist of three α-helices and three antiparallel β-ribbons. The two monomers are interwound, forming a lobster-like structure with two flexible arms consisting of two antiparallel β-ribbons. The two β-ribbons wrap around the DNA through the minor grooves. At the tip of the two flexible arms a hydrophobic proline residue is intercalated between the base pairs, generating two huge kinks. The minor groove is widened and the DNA describes a U-turn with the body of the protein on the inside of the turn. The

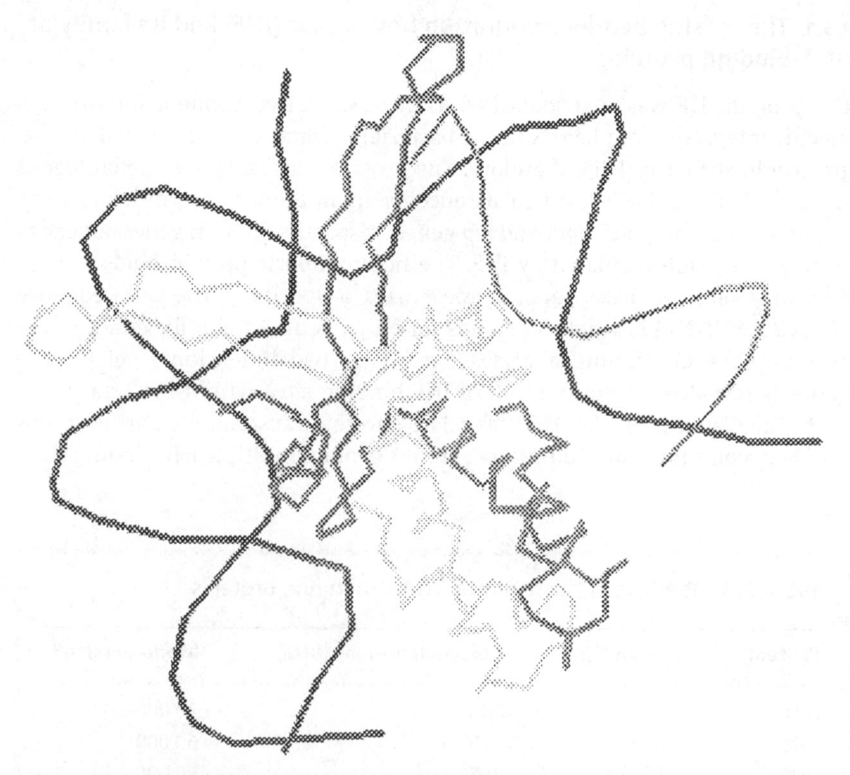

Figure 7.12 Structure of the DNA–IHF complex. The backbone of a dimeric IHF molecule bound to a DNA molecule with the IHF binding sequence is shown. The structural backbones of individual molecules are indicated by different grey lines. The two β-arms of IHF interact with the minor groove of the DNA. The DNA makes a sharp turn (U-turn) of about 180°. The figure is produced by RasMol according to the X-ray diffraction coordinates from the Protein Data Bank. The structural data has been published by Rice *et al.* (1996).

inside of the DNA makes a series of electrostatic interactions with the body of the IHF. The resulting charge neutralization counteracts the normal repulsion of the negatively charged phosphate backbone and allows a considerable narrowing of the DNA grooves on the inside. In this way, pushing on the outside (minor groove widening and kinking the DNA) and pulling on the inside (charge neutralization across the grooves) of the DNA results in the strongly curved U-turn structure. The specificity for the recognition of the IHF consensus sequence resides in the carboxy-terminal α-helix. The structural analysis of the IHF–DNA complex provides a nice example for the recognition model according to **indirect readout** (see Section 6.2.2).

The capacity to bend DNA sharply is probably the major clue to the function of IHF. In many systems the protein is found as an **architectural element**, organizing the structure of nucleoprotein complexes. The same property makes it a versatile protein not only in transcription but also in replication and recombination. Its function in transcription is discussed exclusively here, however. For example, IHF plays an important role in the activation of σ^{54}-directed transcription reactions which involve looped nucleoprotein structures. An example is transcription regulation of the nitrogen-fixation genes. These systems are described in Section 7.6.2. IHF has also been shown to activate several σ^{70} promoters. One example is the expression of the *narGHJI* operon, which encodes the genes for the principal nitrate reductase. The expression of this operon requires two additional regulators, NarL, which is a member of the two-component regulators (see Section 7.6), and FNR, a homologous factor to CRP. The activation of the *narGHJI* operon transcription involves formation of a DNA loop between FNR bound immediately upstream of the *narGHJI* promoter and NarL which binds some 200 base pairs upstream. Apparently, IHF, which binds at a position in between the other two regulators (–120), plays the architectural role and provides the necessary bend of the intervening DNA.

In some cases, however, IHF can inhibit transcription in a similar way to classical repressors. In these cases one or more IHF binding sites are found which overlap with the consensus elements of the core promoters. The *ompB* gene may be taken as an example (Fig. 7.13).

Gel retardation and DNaseI footprinting studies with the *ompB* regulatory region have shown that IHF occupies three sites in the *ompB* promoter region. Two sites are directly overlapping with the –10 and –35 regions. Strong inhibition of transcription by IHF *in vitro* suggests that no other regulatory proteins appear to be involved in repression, and IHF is the sole factor directly responsible for inhibition.

IHF may also act indirectly as a repressor, for instance by obstructing the binding of an activator. The expression of the *ompF* gene may serve as an example in this case. Transcription of the *ompF* gene is controlled by the regulator OmpR which binds to a site upstream of the *ompF* promoter and, depending on the osmolarity, either activates or represses transcription. The OmpR binding sites are flanked by two IHF binding sites. IHF binding has been found to

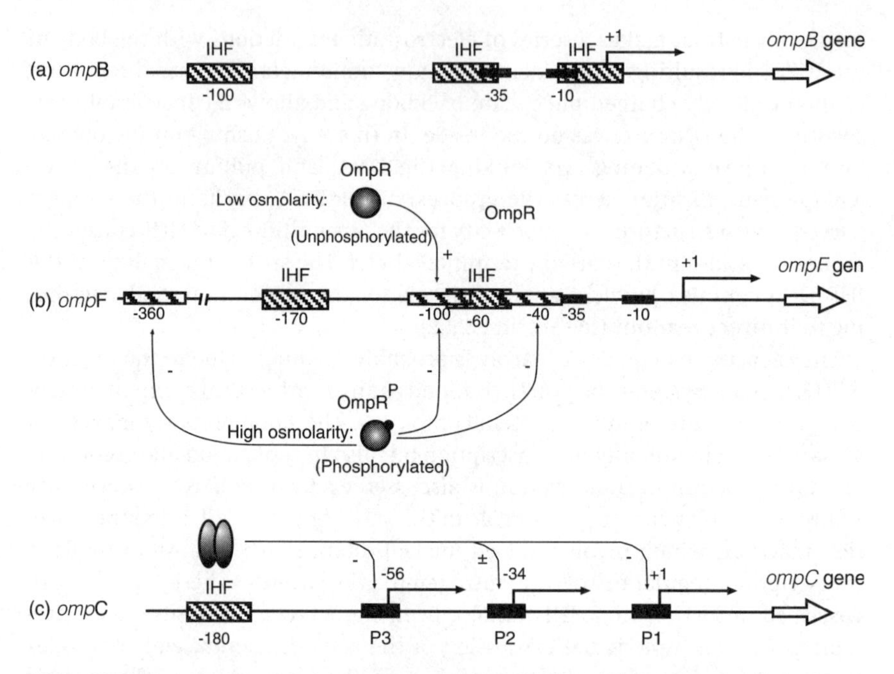

Figure 7.13 Regulation of *omp* gene transcription by IHF. (a) IHF acts a direct repressor for transcription of the *ompB* gene. Three IHF binding sites are found in the *ompB* regulatory region. While one site is located upstream at position −100, the other two sites directly overlap with the −35 and −10 promoter sequences. Binding of IHF thus competes directly with RNA polymerase binding. (b) During expression of the *ompF* gene IHF acts in concert with the regulator OmpR. Two binding sites for IHF are located in the upstream region of the *ompF* promoter with their centres at positions −60 and −170. In addition, three operator sites for the regulator OmpR are found at positions −40, −100 and further distal at −360. The two proximal promoter OmpR sites overlap with the IHF site at −60. The effect of OmpR on transcription depends on its phosphorylation state, which is itself a function of the osmolarity. At low osmolarity OmpR is unphosphorylated and IHF interferes with OmpR binding, causing derepression. At high osmolarity OmpR becomes phosphorylated and binds with high affinity to all three operator sites. Transcription of the *ompF* gene is repressed. (c) Transcription of the *ompC* gene is controlled from three partly overlapping tandem promoters P1, P2 and P3, with their start sites located at positions +1, −34 and −56. A single IHF binding site centres at position −180. Binding of IHF to this site has a differential effect for the activity of the three promoters. Transcription from P1 and P3 is inhibited, while transcription from P2 seems to be unaffected.

interfere with the binding of the regulator OmpR, and thereby to influence *ompF* transcription indirectly .

Interestingly, IHF has also been described as a direct activator. Several operons are known to be activated by IHF without the apparent participation of any other regulator (e.g. the bacteriophage Mu promoter Pe, the λ promoter P_L

and the promoter of the *ilvGMEDA* operon). In all these cases IHF binds just upstream of the promoter and stabilizes closed RNA polymerase-promoter complexes. Activation is thus the result of an increase in K_B. It can be shown that direct activation by IHF is dependent on the orientation of the IHF binding site relative to the core promoter elements. If, for instance, the phasing of the IHF site is altered by varying the distance of the binding site by half a helical turn, activation is lost. This can be taken as an indication that IHF either makes direct contacts with the RNA polymerase or that closed complex formation is stabilized by additional RNA polymerase–DNA interactions to upstream DNA sequences brought into interacting distance through the bending capacity of IHF.

The activity of IHF is not modulated by the interaction of inducer or corepressor molecules; nor is it changed through any form of modification. To explain a physiological role of IHF in transcription regulation one would expect that its cellular concentration might be regulated. This appears not to be the case. Although the absolute concentrations are not precisely known, it is clear that no dramatic change in the concentrations occurs during the cell cycle. How, therefore, can IHF act as a specific regulator of transcription? In those cases where IHF acts primarily as an architectural component, modulation of transcription is certainly brought about by the additional regulators that are also involved. This is the case in the example of the *ompF* gene presented above, where the expression of the regulator OmpR is modulated in response to the osmolarity by phosphorylation. On the other hand, it is believed that other DNA-binding proteins which are sensitive to environmental or cellular conditions influence the binding of IHF. Binding may be affected directly by competition or mediated through the DNA structure in response to changes in the DNA occupancy with other DNA-binding proteins. Several of the proteins listed in Table 7.4 may serve such a function. Binding of the proteins HU or H-NS may compete with IHF binding. In addition, HU binding is known to constrain DNA supercoils, which may lead to a change in the binding behaviour of IHF to special DNA sites. DNA supercoiling may be directly affected by environmental conditions (e.g. ionic strength). The regulatory activity of IHF for specific promoters may, therefore, be tuned through an interplay of DNA supercoiling and the modulation of other DNA-binding proteins.

Several other prokaryotic DNA-binding proteins show strong sequence homology to IHF. These are **HU**, the heterodimer consisting of HUα and HUβ, and the transcription factor **TF1**, which negatively regulates the expression of several SPO1 phage-specific genes. TF1 has a preference for binding to specific sites of *B. subtilis* phage SPO1 DNA, which contains hydroxymethyl uracil instead of thymine. Some eukaryotic transcription factors, such as **TBP** (TATA box binding protein) or the DNA-binding **HMG box proteins** (high mobility group) show strong functional similarities but are structurally completely different. Nevertheless, they may be grouped in the same family of DNA-binding proteins as IHF and HU. They all bind to the minor groove and bend DNA upon binding. While

TBP binds to curved DNA on its convex surface, bending the DNA away from the centre of the protein, IHF is located on the inside of the strongly curved DNA. Both HU and TBP have lobster-like structures very similar to IHF. Binding occurs through flexible arms formed by antiparallel β-sheets which widen the minor groove (HMG domain proteins do not have these β arms). Hydrophobic amino acids at the tip of the flexible arms intercalate on the outside of the DNA minor grooves, causing kinks. The electrostatic repulsion of phosphates on the inside of the bent DNA is minimized through charge neutralization by electrostatic interactions with the body of the proteins. It is not surprising that in several cases, HMG domain proteins or HU and IHF are functionally interchangeable. This functional redundancy might also explain that neither HU nor IHF are essential for the cell. If they are deleted they may functionally be substituted by other DNA-binding proteins of the same family.

It is known that HU is able to modulate the binding of other regulatory proteins, e.g. IHF, as indicated above, but also the binding of CRP or the repressor LexA. Thus, HU can indirectly modulate transcription normally governed by other regulators in a positive or negative way.

HU differs from IHF in the number of proteins per cell; this is much higher for HU (about 60 000 dimers per cell). Furthermore, HU does not have a specific recognition sequence but binds to DNA in a sequence-unspecific manner (it also binds to single-stranded DNA, to cruciform structures and to RNA). The two heterologous subunits of HU, HUα and HUβ, are not coordinately regulated, which is surprising for a protein that acts as a heterodimer. The growth phase-dependent existence of homodimers has been proven, and it is possible that the distribution of heterodimers and homodimers affects the regulatory function of HU. It should be noted, however, that in most bacteria HU is composed of two identical subunits, and the heterodimeric situation found in *E. coli* and *S. typhimurium* is rather the exception.

7.5.2 The growth phase-dependent activator factor for inversion stimulation (FIS)

Originally FIS was discovered as a protein that enhances the inversion reactions of several recombinases. The protein was therefore called **factor for inversion stimulation** (FIS). In the meantime FIS has been recognized as a potent transcriptional regulator. The protein is relatively small (11.2 kDa, 98 amino acids), has a basic pI and exists as a homodimer in solution. FIS has a strong tendency to bind to curved DNA and binding induces a bend of between 40° and 90°. FIS does not only recognize curved DNA, however. Specific binding of the protein requires the presence of a degenerated 15 base pair sequence of interrupted dyad symmetry: $(^{G}/_{T})$NNYRNN$(^{A}/_{T})$NNYRNN$(^{C}/_{A})$ where N stands for any nucleotide, Y stands for pyrimidines (T or C), and R stands for purines (A or G). The three-dimensional structure of FIS is well known (at 2 Å resolution) from X-ray crystallographic analysis. According to this structure the core of the molecule is formed by four α-helices. The N-terminal 20 amino acids are highly disordered. Dimerization of two FIS monomers involves hydrophobic interactions between adjacent N-terminal α-helices. The activation domain is also located at the N-terminal half of the molecule. A classical helix-turn-helix motif located at the C-terminal half of the molecule is responsible for binding to the major groove of DNA. The C-terminus contains two clusters of positively charged amino acid residues which are separated by 24 Å. The distance is too close to fit into two adjacent major grooves of straight B-DNA without distortion of the helix axis (this distance is in the order of 32–34 Å). Bending of the DNA in the direction of the protein by approximately 90° will bring the two major grooves on the inside of the FIS–DNA complex into proximity and allows the accommodation of the α-helices into a perfect fit with the DNA. Although at present no structure from FIS–DNA co-crystals is available, the existing evidence for a bent DNA structure in the complex is compelling. However, the structure of CRP, which bends DNA and interacts by a helix-turn-helix motif with the major groove of the DNA in a similar way to FIS, has been shown. The structure is presented in Fig. 7.14.

For most operons known to be affected by FIS it acts as an activator of transcription. A few systems have been documented, however, where FIS functions as a repressor. The *fis* gene itself, for instance, is negatively autoregulated by FIS. Clearly, the most remarkable function of FIS for the cell is that of an activator for transcription of stable RNA operons, (rRNAs and tRNAs) and components of the translational machinery, e.g. translation factors. A second striking property of FIS is its dramatic change in the cellular concentration under different growth conditions. In *E. coli* cells growing at stationary phase there are less than 100 copies of FIS per cell. However, under conditions of nutritional upshift, for instance after the addition of fresh rich medium, the number of FIS molecules increases rapidly by more than 500-fold, and more than 50 000 dimers are present within the first rounds of cell division. The expression pattern is transient, however, and when the cells enter exponential growth FIS synthesis is shut off and the FIS level decreases as a function of dilution by exponential cell division. Hence, proteolysis is apparently not involved in this regulation. The upshift in the FIS concentration is clearly caused on the transcriptional level which responds directly to an increase in growth rates (see Fig 7.15). The fact that FIS concentrations vary tremendously under different growth conditions clearly points to an important regulatory implication of this protein for cell physiology.

The regulatory region of the *fis* gene largely explains the change in FIS expression at different growth conditions. The promoter displays similarities to stable RNA promoters and has sequence elements characteristic for growth rate and stringent regulated genes (see Section 8.5 and 8.6). These regulatory signals may at least in part explain the dependence of FIS expression on the nutritional situation. As indicated above, *fis* expression is subject to negative transcriptional autoregulation. In line with this, FIS binding sites are found at

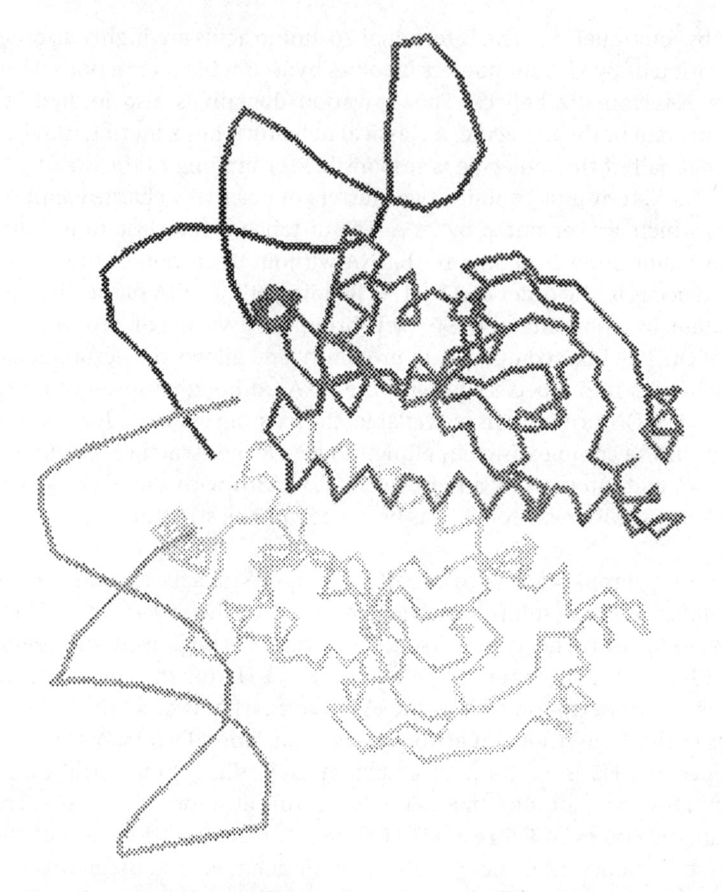

Figure 7.14 Structure of the CRP dimer bound to DNA. The backbone of a CRP dimer that makes contact to two helical DNA segments that contain the CRP binding site is shown. The structural backbones of individual molecules are indicated by different grey lines. The two α-helices of the CRP helix-turn-helix motifs are located within the major groove of the DNA molecules. The DNA contour is clearly bent by more than 60°. The figure is produced by RasMol from a Protein Data Bank file of the crystal coordinates of the CRP–DNA complex. The structure has been published by Parkinson *et al.* (1996).

unusual locations for activation within the *fis* promoter upstream region. Altogether there are six FIS binding sites in the vicinity of the *fis* promoter (Fig. 7.16). One site is located directly overlapping with the site for RNA polymerase binding, and another site is found downstream from the transcription start. Binding to these sites will inhibit transcription from the *fis* promoter at elevated FIS concentrations.

Probably the most information is available for the function of FIS in the activation of rRNA transcription. Three FIS binding sites are present upstream

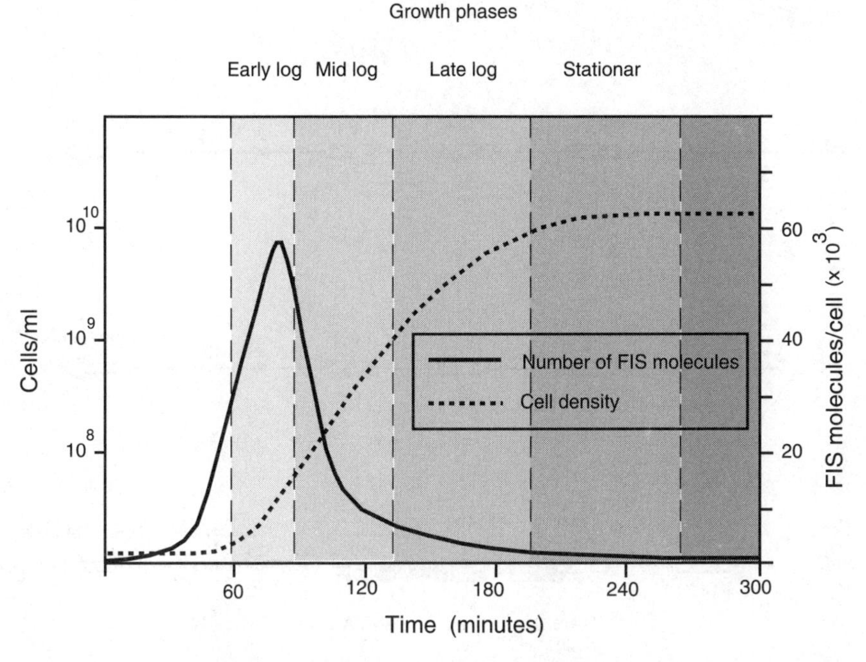

Figure 7.15 Cellular factor for inversion stimulation (FIS) levels during different growth conditions. A growth curve of an *E. coli* culture (cell density plotted as a function of the incubation time) is shown as a dotted line. The different growth phases, early log, mid log, late log and stationary are indicated by highlighting with increasing shading of the background. A solid black curve represents the amount of FIS molecules per cell present at the different growth phases. A sharp maximum is apparent at the early log phase.

of all seven *E. coli* rRNA operons. In addition, FIS binding sites are found upstream for most transcription units coding for tRNAs. Several examples of regulatory sequences containing FIS binding sites are presented schematically in Fig. 7.16. In the case of the upstream region of rRNA genes the FIS site closest to the promoter, usually termed site I, has a rather invariant position (–71) at the different rRNA operons. The other two sites (II and III) have a distance between 20 and 40 base pairs in the upstream direction. Binding is usually very efficient, with dissociation constants in the range of roughly 10^{-9} M. The occupancy of the three sites does not vary greatly and notable cooperativity in FIS binding is not usually observed. The upstream activating sequence (UAS) regions of stable RNA transcription units, which contain the three FIS binding sites, are known to contain strong intrinsic curvature resulting from appropriately phased AT clusters (see Section 6.2.1). One of the best studied examples for the regulation of rRNA transcription is the *rrnB* operon (see Section 8.7). The presence of FIS stimulates rRNA transcription about 5–10-fold from the *rrnB* P1 promoter. The exact effect of FIS is somewhat difficult to assess because the

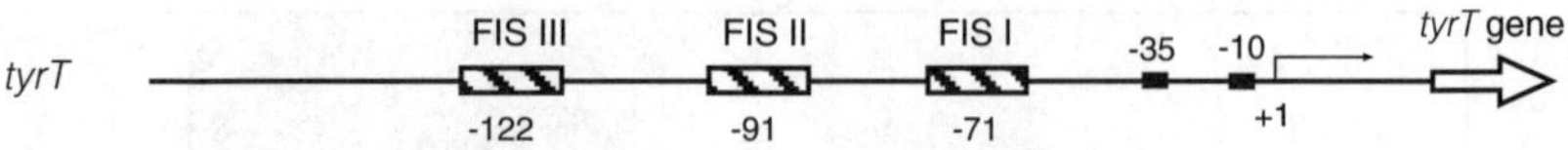
tyrT
FIS III
FIS II
FIS I
-35
-10
tyrT gene
-122
-91
-71
+1
tufB
FIS III
FIS II
FIS I
-35
-10
thrU gene
-123
-91
-71
+1

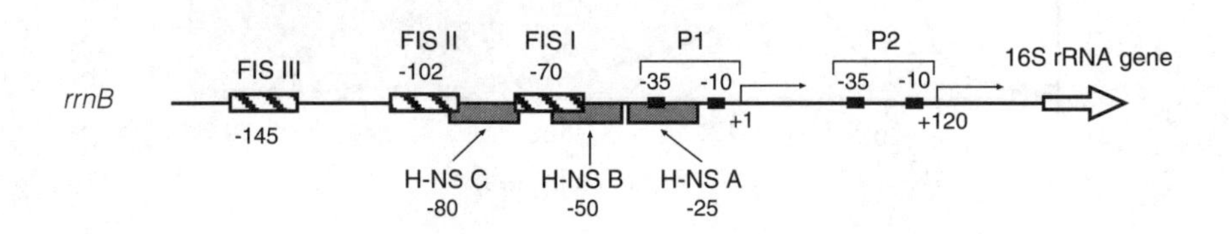
rrnB
FIS III
FIS II
-102
FIS I
-70
P1
P2
-35
-10
-35
-10
16S rRNA gene
-145
+1
+120
H-NS C
-80
H-NS B
-50
H-NS A
-25
fis
FIS VI
FIS V
FIS IV
FIS III
FIS II
-35
-10
FIS I
fis gene
-220
-143
-101
-83
-42
+1
+26
hns
FIS VII
FIS VI
FIS V
-174
FIS IV
FIS III
-71
FIS II
+1
FIS I
hns gene
-35
-10
-282
-243
H-NS III
-185
H-NS II
-110
H-NS I
-50
+28

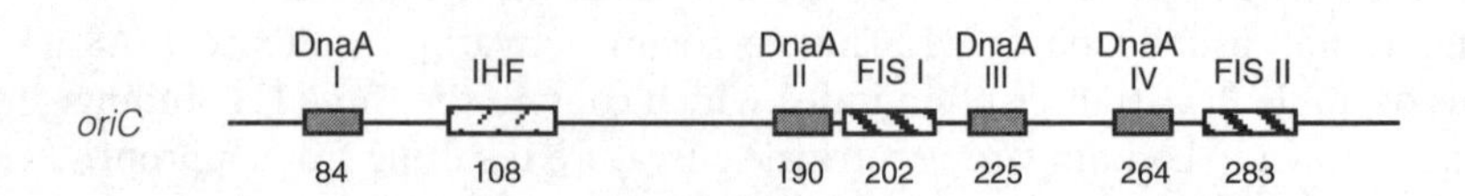
oriC
DnaA
I
IHF
DnaA
II
FIS I
DnaA
III
DnaA
IV
FIS II
84
108
190
202
225
264
283

UAS region of the *rrnB* operon activates through several mechanisms. In addition to the activation by FIS there is clear factor-independent activation of the so called UP element close to the –35 region (see Section 6.2.1). Moreover, the contribution of a second downstream promoter (P2) arranged in tandem is not completely clear (see Chapter 8.7). Interestingly, mutants where the *fis* gene has been disrupted or deleted have no apparent phenotype. At first this finding was unexpected because the adequate synthesis of stable RNAs is vital for the cell. The overall synthesis of stable RNAs is only slightly affected in *fis*$^-$ cells, however. A set of compensatory regulatory mechanisms and redundant control systems involving additional regulators as well as the special arrangement of tandem promoter structures is responsible for the observed phenomenon (note that only P1, the first of the tandem rRNA promoters, is FIS-sensitive). Nevertheless, the presence of an active *fis* gene gives a notable advantage to cells at very rapid growth in rich medium since *fis*$^-$ cells cannot achieve very high growth rates. Moreover, *fis*$^-$ cells have a selective disadvantage when growth conditions change rapidly. The presence of FIS enables a much faster adaptation to high growth rates after a nutritional upshift compared to *fis*$^-$ cells.

The mechanism by which FIS-dependent activation is brought about has been the subject of intensive studies. Several plausible models have emerged from these studies but a complete picture is still pending. Clearly, activation by FIS requires DNA bending. It is believed that bent DNA allows more favourable RNA

Figure 7.16 Location of FIS binding sites at different operons. The complex arrangement of FIS binding sites in the regulatory regions of the following operons is shown: *tyrT*, *tufB*, *rrnB*, *fis*, *hns* and *oriC*. The centres of the FIS binding sites are given by numbers with respect to the transcription start of the FIS-dependent promoters. A similar arrangement of FIS sites is found for the *tyrT*, *tufB* and *rrnB* operons. In each case three binding sites are centred at similar positions upstream of the respective promoter. Note that all three operons encode stable RNAs. The *rrnB* operon is additionally controlled by H-NS, for which three binding sites (A, B and C) are known which partially overlap with the FIS binding sites I and II, and partly with the core promoter. H-NS and FIS thus act as antagonists in the expression of rRNA transcription. Like all rRNA operons, the *rrnB* operon has a second promoter (P2) about 120 base pairs downstream from P1. This promoter is neither FIS nor H-NS dependent. The regulatory region of the *fis* gene itself contains six FIS binding sites. FIS site I is located downstream of the promoter at position +26. The remaining sites II–VI are found upstream between positions –42 and –220. The *fis* gene is thus autoregulated. Transcription of the *hns* gene is affected by FIS and H-NS. Three H-NS binding sites are located at positions –50, –110 and –185. They are obviously implicated in autoregulation. In addition, seven FIS binding sites have been mapped. Site I, similar to the *fis* gene, is located downstream of the promoter at position +28. Site II overlaps directly with the core promoter. Sites III–V partly overlap with the H-NS sites I, II and III. The two sites VI and VII are found further distal at positions –243 and –282. A balanced regulation owing to different affinities of the regulators for the different sites is thus feasible. In the case of *oriC* FIS functions in combination with a set of different factors involving DnaA and IHF in the regulation of replication. All three factors are known to affect DNA conformation.

polymerase contacts. However, amino acid substitution mutations within FIS have been obtained which are still able to bind and bend DNA but which have lost the ability to activate transcription (and recombination). This finding indicates that bending alone is not sufficient and direct contacts between FIS and RNA polymerase are likely. The amino acids in FIS responsible for reduced activity (Arg71, Gly72 and Gln74) are located at the N-terminal end of the helix-turn-helix motif in a surface-exposed loop which might be directly contacted by RNA polymerase. Binding of FIS *and* RNA polymerase to the *rrnB* UAS region occurs cooperatively. The argument that FIS binds directly to RNA polymerase is further supported by the observation that activation requires a precise positioning of the FIS binding sites relative to the polymerase binding site. As shown for several other transcription factors which directly interact with RNA polymerase, activation by FIS is 'face-of-the-helix-dependent'. Distance variations of FIS binding site I by half helical turns abolish activation while insertion of a full turn of DNA partly restores activity. Evidence has been presented for both the α and the σ subunits of RNA polymerase as sites to which FIS interaction may occur. The fact that FIS recruits RNA polymerase into the closed complex has been verified by K_B measurements at the *rrnB* P1 promoter. Increasing the primary RNA polymerase promoter binding is not the only way in which FIS exerts its activating effect, however. It is known that FIS is able to activate sequential steps of transcription initiation. Facilitated promoter melting at the transcription start site as well as enhancement in the escape of RNA polymerase from the promoter have been reported as a result of FIS-mediated activation. These activities of FIS require the binding of the activator to all three FIS sites I, II and III in helical register. This observation has led to a model which suggests that FIS activates stable RNA promoters by the formation of DNA microloops. Such a microloop, once formed, constitutes a separate topological domain. Rotation of the RNA polymerase during intermediate complex formation causes right-handed writhing of the DNA and generates torsional strain within the microloop. Bound FIS molecules are able to revert the torsion into left handed twist. This motion is then transmitted to the initiation site where it drives the transcription bubble to untwist. Activation may thus be the result of topological transitions within a microloop mediated by FIS and directly transmitted to the initiation complex. It is important to note, in this regard, that stable RNA promoters are twist-sensitive because of a 16-base pair suboptimal spacing of the –35 and –10 promoter hexamer sequences (see Section 8.7).

7.5.3 Histone-like nucleoid structuring protein (H-NS), a versatile regulator through DNA conformations

The **histone-like nucleoid structuring** protein (H-NS) is a 15.6-kDa protein of 136 amino acids. In contrast to the other nucleoid associated proteins in *E. coli* H-NS has a neutral pI. The protein is highly conserved among enteric bacteria but a homologous protein has not been found in Gram-positive species so far.

The amino acid sequence of H-NS does not show a convincing similarity to any known DNA-binding protein and no sequence motif characteristic for DNA-binding can be recognized. The active form of the protein is very likely to be a homodimer; however, higher aggregates are also known to exist in solution. H-NS has originally been discovered as a protein that is able to compact DNA and to constrain supercoils. It was therefore believed to serve predominantly as a structural component of the bacterial nucleoid. Today it is clear that H-NS can also act as a direct regulator of transcription for a large number of genes. Moreover, H-NS is now considered to be a global control element in bacterial gene expression. About 30 different genes are affected in *hns* mutants. Most of the genes are derepressed in the absence of a functional *hns* gene, indicating that H-NS mainly functions as a repressor. For some genes, however, H-NS has also been shown to act as a transcriptional activator. Examples where transcription is activated are the *flhD* and *fliA* genes, which are both needed for the synthesis of flagella in *E. coli*. This activation is independent of cAMP–CRP which is also involved in the control of flagella synthesis (some genes are also affected at the post-transcriptional level, e.g. the *rpoS* gene product, see Section 8.3).

The genes regulated by H-NS are very **pleiotropic** and appear to be unrelated at first sight. However, a common denominator may be recognized on closer comparison. The list of genes regulated through H-NS shares the common feature that their expression is related to stress conditions or environmental changes, such as osmolarity, temperature, pH, oxygen availability or nutritional deprivation. Often the genes regulated by H-NS are linked to the stringent or growth rate control (see Sections 8.5 and 8.6). Table 7.5 gives a list of genes for which H-NS has been characterized as a transcriptional regulator.

No high resolution structural information is yet available for the complete H-NS protein. However, a 47-amino acid C-terminal fragment (amino acids 90–136) has been studied by nuclear magnetic resonance (NMR) spectroscopy. A structure consisting of a stretch of antiparallel β-sheets linked to short helical regions has been deduced from this study. Additional information about the secondary structure has been deduced from circular dichroism (CD) and fluorescence measurements. Moreover, the location of several functional domains has been identified by systematic mutational analyses. According to these studies the DNA-binding domain of H-NS resides in the C-terminal half of the molecule. The N-terminal domain of H-NS probably forms an amphipathic α-helix which is involved in transcriptional repression and protein–protein interaction. The central region seems to be important for oligomerization of the protein.

The occurrence of three isoforms of the protein, which differ in their isoelectric points, has been reported. Their significance with respect to the structure and function of H-NS is not known. It seems clear, however, that phosphorylation is not responsible for the three isoforms.

H-NS is encoded by a single copy gene which is under transcriptional autoregulation through H-NS. Transcription of the *hns* gene is activated by the

Table 7.5 Genes regulated by H–NS

Gene	Function	Controlling σ factor	Regulatory level	Effect
appY	Anaerobic growth phase activator	σ^s, σ^{70}	Transcription	Repression
bolA	Cell shape regulator	σ^s	?	
cfaD	Pili formation	σ^{70}	Transcription	Repression
csgA	Curli formation	σ^s	Transcription	Repression
fimA	Type 1 fimbrial subunit	σ^{70}	Transcription	Activation
flhD	Flagellar control protein	σ^{70}	Transcription	Activation
fliA	Flagellar-specific σ factor	σ^{70}	Transcription	Activation
hns	DNA structuring protein, transcription factor	σ^{70}	Transcription	Repression
lysU	Lysyl–tRNA synthetase	σ^{70}	Transcription	Repression
pap	Pili formation	σ^{70}	Transcription	Repression
proU	Proline/glycine betaine transport system	σ^s, σ^{70}	DNA topology, transcription	Repression
rpoS	Stationary σ factor	σ^{70}	Post-transcriptional	Repression
rrn	Ribosomal RNA	σ^{70}	Transcription	Repression
stpA	Molecular backup for H–NS, RNA chaperone activity	σ^{70}	Transcription	Repression
virB	Invasion regulatory gene (*Shigella flexneri*)	σ^{70}	Transcription	Repression

regulator FIS. In addition, expression is regulated by the cold-shock protein **CspA**. H-NS concentrations have been estimated to be between 20 000 and 60 000 molecules per cell. The expression of H-NS does not vary dramatically during cellular growth. There is, however, a measurable increase in the H-NS concentration at the end of the exponential growth phase or at reduced growth rates.

The DNA-binding properties of H-NS deserve some special note. H-NS binds to any kind of double-stranded DNA but also, with less affinity however, to single-stranded DNA or to RNA. There is no apparent DNA sequence requirement for binding but a marked preference to curved DNA. Strong and specific binding occurs to regions of intrinsic curvature. In addition to a high affinity for curved DNA, binding of H-NS is able to induce bends into non-curved DNA. Specific binding of H-NS thus requires either intrinsically curved DNA or DNA which has the appropriate flexibility to adopt a favourable bend during binding. Two types of interactions between H-NS and DNA can be distinguished. *Specific bind-*

ing can occur to curved DNA, and *unspecific binding* to DNA with random structure. The two modes of binding can be clearly distinguished in terms of their salt dependence, and also by fluorescence spectroscopy of a single tryptophan residue present in the amino acid sequence of H-NS. Sites for high affinity H-NS binding usually contain AT tracts in helical register, which are known to occur in intrinsically curved DNA. Binding to random DNA, which is observed at much higher H-NS concentrations, is inhibited at high salt concentrations, and probably depends predominantly on electrostatic interactions. Under conditions of high H-NS concentrations and low ionic strength, cooperative protein oligomerization has also been observed. In the case of specific binding to appropriately curved DNA, amino acids in the vicinity of Trp108 in the C-terminal region of H-NS are involved. This specific binding is distinctly different from the unspecific electrostatic interactions with random DNA and involves hydrophobic interactions and/or hydrogen bonding. Although the binding site for specific DNA recognition can be localized within the C-terminal domain the responsible amino acid sequence elements have not yet been identified.

What is known about the molecular nature of H-NS–DNA recognition? While helix-turn-helix binding proteins recognize the major groove of DNA the typical DNA structuring proteins IHF and HU described above bind to the minor groove. H-NS, which, like HU, does not require a specific recognition sequence may possibly obey the same recognition principles. From the analysis of H-NS–DNA complexes a satisfying answer as to whether H-NS is a major or minor groove binder can not yet be given. Binding to curved DNA is efficiently inhibited by the drug distamycin, for instance. Distamycin is known to bind into the minor groove of AT tracts within curved DNA. By widening the compressed minor groove of curved DNA distamycin is able to straighten out bends. The efficient inhibition of H-NS–DNA binding, that is observed in the presence of distamycin, could mean that H-NS binds to the minor groove. However, the result is also consistent with the view that H-NS interacts with the major groove. Distamycin may obstruct the necessary curvature required for H-NS binding to the major groove and thereby inhibit protein–DNA interaction. In fact, DNaseI footprinting studies of specific H-NS–DNA complexes support the latter assumption. Hyperreactive DNaseI cuts are observed within the H-NS–DNA binding site at about two base pairs distance across the antiparallel DNA strands. This pattern of DNA cleavage is similar to that obtained for FIS–DNA complexes for which binding to the major groove has been firmly established. More high resolution structural information about the H-NS–DNA complexes is clearly required before this important question can be answered unequivocally. There is no doubt at present that the H-NS binding affinity to the target DNA depends on the correct DNA conformation. Strong and specific binding will only occur if the target DNA is appropriately curved. Every reaction mediated through the binding of other transcription factors or environmental stimuli which alter the DNA conformation can potentially influence the extent of H-NS binding to specific sites. This is an important notion because it already provides some mechanistic clues as to how H-NS-dependent regulation of transcription may be modulated.

The precise mechanism as to how H-NS exerts its regulatory function is not yet known and it may actually be very different for different promoters. For many genes under H-NS control, repression could in the first instance be explained by the classical mechanism of sterical hindrance. In these cases H-NS binding sites directly overlap with the RNA polymerase binding site at the promoter. Comparison of the H-NS binding sites which have been mapped so far indicates, however, that such a simple mechanism cannot explain the situation for all genes regulated by H-NS. H-NS binding sites are known to occur at sites too far downstream or upstream of the promoter to prevent RNA polymerase binding directly. In most cases H-NS binding is not restricted to a single site. Usually several adjacent binding sites exist and binding occurs in a highly cooperative way. This observation has led to the hypothesis of regulation by *cooperative protein condensation*, which would finally cover the complete transcription initiation region and thereby prevent any transcriptional activity. There are several observations which indicate that repression is not that simple, and at least in several cases involve more dynamic changes of the DNA conformation. First, the very large mobility shifts observed in gel retardation studies of H-NS–DNA complexes performed with specifically regulated promoter DNA fragments cannot be explained by protein condensation but clearly indicate the formation of strongly curved or looped DNA structures. In addition, for most of the genes regulated by H-NS, a second regulatory protein, usually an activator, is known to bind at sites overlapping the H-NS binding sites. Inhibition of transcription could thus be indirect through obstruction of the binding sites for activators. Most interestingly, binding of the activator and H-NS is not always mutually exclusive. This has been observed in several cases where FIS is the activator and H-NS the corresponding antagonist. For instance, binding competition and detailed footprinting studies between FIS and H-NS at the regulatory region of the rRNA P1 promoter clearly indicate that the two transcription factors can bind simultaneously, although their binding sites show a considerable spatial overlap (see Section 8.7.4). Each factor has three distinct binding sites within the *rrnB* UAS region. Two of the FIS and H-NS sites overlap; however, they are apparently orientated on opposite faces of the DNA helix. Therefore, the two proteins can bind simultaneously to the overlapping DNA regions without displacing each other. One of the H-NS binding sites extends into the *rrnB* P1 promoter core region, where it clearly overlaps with the RNA polymerase binding site. Interestingly, fully occupied H-NS sites do not prevent RNA polymerase binding in this case, although transcription initiation is clearly inhibited. Ternary complexes between promoter DNA, RNA polymerase and H-NS have also been described in other cases, e.g. the non-specific *lac* UV5 promoter. Obviously, H-NS binding induces a DNA conformation which still allows RNA polymerase to interact with the promoter DNA. The geometry of the complex is inadequate, however, for efficient transcription initiation. In particular, the activating effect of the FIS protein can no longer be exerted at the *rrn* P1 promoter occupied by H-NS.

The antagonism between the two regulators FIS and H-NS occurs at other promoter systems as well (e.g. regulation of the *hns* operon: activation by FIS, autorepression by H-NS). As in the case of rRNA promoters the antagonism can be explained by models in which the geometry of the DNA in the transcription initiation complex changes from a transcription-proficient structure to a conformation which does not support efficient transcription. The change between activation and repression is obviously mediated by two alternate proteins which are capable of altering the DNA structure in opposing ways. Other examples of factors showing functional antagonism to H-NS include the activator CfaD at the *cfaAB* genes encoding *E. coli* fimbriae or the VirB-controlled expression of the *Shigella flexneri* invasion genes. Transcription of VirB, for instance, is repressed by H-NS. The protein VirF counteracts the H-NS-dependent VirB transcription.

It is very likely that protein–protein interactions are involved in the conformational transition between the activating and repressing complexes. There is circumstantial evidence for heterologous interactions between H-NS and FIS which may serve as early intermediates in the architectural switch from activation to repression-proficient complexes. Direct evidence for heterologous dimer formations has been presented for several other regulatory proteins. For example, the bacteriophage T4 gene product gp5.5 is known to inactivate H-NS by direct interaction and thereby supports T4 phage propagation. Similarly, the host factor for phage Qβ, HF-I, appears to bind to H-NS. This protein is able to complement an *hns* phenotype. In addition, heterologous interaction between H-NS and **FliG**, one of the bacterial motor proteins, has been demonstrated with the **two hybrid system** (Box 7.3).

Recently an H-NS analogue, the protein **StpA**, has been identified. This protein was first characterized as a splicing suppressor of the phage T4 *td* intron. The protein is 134 amino acids in length and shares 58% identity with the sequence of *E. coli* H-NS. The protein is not only structurally very similar to H-NS; the two proteins can substitute for each other in many but not all functions. Both proteins bind to curved DNA and share the ability to inhibit transcription from several promoters as well as to constrain DNA supercoils. There are, on the other hand, several disparate functions not shared by both proteins. StpA, for instance, has clear functions as an RNA chaperone, which are not efficiently carried out by H-NS. The two proteins are cross regulated by a feedback loop, inhibiting their own and each other's synthesis. For instance, StpA is normally expressed at very low levels. The expression of StpA is efficient, however, in *hns* mutants because of a lack of autorepression. Likewise, overexpression of StpA inhibits the synthesis of H-NS. The expression of StpA is furthermore dependent on the regulator protein Lrp (see Section 8.4) as well as on the temperature and the growth phase. The two proteins are able to interact with

Box 7.3 The two-hybrid system

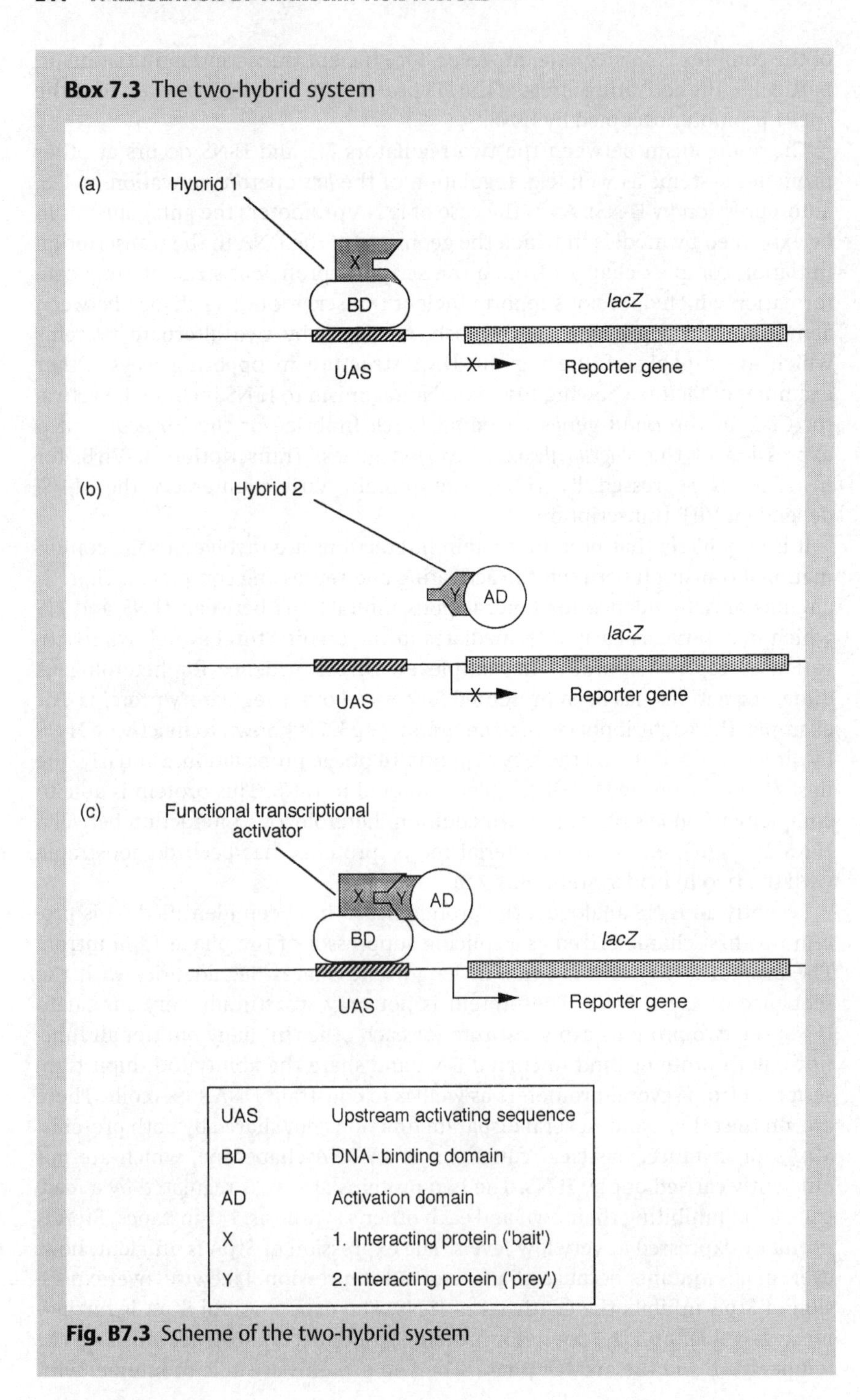

Fig. B7.3 Scheme of the two-hybrid system

The two-hybrid system has been developed as a powerful genetic method to identify proteins that interact *in vivo*. The method relies on the fact that active transcription factors can be assembled from separable functional domains, e.g. a DNA-binding and an activation domain. A functional transcription complex can be reconstituted by non-covalent interactions of two hybrid proteins, which either contain a DNA-binding site or an activator domain. The separable domains are fused to a pair of interacting proteins where they can reconstitute a site-specific transcriptional activator. The hybrid system is composed such that the isolated proteins do not influence transcription, but when a complex of the two interacting proteins can be reconstituted transcription of a flanking reporter gene is initiated. The system was first been developed for yeast but similar strategies are now available for mammalian cells or *E. coli.* The principal two-hybrid system Is schematically depicted in Fig. B7.3.

Common DNA-binding domains (the 'bait') are derived from either the yeast Gal4 or the *E. coli* LexA proteins. The activation domains (the 'prey') are either taken from the Gal4 protein or the Herpes simplex virus activator VP16. Frequently used reporter genes are the *lacZ* gene, which allows a simple blue-white screen if the chromogenic substrate X-gal is employed. The two-hybrid system has been used with great success to identify genes which associate in the cell either permanently or transiently. The analysis can be refined to the sites of interaction if deletion or systematic mutations are introduced into the pair of interacting proteins. The versatility of the system has been extended considerably in the past and different variants of the system, such as the three-hybrid system where a heterodimeric ligand or an RNA molecule are employed to link the DNA binding and activation proteins, have been described.

each other, forming heterodimers. The biological significance of such heterodimers is not known.

There are several additional interesting functions which have been observed for H-NS for which there is no clear mechanistic explanation yet. For instance, H-NS appears to be able to discriminate between the different RNA polymerase holoenzymes $E\sigma^{70}$ and $E\sigma^{s}$. Note that H-NS is involved in the expression of σ^{s}, however, at the level of *rpoS* mRNA translation and protein stability (see also Section 8.3). Promoters which are normally recognized by $E\sigma^{s}$ will be transcribed by $E\sigma^{70}$ in the absence of H-NS. It is speculated that the promoter UP element (see Section 2.1.2 and 6.2.1) plays a role in this change of specificity. The UP element is the preferred site of interaction of H-NS *and* it is recognized by the αCTD of RNA polymerase. Competition between these two elements may therefore partly explain the change in specificity.

The involvement of a small RNA molecule has been described which is able to antagonize H-NS-mediated repression. It has been observed that H-NS-dependent repression of the *rcsA* gene, encoding a regulator of capsular polysaccharides, is reversed by the presence of a small untranslated RNA, **DsrA**. The mechanism of this interesting regulation is still elusive. In a similar way, H-NS-dependent repression of the stationary phase-specific σ factor, σ^s, is released by the DsrA RNA. This control of σ^s synthesis is mediated at the post-transcriptional level (see Section 8.3).

It will not be a surprise when more regulatory contexts and examples are uncovered in near future which demonstrate the importance of H-NS as a global regulator of bacterial gene expression.

7.6 Two-component regulatory systems

Generally, bacteria live in an environment that frequently undergoes drastic changes in the chemical composition or physical parameters. It is extremely important for the survival of bacteria that they are able to respond rapidly to such changes, either by adaptation or by movement. The fact that bacteria have evolved very efficient mechanisms for sensing external stimuli and transferring this information to cellular sites where an appropriate response is exerted, often involving an altered pattern of gene expression, has stimulated a fascinating subject of research. Considerable progress has been made during the past years, and today there is some insight into the general mechanisms on how bacteria sense external stimuli, and how these signals are transferred to the transcriptional apparatus. Signal transduction generally involves the transfer of phosphoryl groups between proteins. Many of these signals are sensed and transmitted to the transcription apparatus by pairs of proteins which belong to the family of **two-component regulatory proteins**. The stimulus is normally transferred between a group of proteins termed **histidine kinases** and another group of proteins called **response regulators**. The following section summarizes structural and functional data on these important groups of bacterial regulators and presents some selected examples as to how an external stimulus is transferred to the transcription machinery by members of the two-component regulatory system.

7.6.1 Sensor kinases and response regulators

Today, more than 50 pairs of two-component systems are known to function in the transfer of environmental stimuli in bacteria. The cognate pairs consist of **sensor kinases** and **response regulators**, both of which are characterized by virtue of extensive amino acid homology of subdomains within the two protein families.

Response regulators contain a conserved domain of about 125 amino acids at their N-terminus with 20–30% identity between each two proteins and a considerable number of invariant amino acid residues. The conserved N-terminus of response regulators is termed the **receiver domain**. Among the highly conserved residues of the receiver domain are three aspartate residues near the N-terminal border and the centre of the domain as well as a lysine residue near the C-terminal border. The three-dimensional structure of the receiver domain of **CheY** has been resolved by X-ray crystallography and NMR analysis. CheY is involved in **chemotaxis**, where it affects the flagellar rotation in either a clockwise or counterclockwise direction. It is not a transcriptional regulator but because of the outstanding sequence homology it is considered to be a structural prototype for receiver domains. According to this structural analysis the receiver domain is composed of five parallel β strands alternating with α-helices *via* short loops which form a roughly cylindrical structure (α/β barrel). The N- and C-terminal residues are both on one end of the cylinder and the invariant aspartate residues, the site of phosphorylation, form an acidic pocket at the opposite site. The side chain of the invariant lysine residue protrudes into this pocket.

In addition to the conserved receiver domain, response regulators usually contain different **output domains** at the C-terminus. These output domains interact with the transcription complex and mediate transcriptional regulation by either altered DNA and/or RNA polymerase binding or other mechanisms of transcriptional regulation. The change in activity of the output domain is brought about through the reversible transfer of phosphoryl groups to the invariant aspartate residues at the receiver domain. According to the different structural and functional organization of the output domains, response regulators are generally divided into three groups. These groups are termed after the prototype regulators OmpR, FixJ and NtrC, each representing a distinct subfamily (Fig. 7.17).

The members of **OmpR like response regulators** contain a C-terminal DNA-binding domain of roughly 150 amino acids in length linked to the receiver domain. DNA-binding probably involves helix-turn-helix motifs which are present in the C-terminal domain of OmpR like regulators. The capability to bind DNA is modulated by phosphorylation of the receiver domain. Strong binding to DNA is observed in the phosphorylated state. In the dephosphorylated state the receiver domain appears to interfere with DNA-binding. The OmpR type of regulators act at promoters which are recognized by the $E\sigma^{70}$ holoenzyme of RNA polymerase. Transcription can either be activated or repressed by OmpR like regulators depending on the type of promoter. For several of the regulators direct contacts with the C-terminal domain of the α subunits (αCTD) has been demonstrated (e.g. OmpR; see Section 2.2.1).

The **FixJ type of response regulators** are characterized by a conserved C-terminal DNA-binding domain of around 100 amino acids. This domain does not share any convincing homology with the OmpR like proteins, although a helix-turn-helix motif near the C-terminus probably represents the DNA-binding site. As described for the OmpR like family, the FixJ like regulators act in combination with the $E\sigma^{70}$ holoenzyme of RNA polymerase. Interestingly,

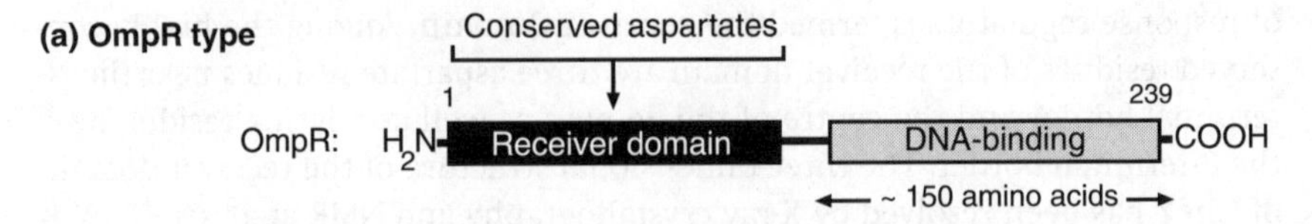

Members comprise: AraC, CopR,CpxR, OmpR, PhoB,VirG, etc.
Some members occur without receiver domain, e. g. GlnR

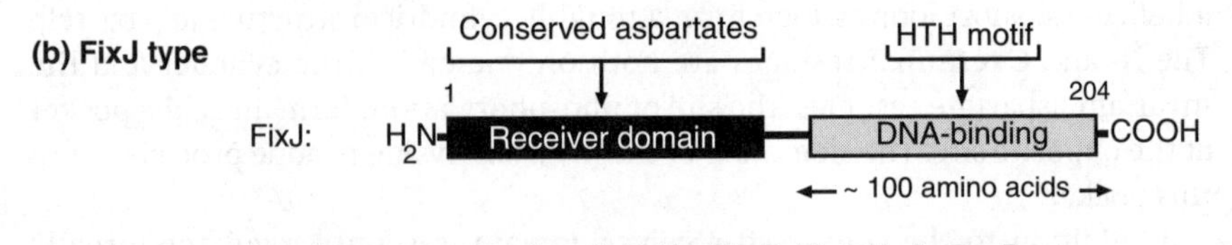

Members comprise: ComA, DctR, FimZ, FixJ, NarL, RcsB, etc.
Some members occur without receiver domain, e. g. MalT

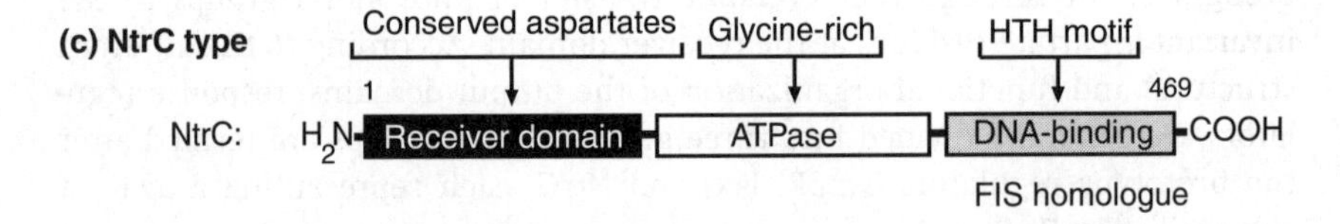

Members comprise: DctD, HoxA, HupR, NtrC, PilR, etc.
Some members occur without receiver domain, e. g. NifA, FliD

Figure 7.17 Families of response regulators. Based on sequence homology three types of response regulators can be defined. (a) OmpR type regulators contain an N-terminal receiver domain with invariant aspartate positions which function as phosphate acceptors and a C-terminal DNA-binding domain of roughly 150 amino acids in length. (b) The FixJ type of regulators are composed of an N-terminal receiver domain and a C-terminal DNA-binding domain of about 100 amino acids. A helix-turn-helix motif (HTH motif) is responsible for DNA-binding. (c) The NtrC type of regulators have, in addition to the N-terminal receiver domain and the C-terminal DNA-binding domain, a central domain which is glycine-rich and contains an ATPase activity. The DNA-binding domain contains an HTH motif and has strong homology with the FIS protein.

many homologues to the FixJ like regulators are known which lack a receiver domain.

The **NtrC family of response regulators** usually consists of three separate domains. The receiver domain is localized at the N-terminus. A helix-turn-helix motif within the C-terminal domain is responsible for DNA-binding. The DNA-binding motif is similar to that found for the transcription factor FIS. In contrast to the other families of regulators, members of the NtrC family contain within their central domain an ATPase site, typically found in many ATP binding proteins. The central domain is additionally characterized by a glycine-rich sequence. Members of the NtrC family act at promoters recognized by the $E\sigma^{54}$ form of RNA polymerase. In contrast to the regulators from the OmpR or FixJ family, which act by improving RNA polymerase–DNA interaction (RNA polymerase recruitment), NtrC-like regulators exert their function by facilitating promoter melting and the transition from the closed to the open transcription initiation complex. This reaction involves the ATPase function in the central domain of the regulator. The ATPase activity itself is dependent on the prior phosphorylation of the receiver domain which enables the formation of tetramers or higher aggregates of NtrC-like proteins (see Fig. 7.19). In contrast to the dimers which prevail in the dephosphorylated state, the tetramers or higher oligomers of the NtrC-like proteins represent the active regulator species.

The transfer of phosphoryl groups to the receiver domains of response regulators is maintained by a cognate kinase. Note, however, that the catalytic activity to phosphorylate resides in the response regulator itself, and autophosphorylation by low molecular weight phosphate donors, e.g. phosphoramidate or acetyl phosphate, has been observed. Nevertheless, for each response regulator a specific kinase can be assigned. These kinases are termed **histidine kinases** because the source of the phosphoryl group is a phosphohistidine residue. Histidine kinases are also characterized by their modular structure. Different kinase domains comprise a conserved stretch of approximately 250 amino acids, with more than 20% identical residues. The domains can be subdivided into five characteristic segments termed H-, N-, D-, F- and G-boxes. A conserved histidine within the H-box represents the active site which catalyses the transfer of a γ phosphoryl group from ATP to the N3 position of the histidine. The ATP binding site is probably provided by the highly conserved residues of the G-box. Histidine kinases are generally homodimeric molecules. Phosphorylation occurs mutually from one subunit to the other. This autophosphorylation is dependent on a specific signal. The nature of this signal and the mechanism by which it is sensed is not known for most histidine kinases, except for a few examples (see below). The N-terminus of histidine kinases generally contains variable sequences which provide unique specificity for a given pair of two component regulators. Most histidine kinases contain at their N-terminus several hydrophobic membrane-spanning sequences and can be regarded as integral membrane proteins which function as transmembrane receptors. In this way, extracytoplasmic stimuli are assumed to be transferred to the cytoplasm by conformational changes of the histidine kinase after ligand-binding to the outside sensing domain. In case of the kinase FixL the nature of the sensor domain is known. The FixL histidine kinase is involved in regulating the expression of the nitrogen-fixation genes. The N-terminal domain contains a heme group which senses oxygen. The kinase domain of

FixL is stimulated when no oxygen is bound to the heme group. In other cases, the situation is more complex and separate components are involved in regulating the activity of the histidine kinase. The histidine kinase PhoR may be taken as a typical example. PhoR activates the corresponding response regulator PhoB, which regulates the expression of the Pho regulon involved in the metabolism and uptake of phosphate. The PhoR kinase is a membrane-spanning protein with an extracellular sensor domain and an intracellular kinase activity, which senses phosphate levels. One might expect that periplasmic phosphate binds directly or indirectly via the phosphate binding protein PstS to the extracytoplasmic domain of PhoR and thus activates the kinase. Yet this is not the case. The PhoR activity is regulated indirectly instead by the phosphate-specific transporter system PTS, which belongs to the family of ABC transporters. It is not yet known how the signal from the PTS system is transferred to PhoR.

A general scheme for the signal transduction involving two-component regulatory systems is presented in Fig. 7.18 and Table 7.6 gives some representative examples of two-component regulators.

7.6.2 Regulation of nitrogen-fixation genes

The most important (sometimes the only) way to utilize nitrogen for many bacteria is the synthesis of glutamine from glutamate and ammonia catalysed

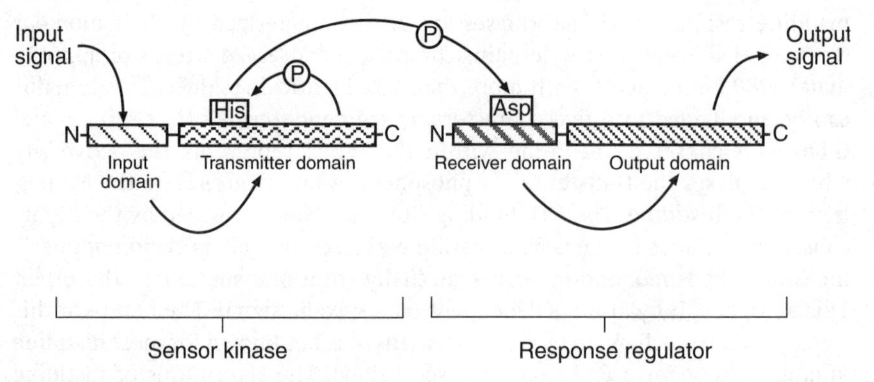

Figure 7.18 Two-component signal transduction scheme. The path of an arbitrary environmental signal involving the two-component sensor kinase and response regulator is schematically illustrated. The sensor is often a membrane protein able to monitor environmental signals. The signal is transmitted to a cognate response regulator, usually a cytoplasmic protein, which mediates a change in gene expression. A signal or stimulus causes a change in the input domain of a sensor kinase, which induces an autophosphorylation reaction transferring a phosphate from ATP to a histidine residue within the transmitter domain. The phosphate is subsequently transferred to an aspartate residue within the receiver domain of the response regulator. This causes a change in the output domain, usually leading to an altered effect on transcriptional activity. The scheme is adapted from Parkinson and Kofoid (1992).

Table 7.6 Typical two-component regulatory systems

Sensor/ transmitter	Regulator/ receiver	Regulated genes	Stimulus	Organism
CheA	CheB/CheY	Chemotaxis	Repellents, attractants	*E. coli, B. subtilis, S. typhimurium*, etc.
EnvZ	OmpR	*ompC, ompF*	Change in osmolarity	*E. coli, S. typhimurium*
NtrB	NtrC	*glnA* etc.	Ammonia limitation	*E. coli, S. typhimurium* etc.
PhoR	PhoB	*phoA, phoE*, etc.	Phosphate limitation	*E. coli, B. subtilis*, etc.
NarX	NarL	*narGHJI*	Nitrate concentration	*E. coli*
FixL	FixJ	*nifA*	Nitrogen limitation	*R. meliloti, A. vinelandii*, etc.
VirA	VirG	*virB, virC*, etc.	Plant wound exudate	*A. tumefaciens*
DegS	DegU	Degradative enzymes	Starvation	*B. subtilis*

Part of the data is adapted from Gross *et al.* (1989)

by **glutamine synthetase** at the expense of ATP hydrolysis. A drop in the intracellular concentration of glutamine provides a signal for the activation of transcription of nitrogen-regulated genes. The activity of glutamine synthetase may thus be regarded as a cellular ammonia sensor. This explains why the expression of glutamine synthetase is precisely regulated in response to the nitrogen demands of the cell. Glutamine synthetase is the product of the *glnA* gene, which is encoded in the *glnALG* operon (Fig. 7.19).

In addition to glutamine synthetase the *glnALG* operon encodes the structural genes *glnL* and *glnG* whose products are the regulatory proteins **NtrB** and **NtrC** respectively. (A different nomenclature for the two proteins is often used, with NtrC termed NR_I and NtrB termed NR_{II}.) The *glnALG* operon is characterized by the presence of three promoters, Ap1, Ap2 and Lp. There is a Rho-independent terminator upstream of the Lp promoter near the end of the *glnA* gene which significantly reduces readthrough from *glnA* into the genes *glnL* and *glnG*. The promoters Ap1 and Lp are σ^{70}-dependent promoters which are located 180 base pairs upstream of the *glnA* gene or 32 base pairs upstream of the *glnL* gene, respectively. In contrast, the promoter Ap2, located 73 base pairs upstream of glnA, is a typical σ^{54}-dependent promoter (see Section 2.3.4). Instead of the standard σ^{70} –10 and –35 consensus elements, σ^{54} promoters contain conserved

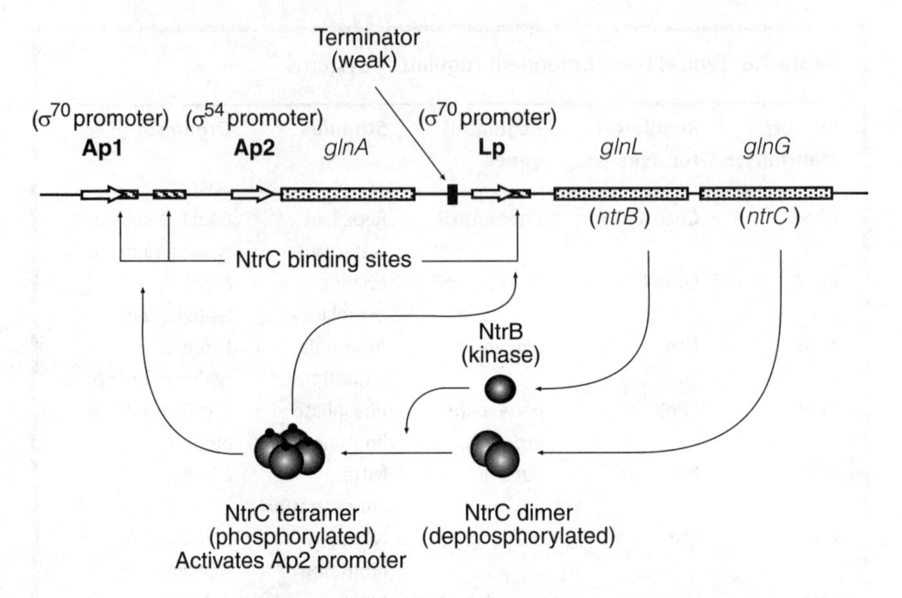

Figure 7.19 Structure of the *glnALG* operon. The *glnALG* operon encodes the genes for glutamine synthetase (*glnA*) and for the two regulatory proteins NtrB (*glnL*) and NtrC (*glnG*). Transcription of the operon is directed from three promoters, the Eσ^{70}-dependent promoters Ap1 and Lp and the Eσ^{54}-dependent promoter Ap2. The promoters are depicted as open arrows. A Rho-independent terminator (black bar) at the end of the *glnA* gene prevents efficient readthrough into the *glnL* and *glnG* genes. Three binding sites for the oligomeric NtrC regulator are found which overlap with the promoters Ap1 and Lp (shaded boxes). The ability of NtrC to activate transcription from the Ap2 promoter depends on the prior phosphorylation by NtrB (phosphate residues are marked as small black dots). NtrC and NtrB are typical members of two-component regulatory proteins, with NtrB acting as sensor kinase and NtrC as response regulator.

sequences at –12 and –24. Moreover, as opposed to σ^{70} promoters, transcription from σ^{54}-dependent promoters is characterized by a number of features typical of eukaryotic transcription. First, σ^{54} promoters require ATP hydrolysis for the isomerization from the closed to the open initiation complexes. Second, σ^{54} promoters need activator proteins to form productive initiation complexes and, third, σ^{54}-dependent promoters are preceded by upstream sequences which function as activator binding sites. These binding sequences resemble very much eukaryotic **enhancer sequences**. They can be moved upstream or downstream of the promoter by more than 1000 base pairs without loss of the activating function. Hence, the upstream sequences of σ^{54} promoters are termed **enhancer-like elements**. In case of the *glnALG* operon the activator binding to the enhancer-like elements is NtrC. Binding of NtrC occurs to three sites with the sequence **GCACN$_5$TGGTGC**. NtrC binding sites overlap the –35 region, the transcriptional start site of the Ap1 promoter, and in addition, the –10 element of the Lp promoter.

NtrC and NtrB are typical members of the two-component regulatory proteins. NtrC is the response regulator and NtrB functions as sensor kinase which can both phosphorylate and dephosphorylate NtrC. The phosphate is transferred from a histidine at position 139 to an aspartate of NtrC. NtrC has the characteristic modular structure of regulators with a central domain of approximately 240 amino acid residues which contains an ATP binding motif and comprises the activation region. The roughly 120 residue long N-terminal domain contains the phosphorylation site (receiver domain), including a characteristic aspartate residue at position 54. The C-terminal domain (about 60 residues) comprises the DNA-binding region which is characterized by a helix-turn-helix motif similar to that found in the FIS protein (see Fig. 7.17 and Section 7.5.2). The components for the expression and regulation of the *glnALG* operon are summarized in Table 7.7.

How is transcription of the *glnALG* operon modulated in response to the availability of nitrogen? In case of excess ammonia, transcription of *glnA* is initiated at promoter Ap1, and transcription of the *glnLG* genes starts at the Lp promoter. Transcription from these promoters is limited by NtrC bound to the overlapping sites. Under conditions of limiting ammonia immediate activation of the σ^{54}-controlled Ap2 promoter occurs. Activation is mediated by NtrC which is controlled by NtrB through reversible phosphorylation. Only a few molecules of NtrC are present in the cell in an excess of ammonia. This number increases significantly under nitrogen deprivation. Phosphorylation of NtrC by NtrB is a strict requirement for the ATPase activity of NtrC which mediates open complex formation. Phosphorylation is not necessary for the binding of NtrC to the enhancer-like sequence, however.

Table 7.7 Components for the expression and regulation of the *glnALG* operon

Gene	Alternative designation	Protein	Alternative designation	Function
glnA		GlnA	Glutamine synthetase	Sensor
glnB		P_{II}		Signal transducer
glnD		GlnD	UTase/UR	Signal transducer
glnE		GlnE	AT	Adenylyl transferase
glnF	*ntrA* *rpoN*	RpoN	NtrA/σ^{54}	Alternative sigma factor
glnG	*ntrC*	NtrC	NR_I	Transcriptional regulator
glnL	*ntrB*	NtrB	NR_{II}	Kinase/phosphatase

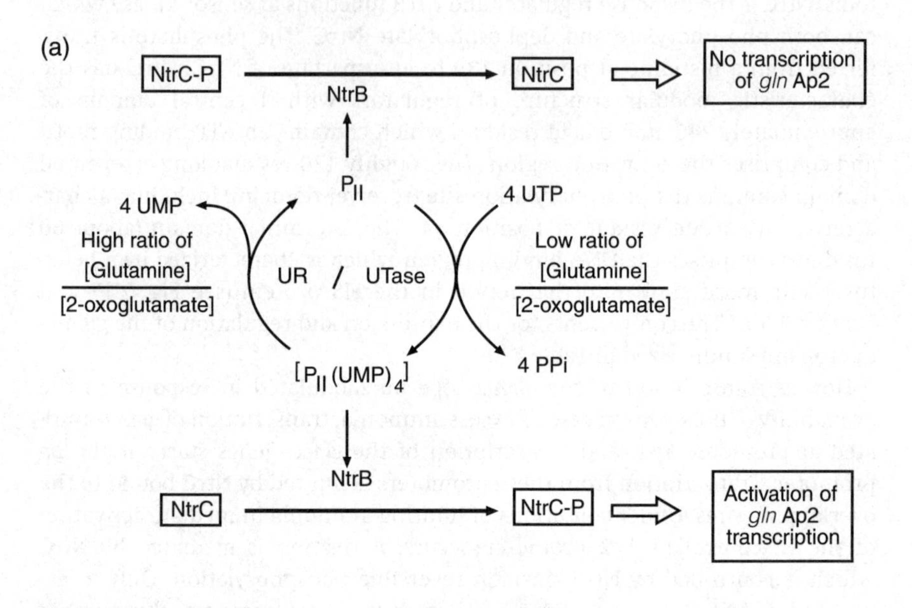
(a)
NtrC-P
NtrC
No transcription of *gln* Ap2
NtrB
P_{II}
4 UMP
4 UTP
High ratio of
[Glutamine]
[2-oxoglutamate]
UR / UTase
Low ratio of
[Glutamine]
[2-oxoglutamate]
$[P_{II}(UMP)_4]$
4 PPi
NtrB
NtrC
NtrC-P
Activation of *gln* Ap2 transcription

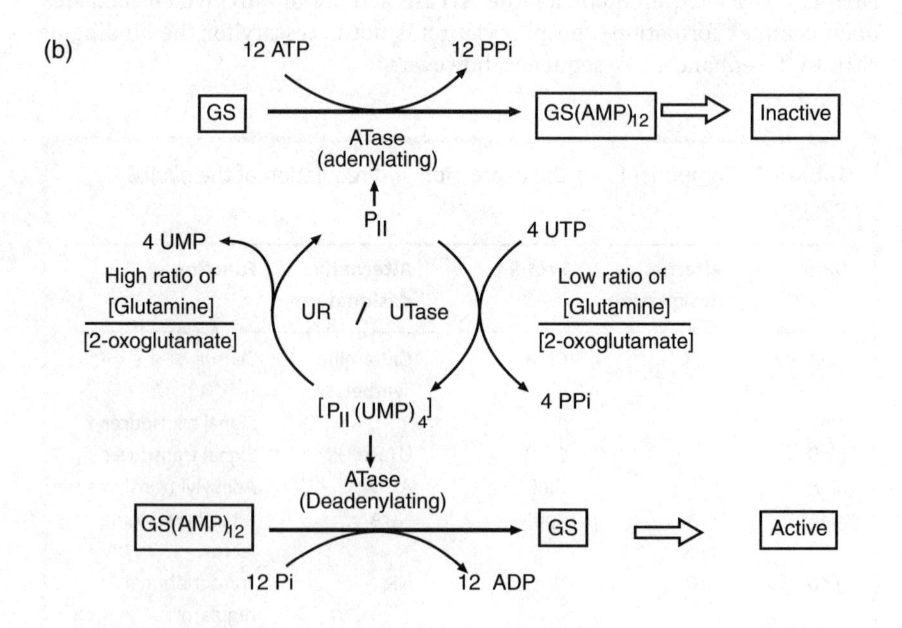
(b)
12 ATP
12 PPi
GS
$GS(AMP)_{12}$
Inactive
ATase
(adenylating)
P_{II}
4 UMP
4 UTP
High ratio of
[Glutamine]
[2-oxoglutamate]
UR / UTase
Low ratio of
[Glutamine]
[2-oxoglutamate]
$[P_{II}(UMP)_4]$
4 PPi
ATase
(Deadenylating)
$GS(AMP)_{12}$
GS
Active
12 Pi
12 ADP

Phosphorylated NtrC (P-NtrC) binds specifically to the Eσ^{54} RNA polymerase. The two NtrC binding sites upstream of the Ap2 promoter are too far away from the polymerase at the Ap2 promoter to allow direct contact between the activators and the polymerase without considerable deformation of the intervening DNA. The mechanism of activation of the *gln* Ap2 promoter by NtrC thus involves **DNA loop** formation, which increases the local concentration of the activators near the RNA polymerase (see Section 7.4). Such loops can be visualized in electron micrographs. Displacement of the activator binding sites by as much as 1000 base pairs upstream or downstream of the Ap2 promoter still allows specific P-NtrC–RNA polymerase contact and therefore does not notably diminish the activating function. The upstream region of the *gln* Ap2 promoter can therefore be considered as a true *enhancer-like* sequence. Although phosphorylation of NtrC does not increase the affinity of the protein for the single binding site, it induces cooperative interaction between the phosphorylated dimers resulting in the binding of tetramers or higher oligomers which may enhance activation.

How is the signal for the regulation of the synthesis of the genes necessary for nitrogen utilization generated and transferred? Clearly, the list of regulatory

Figure 7.20 Scheme for the regulation of nitrogen-fixation genes. (a) Transcription of the *gln* Ap2 promoter depends on the phosphorylation of NtrC by NtrB. NtrB is also responsible for the dephosphorylation of NtrC. This reaction requires the additional activity of the protein P_{II}. The activity of P_{II} is modulated in turn by the reversible addition of four UMP residues, catalysed by a uridylyl transferase (UTase), which also acts as uridylyl-removing enzyme (UR). Only the deuridylated P_{II} enzyme supports NtrB-dependent dephosphorylation of NtrC. Removal or addition of uridylyl residues depends on the ratio of glutamine to 2-oxoglutamate in the cell, which represents the nitrogen sensor. A high ratio of glutamine to 2-oxoglutamate activates the uridylyl removing activity and thereby causes P_{II} to support NtrB-dependent dephosphorylation of NtrC. As a consequence, transcription from the *gln* Ap2 promoter is inhibited. In contrast, a low ratio of glutamine to 2-oxoglutamate activates the UTase, which results in uridylated P_{II} enzyme, inactive to support NtrB-dependent dephosphorylation. As a consequence NtrC is phosphorylated by NtrB and transcription from the *gln* Ap2 promoter is activated. (b) Additional regulation occurs at the level of glutamine synthetase (GS) activity. The addition of 12 adenylyl groups (AMP) by adenylyltransferase (ATase) inactivates the enzyme. Again, the P_{II} protein is a modulator of the ATase activity. Uridylated P_{II} protein activates the deadenylating activity of ATase. AMP residues are removed from glutamine synthetase which renders the enzyme active. A low ratio of glutamine to 2-oxoglutamate is required for this reaction. At a high ratio of glutamine to 2-oxoglutamate, however, deuridylation occurs and the adenylating activity of P_{II} is supported. Transfer of 12 AMP residues to glutamine synthetase renders the enzyme inactive. The scheme is adapted from the article by Magasanik (1987).

components outlined above is incomplete and additional regulators are necessary to understand the full regulatory cycle. One important aspect concerns the removal of the phosphate group from P-NtrC. Although NtrB is responsible for the removal of the phosphate it requires an additional small protein for rapid dephosphorylation. This protein is termed P_{II}, the gene product of the *glnB* gene. The activating effect of P-NtrC is rapidly lost because of dephosphorylation caused by NtrB in the presence of P_{II}. P_{II} itself is modified by the addition or removal of uridylyl groups. A **uridylyl transferase**, the product of the *glnD* gene, causes the addition of four UMP residues from UTP to P_{II} (UTase activity). The resulting P_{II}-UMP will not catalyse dephosphorylation of NtrC and the nitrogen regulated promoters can be activated by P-NtrC. The UMP residues can alternatively be removed from P_{II} by a uridylyl group-removing activity (UR activity). The unmodified P_{II} in turn causes a rapid dephosphorylation of NtrC and the nitrogen-regulated promoters are no longer activated. Whether P_{II} is uridylylated or deuridylylated depends on the ratio of two metabolites, namely **glutamine** and **2-oxoglutarate**. Excess 2-oxoglutarate activates uridylylation and excess glutamine activates deuridylylation. Thus, if the cellular concentration of 2-oxoglutarate is high, P_{II}–UMP is formed. As a consequence, P-NtrC will not be dephosphorylated and can activate nitrogen regulated promoters. If, on the other hand, the concentration of glutamine is high, deuridylylated P_{II} will cause NtrB to remove the phosphate from P-NtrC, which causes cessation of transcription from the nitrogen regulated promoters (Fig. 7.20).

Moreover, regulation occurs also at the level of the activity of glutamine synthetase. Some of the subunits of the multimeric enzyme can be modified by the addition of adenylyl groups to tyrosines via an adenylyl transferase. Transfer of the adenylyl groups renders the glutamine synthetase inactive. The adenylyl transferase in turn is activated by P_{II}. Hence, the presence of P_{II} affects both the activity and the synthesis of glutamine synthetase.

It is not only the *glnALG* operon that is activated by NtrC. Regulation by NtrC occurs at several other promoters as well. These include, for instance, the operons for the uptake of histidine (*hisJQMP*) or glutamine (*glnHPQ*), respectively. Interestingly, the *glnH* p2 promoter requires a second regulatory protein in addition to P-NtrC for efficient transcription initiation. This regulator is the IHF protein (see Section 7.5.1). A binding site for the DNA-bending protein IHF is present between the NtrC site and the *glnH* p2 promoter. Binding of IHF is obligatory for activation and obviously helps to bend the DNA in a suitable conformation, allowing contact between NtrC and the RNA polymerase. If the binding sites for the activator, the RNA polymerase or IHF are changed within that system by including or deleting a number of intervening base pairs corresponding to half a helical turn, transcription will no longer be activated but will rather be inhibited. As observed in many other cases, the efficiency of regulators which exert their activity by DNA bending depends strongly on the position in which the binding sites are located with respect to the promoter (see Fig 7.21).

(a) *glnH* p2 regulatory region

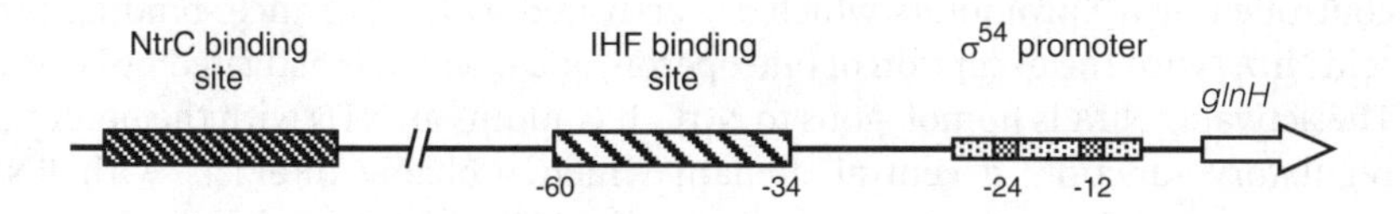

(b)

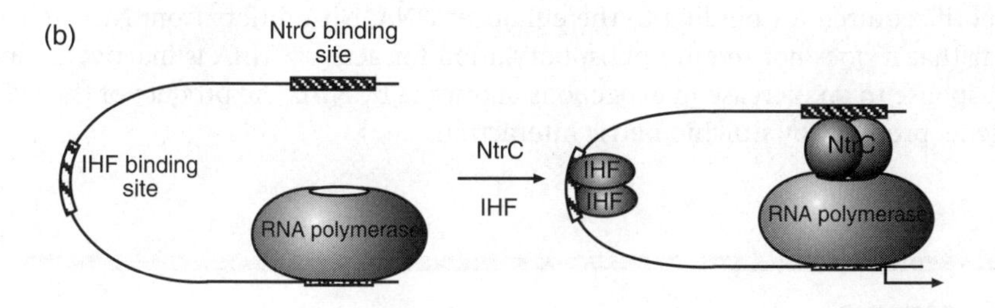

Displacement of IHF or NtrC binding sites by half helical turn (5 bp)

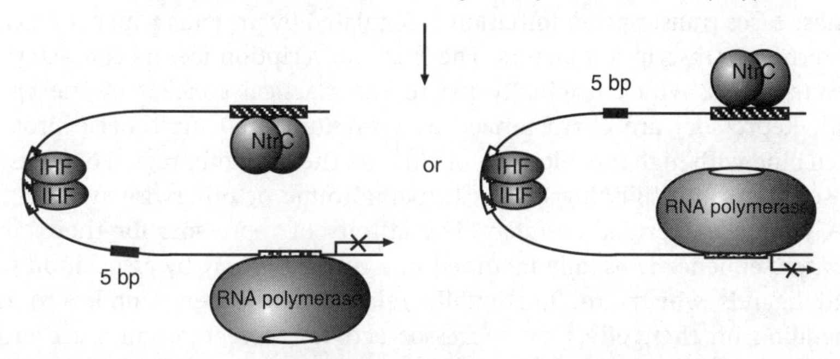

Figure 7.21 Regulation of the *glnH* p2 promoter. (a) The regulatory region of the *glnH* gene is presented. The σ^{54}-dependent promoter with the consensus sequences –12 and –24 is indicated by stippled bars. A binding site for IHF, located between positions –34 and –60, is denoted by a shaded box. An NtrC binding site further upstream is indicated by a dark shaded box. (b) Formation of an active transcription complex through the action of IHF and NtrC is schematically presented. RNA polymerase (large grey sphere) binds to the *glnH* p2 promoter. An NtrC oligomer bound to its DNA-binding site makes protein–protein contacts with a region of RNA polymerase (white oval), causing the DNA to form a loop. Binding of IHF to its binding site in the centre of the DNA curvature enables the strongly bent DNA conformation necessary for loop formation. The lower part of the figures indicates that introduction of five base pairs (half a helical turn) between the IHF binding site and the promoter or between the binding sites for IHF and NtrC, changes the orientation of the DNA-associated macromolecules such that stable loop formation and concomitant promoter activation is lost.

IHF is also involved in supporting a proper DNA loop of the enhancer-like sequence upstream of the *nif* promoters which control the expression of the nitrogen-fixation genes. For instance in *Klebsiella pneumoniae*, there are altogether 19 *nif* genes encoded in seven continuous operons. All operons are controlled by σ^{54} promoters which are activated by the enhancer-binding protein **NifA**, with the exception of one operon, *nifLA*, which is activated by P-NtrC. The activator NifA is homologous to NtrC. It contains an NTD with the potential regulatory function, a central domain which probably interacts with RNA polymerase and comprises a sequence for ATP binding and hydrolysis. As expected for activators of σ^{54} promoters, ATP hydrolysis is required for open complex formation. The CTD, as in the case of NtrC, contains a helix-turn-helix motif required for binding to the enhancer DNA. NifA differs from NtrC in so far that it does not require phosphorylation for activity. NifA is inactivated in response to an increase in exogenous ammonia by **NifL**, the product of the *nifL* gene, probably by stoichiometric interaction.

Summary

In most cases transcription initiation is regulated by proteins which are generally termed transcription factors. The first transcription factors characterized were *repressors*, which originally led to the classical concept of the *operon model*. Repressors are characterized as even-numbered multimeric proteins which bind with high specificity at, or close to, the promoter region of a specific transcription unit. Binding occurs to palindromic or otherwise symmetrical DNA sequences, termed *operators*. The affinity of repressors for their target operator sequence is usually modified in a reversible way by association with small ligands which are functionally related to the genes under control. Depending on their effect on repressor activity these molecules are either termed *inducers* or *co-repressors*. In some cases repressors are also inactivated by proteolysis. According to the classical concept the function of repressors is explained by sterical hindrance of RNA polymerase binding to the promoter (binding competition). The efficiency of repression can vary from a complete shut-down to a moderate reduction and depends largely on the affinity of the repressors to their respective operator sites. Generally, transcription factors contain different *functional domains* which are reflected in a *modular organization* of the molecules. They contain specialized domains responsible for DNA-binding, oligomerization, activation or repression, and usually an allosteric binding site for a corepressor or an inducer compound. In many cases functional domains are interchangeable between different transcription factors, resulting in hybrid molecules with mixed properties. Generally, transcription factors have dual functions and can act as repressors at some promoters and as activators at others. The property of transcription factors to act as repressor or activator depends to a large extent on the location of the binding site relative to the transcription start position. Preferred sites for activation are between nucleotide positions −80 and −30, while repressors bind preferentially downstream from nucleotide position −30. Generally transcription factors exert their function by changing the affinity of RNA polymerase for the respective promoter (K_B) or, through conformational changes, they alter later steps of the transcription cycle, such as isomerization (k_2) or promoter escape. Many activators function by enhancing the affinity of RNA polymerase binding to the promoter. This activation through *recruitment* can be achieved by activator contacts to either the σ or the α subunit (αCTD) of RNA polymerase.

According to common structural and functional properties, transcription factors can be grouped into several different families. Examples are the LacI/GalR or the LysR families, which comprise a large number of regulators, both activators and repressors. Some regulator proteins, e.g. the transcription factor TyrR, control several different transcription units which are functionally related and collectively describe a regulon. The activity of TyrR to function as an activator or repressor is modified by different cofactors (aromatic amino acids) and the presence or absence of ATP.

A large number of genes involved in catabolism are regulated by a unique protein, the *catabolite regulatory protein* (*CRP*). This regulator requires the presence of the obligatory coregulator *cyclic adenosine monophosphate* (*cAMP*). Transcription units under the control of CRP have been grouped into three different classes, depending on the sites of CRP–cAMP interaction with the upstream promoter regions. Class I CRP-dependent promoters are characterized by binding sites that centre around position −61, Class II CRP-dependent promoters have sites around −41, and at Class III CRP-dependent promoters the regulator binds around position −100. Activation at Class I sites involves a direct interaction between CRP and the C-terminal domain of the RNA polymerase α subunit (αCTD). At Class II promoters one subunit of the CRP dimer is bound to the αCTD and the other, more promoter proximal CRP subunit, interacts with the N-terminal domain of α (αNTD). Regulation at Class III sites requires the activity of at least one additional transcription factor. This type of regulation, with two transcription factors interacting at distant DNA sites, usually involves the formation of *DNA loops* and can be characterized as *action-at-a-distance*. The *galETK* and the *araBAD* operons, with the *GalR* and the *AraC* proteins as additional transcription factors, are well studied examples of the formation of DNA loop structures during regulation.

A group of small abundant proteins is known to associate with the bacterial nucleoid and to change the overall structure and topology of DNA. These proteins have often been designated as histone-like, although they do not share structural or functional properties with eukaryotic histones. They may better be characterized as *DNA structuring* proteins. Notable representatives are the proteins IHF, HU, FIS and H-NS. These DNA structuring proteins participate in

DNA rearrangement or recombination reactions but they are also known to function as specific transcription factors in many cases. Among all the DNA-binding proteins the heterodimeric IHF causes the strongest bend in the target DNA (U-turn). It binds to the minor groove of DNA, shows considerable functional homology to the eukaryotic HMG-box or TATA box binding proteins (TBP). IHF can often be functionally replaced by HU, which has a similar structure. FIS is known to be involved in regulation of the growth phase. Its cellular concentration strongly varies with the growth phase and its most outstanding function is to activate stable RNA synthesis. FIS binds to the major groove of DNA and usually has several regularly spaced binding sites. The protein H-NS is involved in the transcriptional regulation of a large number of different genes. The activity of H-NS is often linked to stationary phase expression or situations of cellular stress. H-NS does not recognize a specific DNA primary structure but requires curved DNA for specific binding. Like the other DNA structuring proteins H-NS is considered to be an *architectural component* of many transcription complexes. It is involved in the shutdown of stable RNA transcription at slow growth or under stress conditions. H-NS, which predominantly acts as a repressor, functions in several cases as a direct antagonist to the activator FIS.

A large group of transcriptional regulators is known to function as *two-component regulatory systems*. These proteins are involved in signal transduction through the transfer of phosphoryl groups. Two component regulatory systems generally consist of a pair of proteins, a *sensor kinase* (histidine kinase) and a suitable *response regulator*. Response regulators are grouped into three families of similar structure: the OmpR like, the FixJ like and the NtrC like regulators. Response regulators are composed of a *receiver domain*, with an invariant aspartate residue to which a phosphate is transferred from the sensor kinase, and an output domain, which normally interacts with the transcription complex.

Many genes necessary for the fixation of nitrogen are controlled by the alternative σ factor σ^{54}. The regulation of σ^{54}-dependent promoters differs from σ^{70}-dependent promoters and requires specific *activator proteins* and the hydrolysis of *ATP* to form open complexes. The activators bind to DNA regions, usually located upstream of σ^{54}-dependent promoters. These binding sites resemble eukaryotic *enhancer elements* and can be moved several hundred base pairs upstream or downstream without loss of activity. The regulation of the *glnALG* operon, for example, involves the activator *NtrC*, which is a member of the two-component regulatory systems. NtrC must be phosphorylated by *NtrB* to activate the Ap2 promoter. Phosphorylated NtrC binds to two sites far distant from the promoter. Activation is brought about by DNA loop formation between the activators and the transcription complex. The activation of the σ^{54}-dependent *glnHPQ* promoter, which controls the genes necessary for the uptake of glutamine, requires a second regulatory protein in addition to phosphorylated NtrC. In this case DNA loop formation is facilitated through the interaction of IHF, which helps to bend the DNA in a favourable orientation for NtrC–RNA polymerase interaction.

References

Busby, S. and Ebright, R. H. (1997) Transcription activation at Class II CAP dependent promoters. *Molecular Microbiology* **23**: 853–9.

Gonzáles-Gil, G., Kahmann, R. and Muskhelishvili, G. (1998) Regulation of *crp* transcription between distinct nucleoprotein complexes. *EMBO Journal* **17**: 2877–85.

Gralla, J. D. and Collado-Vides, J. (1996) Organization and function of transcription regulatory elements. In: Neidhard, F. C., Curtiss III, R., Ingraham, J. L. *et al.* (eds) *Escherichia coli and Salmonella Cellular and Molecular Biology*, Vol. 1. Washington DC: ASM Press, pp. 1232–45.

Gross, R., Aricò, B. and Rappuoli, R. (1989) Families of bacterial signal-transducing proteins. *Molecular Microbiology* **3**: 1661–7.

Hanamura, A. and Aiba, H. (1992) A new aspect of transcriptional control of the *Escherichia coli crp* gene: positive autoregulation. *Molecular Microbiology* **6**: 2489–97.

Jacob, F. and Monod, J. (1961). Genetic regulatory mechanisms in the synthesis of proteins. *Journal of Molecular Biology* **3**: 318–56.

Magasanik, B. (1988) Reversible phosphorylation of an enhancer binding protein regulates the transcription of bacterial nitrogen utilization genes. *Trends in Biochemical Sciences* **13**: 475–9.

Parkinson, J. S. and Kofoid, E. C. (1992) Communication modules in bacterial signaling proteins. *Annual Reviews of Genetics* **26**: 71–112.

Parkinson, G., Gunaserka, A., Vojtechovsky, J., Zhang, X., Kunkel, T. A. and Ebright, R. H. (1996) Aromatic hydrogen bond in sequence-specific protein DNA recognition. *Nature Structural Biology* **3**: 837–41.

Pittard, A. J. and Davidson, B. E. (1991) TyrR protein of *Escherichia coli* and its role as repressor and activator. *Molecular Microbiology* **5**: 1585–92.

Rice, P. A., Yang, S.-W., Mizuuchi, K. and Nash, H. (1996) Crystal structure of IHF–DNA complex: a protein-induced DNA U-turn. *Cell* **27**: 1295–1306.

Schell, M. A. (1993) Molecular biology of the LysR family of transcriptional regulators. *Annual Reviews of Microbiology* **47**: 597–626.

Further reading

Adhya, S. (1989) Multipartite genetic control elements: communication by DNA loop. *Annual Reviews of Genetics* **23**: 227–50.

Albright, L. M., Huala, E. and Ausubel, F. M. (1989) Prokaryotic signal transduction mediated by sensor and regulator protein pairs. *Annual Reviews of Genetics* **23**: 311–36.

Atlung, T. and Ingmer, H. (1997) H-NS: a modulator of environmentally regulated gene expression. *Molecular Microbiology* **24**: 7–17.

Bellomy, G. R. and Record, M. T. (1990) Stable DNA loops *in vivo* and *in vitro*: roles in gene regulation at a distance and biophysical characterization of DNA. *Progress in Nucleic Acids Research and Molecular Biology* **39**: 81–128.

Bewley, C. A., Gronenborn, A. M. and Clore, G. M. (1998) Minor groove-binding architectural proteins: structure, function, and DNA recognition. *Annual Reviews of Biophysics and Biomolecular Structures* **27**: 105–31.

Bourret, R. B., Borkovich, K. A. and Simon, M. I. (1991) Signal transduction pathways involving protein phosphorylation in prokaryotes. *Annual Reviews of Biochenistry* **60**: 401–41.

Busby, S. and Ebright, R. H. (1997) Transcription activation at class II CAP-dependent promoters. *Molecular Microbiology* **23**: 853–859.

Choy, H. and Adhya, S. (1996) Negative control. Vol. 1. Neidhard, F. C., Curtiss III, R., Ingraham, J. L., *et al.* (eds) *Escherichia coli and Salmonella Cellular and Molecular Biology*, Washington DC: ASM Press. pp. 1287–99.

Collado-Vides, J., Magasanik, B. and Gralla, J. D. (1991) Control site location and transcriptional regulation in *E. coli*. *Microbiological Reviews* **55**: 371–94.

Dove, S. L., Young, K. J. and Hochschild, A. (1997) Activation of prokaryotic transcription through arbitrary protein–protein contacts. *Nature* **386**: 627–30.

Ebright, R. H. (1993) Transcription activation at class I CAP-dependent promoters. *Molecular Microbiology* **8**: 797–802.

Finkel, S. E. and Johnson, R. C. (1992) The FIS protein: it's not just for DNA inversion anymore. *Molecular Microbiology* **6**: 3257–65.

Freundlich, M., Ramani, N., Mathew, E., Sirko, A. and Tsui, P. (1992) The role of integration host factor in gene expression in *Escherichia coli*. *Molecular Microbiology* **6**: 2557–63.

Gallegos, M., Michàn, C. and Ramos, J. (1993) The XylS/AraC family of regulators. *Nucleic Acids Research* **21**: 807–10.

Garges, S. (1994). Activation of transcription in *Escherichia coli*. The cyclic AMP receptor protein. In: Conaway, R. C. and Conaway, J. W. (eds) *Transcription: Mechanisms and Regulation*, New York, Raven Press, pp. 343–52.

Goosen, N. and van de Putte, P. (1995) The regulation of transcription initiation by integration host factor. *Molecular Microbiology* **16**: 1–7.

Gralla, J. D. and Collado-Vides, J. (1996) Organization and function of transcription regulatory elements. Neidhard, F. C., Curtiss III, R., Ingraham, J. L. *et al.* (eds) *Escherichia coli and Salmonella Cellular and Molecular Biology*. Vol. 1. Washington DC: ASM Press, pp. 1232–45.

Gross, R., Aricò, B. and Rappuoli, R. (1989) Families of bacterial signal-transducing proteins. *Molecular Microbiology* **3**: 1661–7.

Hakenbeck, R. and Stock, J. B. (1996) Analysis of two-component signal transduction systems involved in transcriptional regulation. *Methods in Enzymology* **273**: 281–300.

Hilchey, S., Xu, J. and Koudelka, G. B. (1997) Indirect effects of DNA sequence on transcriptional activation by prokaryotic DNA binding proteins. In Eckstein, F. and Lilley, D. M. J. (eds) *Nucleic Acids and Molecular Biology*. Vol. II. Berlin–Heidelberg, Springer Verlag. pp. 115–34.

Hu, J. C. (1995) Repressor fusion as a tool to study protein–protein interactions. *Structure* **3**: 431–3.

Kolb, A., Busby, S., Buc, H., Garges, S. and Adhya, S. (1993) Transcriptional regulation by cAMP and its receptor protein. *Annual Reviews of Biochemistry* **62**: 749–95.

Kustu, S., North, A. K. and Weiss, D. S. (1991) Prokaryotic transcriptional enhancers and enhancer-binding proteins. *Trends in Biochemical Sciences* **16**: 397–402.

Magasanik, B. (1988) Reversible phosphorylation of an enhancer binding protein regulates the transcription of bacterial nitrogen utilization genes. *Trends in Biochemical Sciences* **13**: 475–9.

Matthews, K. S. and Nicols, J. C. (1998) Lactose repressor protein: functional properties and structure. *Progress in Nucleic Acid Research and Molecular Biology* **55**: 127–64.

Ninfa, A. J. (1996) Regulation of gene transcription by extracellular stimuli. In: Neidhard, F. C., Curtiss III, R., Ingraham, J. L *et al.* (eds). *Escherichia coli and Salmonella Cellular and Molecular Biology*. Vol. 1. Washington DC, ASM Press, pp. 1246–62.

Pabo, C. O. and Sauer, R. T. (1992) Transcription factors: structural families and principles of DNA recognition. *Annual Reviews of Biochemistry* **61**: 1053–95.

Parkinson, J. S. (1993) Signal transduction schemes of bacteria. *Cell* **73**: 857–71.

Parkinson, J. S. and Kofoid, E. C. (1992) Communication modules in bacterial signaling proteins. *Annual Reviews of Genetics* **26**: 71–112.

Pittard, A. J. and Davidson, B. E. (1991) TyrR protein of *Escherichia coli* and its role as repressor and activator. *Molecular Microbiology* **5**: 1585–92.

Ptashne, M. and Gann, A. (1997) Transcriptional activation by recruitment. *Nature* **386**: 569–77.

Rippe, K., von Hippel, P. and Langowski, J. (1995) Action at a distance: DNA-looping and initiation of transcription. *Trends in Biochemical Sciences* **20**: 500–6.

Schell, M. A. (1993) Molecular biology of the LysR family of transcriptional regulators. *Annual Reviews of Microbiology* **47**: 597–626.

Schleif, R. (1992) DNA looping. *Annual Reviews of Biochemistry* **61**: 199–223.

Schmid, M. (1990) More than just "histone-like" proteins. *Cell* **63**: 451–3.

Stock, J. B., Ninfa, A. J. and Stock, A. M. (1989) Protein phosphorylation and regulation of adaptive responses in bacteria. *Microbiological Reviews* **53**: 450–90.

Travers, A. A. and Muskhelishvili, G. (1998) DNA microloops and microdomains: a general mechanism for transcription activation by torsional transmission. *Journal of Molecular Biology* **279**: 1027–43.

Ussery, D. W., Hinton, J. C. D., Jordi, B. J. A. M., *et al.* (1994) The chromatin-associated protein H-NS. *Biochimie* **76**: 968–80.

Werner, M. H. and Burley, S. K. (1997) Architectural transcription factors: proteins that remodel DNA. *Cell* **88**: 733–6.

8 Regulatory networks

This chapter presents a collection of regulatory circles and interlinked control systems rather than individual regulatory mechanisms affecting a single operon. The examples given comprise the physiological adaptation of the cell to internal growth signals but also the response to external changes, for instance nutritional deprivation or during stress or damaging conditions. Some of the networks described below control genes that are related in function while in other cases control over a large number of apparently unlinked genes is involved. It should be emphasized, however, that the networks presented are generally interlinked regulatory units which are not only controlled by a single stimulus. Thus, they should not be understood as independent control circuits; for many of these networks the term **global network** is justified. Although many of the interrelations between the individual regulatory systems may already appear rather complicated, it is obvious that the real degree of complexity and the intricate coupling between the networks is far more complex than it might appear in this description. The unexpected coupling in the expression of apparently functionally unrelated operons indicates that almost nothing happens in the cell without affecting the remaining cellular activities and reactions. The highly redundant control of many central and important genes is a further indication for the interconnection of different regulatory loops. While this has not been explicitly emphasized in the first six chapters, where networks which were selected according to their different stimuli were described, an attempt to indicate the highly complex and tuned interplay of many different control systems and stimuli is made in this last chapter describing the regulation of rRNA synthesis.

A few explanatory definitions often used in connection with regulatory networks are required although some of these terms have been used frequently in the preceding chapters. Gene regulation can be organized in several steps of increasing complexity. Genes which are regulated may be grouped in **operons**, **regulons**, **modulons** or **stimulons**. These terms, describing regulatory units, are generally used in the following sense.

In an *operon* a number of genes, often related in function, is encoded as a unit (transcription unit), and transcriptionally coregulated. The polycistronic arrangement of genes in operons is typical of prokaryotes. Proteins that belong to the same metabolic pathway are normally organized in operons. In the past chapters many examples in which the subject of regulation is an operon or a transcription unit have been encountered. Typical examples are the *lacZYA*, the *galETK*, or the *araBAD* operons, which have been described in Chapter 7.

A *regulon* describes the organization of several independent operons, which are controlled in a coordinated way. Regulons usually share a common regulatory protein. Coordinated regulation within a regulon does not imply that individual operons are always expressed in a strictly parallel manner. This means the extent of transcription regulation between different operons can vary widely. Moreover, transcription of one operon may be activated while another operon from the same regulon can simultaneously be repressed by the same regulator protein. Examples of regulons comprise genes which are coordinately regulated by alternative σ factors or the TyrR regulon, for instance, which is presented in Section 7.3.1. Further examples, such as the heat-shock response, the SOS response or the Lrp regulon, are outlined below.

Modulons describe independent operons which may belong to different regulons but respond to a common regulator. Thus, modulons characterize a regulatory level above the regulon. Individual control of operons in different regulons is maintained by individual inducing ligands. Modulons therefore describe global regulatory units. The standard example of a modulon is the set of metabolic genes encoding the catabolic enzymes which are regulated by the cAMP–CRP complex, for instance.

The term *stimulon* describes a collection of genes that respond to a common stimulus independent of their regulatory organization or mechanism of activation or inhibition. The definition is therefore *operational* and stimulons are thus composed of heterogeneous members. For example, raising the growth temperature will induce the regulon controlled by the alternative σ factor σ^{34} (see Section 2.3.2). Many more genes not under the control of σ^{34} are also affected by heat, however, and will be activated or repressed. These genes may be members of different operons or regulons, and the mechanism of regulation or the nature of the regulator may be different in each case.

The first network in the chapter below describes the SOS response, which characterizes the coordinated expression of repair functions when bacteria encounter conditions damaging to DNA. The second regulon presented is the heat-shock response. During the heat-shock response alternative σ factors are responsible for the transcription of operons encoding families of proteins to protect the cell from heat damage. A complex change in the metabolism of bacterial cells is observed when exponential growth ceases and cells enter the stationary phase. Stationary phase control therefore comprises the next regulatory network presented below. Much of the stationary phase regulation is brought about by the stationary phase-specific σ factor σ^{s}. There follows a description of the Lrp regulon, whose function in the cell may be summarized as an adaptation of the cell metabolism to the quality of the nutritional growth conditions. Two networks will then be presented which, at least in part, share the same effector molecule although they are physiologically different. One is the global network of the stringent response, which is induced as a consequence of amino acid deprivation and which affects many genes in different ways. The other network is growth rate control, which links the expression of a rather small number of vital genes to the cellular growth rate. In the last section an attempt is made to summarize as many as possible of the different regulatory mechanisms presented so far, describing a very complex example of regulation, namely the synthesis of rRNAs. This chapter aims to demonstrate the many facets of regulation that are involved in the regulation of one group of vital genes and might also be helpful in recalling some of the facts presented in the preceding chapters of the book.

8.1 The SOS response

The **SOS response** was one of the first clear networks of transcriptional regulation identified in bacteria. It comprises a set of coordinated physiological responses induced by DNA damage. The expression of more than 20 genes can be induced when DNA damage occurs. Most of these genes are involved in different mechanisms of DNA repair, like mismatch repair, recombinational repair, excision repair or mutagenesis (error-prone repair) (Table 8.1). Expression of these genes involves the two regulatory proteins **LexA** and **RecA**.

LexA is a typical transcriptional repressor. It has a molecular weight of 22.7 kDa and consists of two domains connected by a flexible hinge. The N-terminal domain (amino acids 1–84) comprises the DNA-binding region, which is slightly different but structurally related to a helix-turn-helix DNA-binding motif. The protein exists as a dimer, and the dimerization domain is located in the C-terminus. A palindromic DNA sequence, termed the SOS box, is recognized by the dimeric repressor. The SOS box has the consensus sequence: TA**CTG**-TATATATATA**CAG**TA. The bold letters indicate more highly conserved nucleotide positions. The SOS box operators are found upstream of the subordinate genes. The number of operators for the different genes can vary between one and three. In addition, the operator positions with respect to the promoters are variable. Some SOS boxes overlap with the -35 hexamer (e.g. *uvrA*), some are located between the –35 and the –10 regions (*recA* or *uvrB*), or only overlap the –10 region (*sulA*, *lexA*), and still others are located downstream of the –10 promoter sequence (*uvrD*). The variability in the number and location of operator sites indicates that the individual genes which are under the control of LexA are not regulated to the same extent and are obviously subject to fine regulation. This is further supported by slight differences in the sequence of the SOS boxes, which cause different affinities of LexA for the individual operators.

Table 8.1 Genes regulated by the SOS response

Gene	Function	Map location (min)	Size (kDa)
lexA	SOS repressor	91	22.7
recA	DNA recombination, co-protease (LexA cleavage), SOS mutagenesis	58	37.8
recN	Repair of DNA double-strand breaks	58	60
recQ	Recombination, resistance to thymineless death	85	68
ruv	Daughter-strand gap repair	41	41
sulA (*sfiA*)	Inhibition of cell division	22	18
umuC	SOS mutagenesis	25	48
umuD	SOS mutagenesis	25	15
uvrA	Excision repair	92	103
uvrB	Excision repair	17	76
uvrD	Excision repair, methyl-directed mismatch repair, DNA helicase	85	75

The second protein involved in the regulation of the SOS response is RecA. This 37.8-kDa protein serves several functions in the cell. RecA is an important protein for homologous recombination. Within the framework of the SOS response, however, RecA mediates the induction of the genes coordinately repressed by LexA. This repression is released by proteolytic cleavage of the LexA repressor protein. The proteolytic activity is mediated by RecA, which is not a classical protease, however. It only facilitates a latent capacity of LexA to autodigest. Cleavage occurs between the amino acid positions Ala111 and Gly112, resulting in two fragments of similar size. In the absence of DNA damage LexA is perfectly stable, although there are about 7000 copies of RecA per cell. The activity of RecA to stimulate LexA cleavage depends on a signal which is related to DNA damage. This signal is almost certainly the presence of single-stranded DNA regions resulting from various forms of DNA damage. The activity of RecA is changed upon binding to such single-stranded DNA sites resulting in an activated conformer designated **RecA***. To mediate LexA cleavage the activated RecA* needs ATP, although hydrolysis of the bound ATP is not required for cleavage. Activation of RecA is reversible and prevails as long as the inducing signal (single-stranded DNA) is abundant. Cleavage of LexA results in derepression of the genes, which are coordinately repressed by LexA (see Table 8.1). Induction is fast, and 3 minutes after a mild ultraviolet (UV) treatment about 80% of the LexA molecules are cleaved. Expression of the coordinately

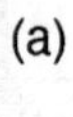
(a)

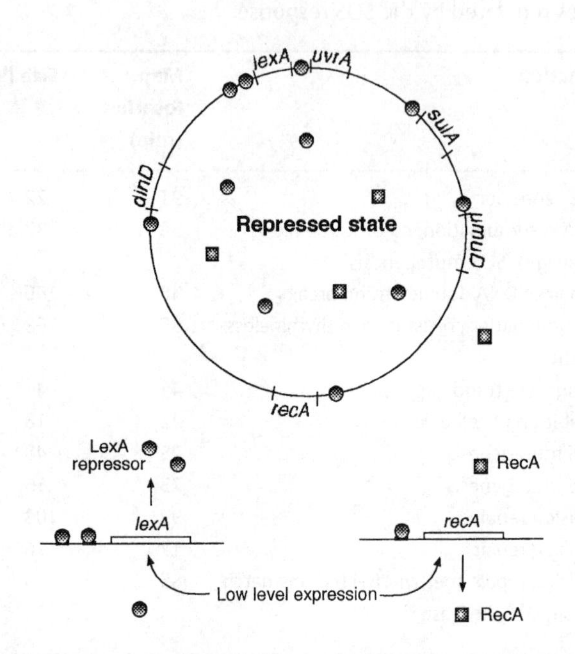
lexA
uvrA
sulA
umuD
recA
dinD
Repressed state
LexA
repressor
RecA
lexA
recA
Low level expression
RecA

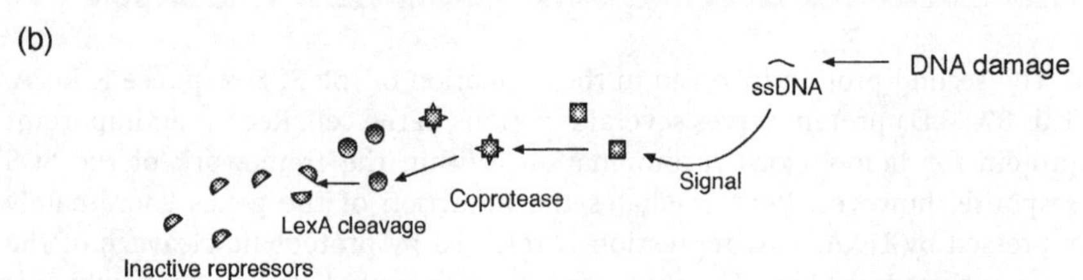
(b)
DNA damage
ssDNA
Signal
Coprotease
LexA cleavage
Inactive repressors

(c)

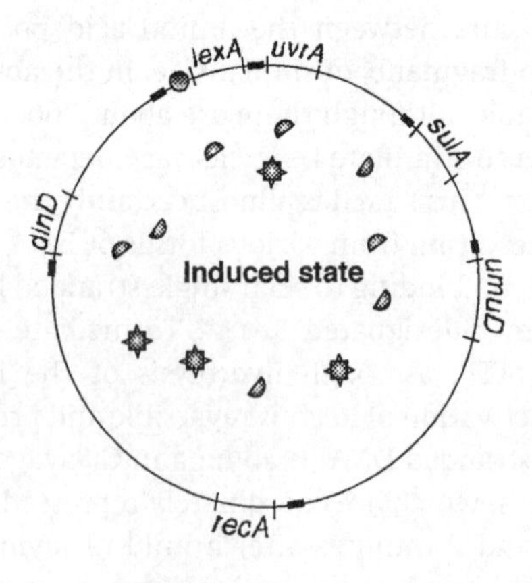
lexA
uvrA
sulA
umuD
recA
dinD
Induced state

regulated SOS genes occurs within 5 minutes of UV irradiation. The induction level for the different SOS genes varies from around fourfold up to 100-fold, and depends on the number of operator sites and location as well as on the affinity of LexA for the respective operators. Moreover, the individual promoter strength or the presence of additional constitutive promoters is important. Not all members of the SOS response are therefore induced in a unique fashion. The response is rather gradual and depends on the amount of intact LexA repressors in the cell. The LexA pool in turn depends on the intensity of the signal induced by the damage. Intermediate induction of the DNA repair genes, between the fully repressed and fully induced situation, can therefore be achieved. For instance, the genes *lexA*, *uvrA*, *uvrB* and *uvrD* are induced first, while the *sulA*, *umuDC* or the *recA* genes are only induced at the full SOS response. The fact that the set of genes regulated by LexA includes the *lexA* and *recA* genes, *lexA* being derepressed first, ensures that the system returns to the repressed state after the signal indicating DNA damage has disappeared from the cell. Within 30–60 minutes after a light UV dose the transcription rate of LexA-controlled genes drops back to the basal levels of the uninduced state.

The activated RecA* is also capable of mediating cleavage of a number of other repressors. For instance, the phage repressors from λ, P22, phage 434 or Φ80 all contain a cleavable internal site within the C-terminal domain. All of these domains share considerable homology with the LexA protein. Phages have thus opportunistically adapted a mechanism of sensing a critical status of the host cell and, by activating their lytic cycle, they may escape before the cell dies.

The inducing signal, which is almost certainly single-stranded DNA, is created as a result of DNA damaging reagents or UV irradiation. Neither UV irradiation nor damaging agents such as intercalating drugs, however, cause

Figure 8.1 The SOS response regulon. (a) The structural arrangement of several genes that are regulated by the SOS response are schematically presented on a circular chromosome (*din* stands for damage-induced genes; see Table 8.1). The figure represents the repressed state with LexA repressor molecules (round grey spheres) bound to the upstream region of the SOS genes. Grey squares denote RecA proteins. The drawing in the lower part indicates that, even in the repressed state, the LexA repressor and the RecA proteins are expressed at a low level. (b) The signal cascade from DNA damage to the inactivation of the LexA repressor is illustrated. DNA damage causes the appearance of small single-stranded DNA fragments in the cell (ssDNA, indicated by a small wavy line). This signal causes a change in the activity of the RecA protein, which is rendered to a coprotease (indicated by spikes on the squares that symbolize the RecA molecules). The RecA coprotease causes cleavage of the LexA repressor molecules (half circles), which are thus inactivated. (c) The induced state is characterised by a high concentration of cleaved LexA repressor molecules and RecA molecules with coprotease activity. No repressor molecules are bound to the promoters of the SOS genes. The first gene to be inactivated when the situation returns to the repressed state is the *lexA* gene, which provides the cell with necessary repressor molecules.

the inducing signal directly. Obviously, the inducing signal is created when damaged DNA undergoes replication. UV irradiation in the absence of replication is therefore not sufficient for induction. The single-stranded DNA segments are produced upon DNA replication when the DNA polymerase encounters a DNA lesion. RecA, in the presence of ATP, apparently binds to such segments and is activated. Interestingly, no activation of RecA occurs from the natural single-stranded regions that are normally present during lagging strand synthesis near the replication fork. These sites are probably covered with the bacterial single-strand binding protein SSB.

RecA-mediated proteolysis has a second effect during the SOS response. One of the subordinate gene products, UmuD, is activated post-translationally by RecA through cleavage. UmuD is involved in the repair of UV damage and chemical mutagenesis. The complete protein is inactive but activity is unmasked in the C-terminal fragment after cleavage. Although cleavage occurs between positions 24 and 25 at a Cys–Ala site, and not as in LexA at a Ala–Gly site, the corresponding domain is homologous with the C-terminal domain of LexA. The regulatory circuit of the SOS response is schematically depicted in Fig. 8.1.

8.2 The heat-shock response

When bacterial cells are exposed to a modest temperature increase (e.g. from 30°C to 40°C) the synthesis of a number of small proteins, so called **heat-shock proteins** (HSP) is induced (Table 8.2). The reaction is termed the **heat-shock response**. The heat-shock response comprises about 20 different proteins whose expression is transiently altered. Interestingly, this response is universal from bacteria to mammals and, surprisingly, many of the induced proteins are also universally conserved. The general function of heat-shock proteins is to monitor the state of protein folding in the cell. They function as **molecular chaperones**, facilitating the folding of partly or fully unfolded proteins. Other members of the family of heat-shock proteins act as **proteases**, which remove misfolded or aggregated proteins from the cell. Expression of the heat-shock genes is regulated at the transcriptional level through the alternative σ factor σ^{32} (see Section 2.3.2). RNA polymerase holoenzyme with the σ^{32} subunit ($E\sigma^{32}$) recognizes specific promoter sequences characterized by the C-rich consensus –35 sequence **TCTCNCCCTTGAA** separated by a 13–17 base pair spacer from the consensus –10 element **CCCATNTA** (see Table 2.5). Initially it was thought that the high GC content of the consensus elements might influence the isomerization step during promoter melting, and thus provide a link to the temperature-dependent induction. This is not the case, however, since it has been shown that the initiation mechanism at σ^{32}-dependent promoters does not have a different temperature requirement compared to σ^{70}-dependent promoters. The heat-shock σ factor σ^{32} is not only responsible for the transcription of the heat-shock genes under conditions of heat stress but also during growth at normal temperature, where it assures a basal level for many of the heat-shock proteins. In addition, transcription of σ^{32}-dependent promoters is sensitive to temperature upshift but also to temperature downshift. The response in both directions is very fast. When the temperature is raised from 30°C to 40°C, for instance, a 10-fold increase in transcription of the heat-shock genes occurs within 5 minutes of temperature upshift. It follows an adaptive phase of variable length, depending on the temperature. During this phase transcription of the heat-shock genes declines and reaches a new steady-state level. Within 15 minutes the concentration of heat-shock proteins reaches about twice the level before heat stress. The expression level depends on the temperature difference. If the temperature is shifted to 46°C the concentration of the heat-shock proteins increases 10-fold, accounting for around 20% of the total cellular proteins. If, on the other hand, a temperature downshift occurs, for instance from 42°C to 30°C, the reverse response can be observed. Within 5 minutes transcription of the heat-shock genes decreases about 10-fold, and the concentration of heat-shock proteins is diluted through growth.

σ^{32}, the product of the *rpoH* gene, is the regulator responsible for the change

Table 8.2 Representative list of proteins induced by the heat-shock response in *E. coli*

Protein	**Function**	**Molecular weight (kDa)**
ClpB	Chaperone	84
ClpP	Protease	24
ClpX	Protease	46
ClpY	Protease	49
DnaJ	Chaperone	39
DnaK	Chaperone	69
FtsH	Protease	70
GapA	Dehydrogenase	35.5
GroEL	Chaperone	60
GroES	Chaperone	16
GrpE	Nucleotide exchange factor	26
HtpE	Chaperone	16.3
HtpG	Chaperone	70
HtpN	Chaperone	15.8
HtrM	Epimerase	34
Lon	Protease	89
RpoD (σ^{70})	σ factor	70

in transcriptional activity. The major regulation occurs through the concentration of the σ^{32} protein. At 30°C only a very small number of σ^{32} proteins, around 10–30 copies are present in the cell. However, because σ^{32}-dependent promoters are very strong this low level of heat-shock-specific σ factors ensures a sufficient basal transcriptional level of the heat-shock genes which are required at normal growth temperature in unstressed cells. It should be noted, however, that some heat-shock genes are additionally controlled by σ^{70}-dependent promoters. Upon temperature upshift the concentration of σ^{32} increases rapidly. About 500 copies of σ^{32} are present in the cell shortly after a temperature upshift from 30°C to 42°C. The increase in σ^{32} concentration is transient. It is the result of both enhanced new synthesis and a change in the stability of the protein. The increase in σ^{32} synthesis is not primarily due to enhancement in transcription, which accounts only for a factor of two. The raise in synthesis is predominantly the result of an *enhanced translational efficiency* of the *rpoH* mRNA. Improved translation accounts for a roughly 12-fold increase in σ^{32} concentration. It has been suggested that a temperature-dependent change in the secondary structure of the *rpoH* mRNA might be responsible for this increase in translation. According to this assumption the secondary structure of a particular segment of the *rpoH* mRNA would function as the temperature sensor.

The increase in the cellular level of σ^{32} is not only the result of enhanced synthesis. In addition, the lifetime of the σ^{32} protein contributes significantly to the number of cellular copies. At 30°C, the lifetime of σ^{32} is very short, with a half-life ($t_{1/2}$) in the range of 1 minute. When the temperature is shifted to 42°C the half-life is increased more than eightfold. Cellular proteases are responsible for the rapid turnover of the σ^{32} protein. The exact nature of these proteases was unknown for long time, and none of the heat-shock induced major proteases, e.g. **Lon** or **Clp**, appeared to be involved. Recently it has been shown, however, that the protease **FtsH** is involved in the degradation of the σ^{32} protein.

Regulation of σ^{32} occurs additionally at the level of activity. The heat-shock σ factor can be activated or inactivated reversibly. The responsible modulators are the heat-shock proteins **DnaK**, **DnaJ** and **GrpE**. These proteins, which normally act as molecular chaperones for the folding of damaged proteins, have been shown to interact directly with σ^{32}. Interaction with the three heat-shock proteins occurs independently and involves ATP, which is hydrolysed after DnaJ joins the ternary σ^{32}–DnaK–ATP complex. Interaction with the heat-shock proteins DnaK and DnaJ renders σ^{32} inactive, possibly by removing it from the RNA polymerase holoenzyme $E\sigma^{32}$. GrpE acts as a nucleotide exchange factor, and appears to reverse this inhibition partially, by initiating dissociation of the σ^{32}-chaperone complex through the release of ADP. Since DnaK, DnaJ and GrpE are involved in the repair of damaged proteins, they are sequestered by binding in the case of heat-induced protein damage. A feedback loop is thus feasible, which causes derepression and stabilization of σ^{32} when the demand for heat-shock proteins DnaK, DnaJ and GrpE increases. Moreover, the inactivation of σ^{32} through DnaK, DnaJ and GrpE may be the first step in the degradation pathway of the heat-shock σ factor. The three heat-shock proteins are therefore likely candidates to initiate the σ^{32} turnover (see above). A schematic representation of the heat-shock response is depicted in Fig. 8.2.

Transcription of the *rpoH* gene is coupled to a second heat-inducible regulon controlled by the minor σ factor σ^{24} (σ^{E}, see Section 2.3.2). The second heat-shock factor, σ^{24}, is induced by the extracytoplasmic accumulation of misfolded proteins. About 10 genes are regulated by σ^{24}, including transcription of the σ^{24} gene (the *rpoE* gene) itself. While the σ^{32} regulon is responsible for the major response to thermal stress, the σ^{24} regulon is necessary for more extreme heat conditions, and controls σ^{32} (*rpoH*) transcription at very high temperatures.

Transcription of the *rpoH* gene starts from four promoters, designated P1, P3, P4 and P5. Promoters P1, P4 and P5 are σ^{70}-dependent, while P3 is σ^{24}-dependent. At normal growth temperatures the σ^{70}-dependent promoters P1 and P4 contribute to more than 90% of the total *rpoH* transcription, while the σ^{24}-dependent promoter P3 accounts for only about 2%. With increasing temperature, however, P3 becomes more and more active, and at very high temperatures (around 50°C) transcription from P1 and P4 is completely suppressed, whereas that from P3 persists at high level. At such temperatures σ^{70} (and probably other σ factors) is inactivated and the cell is only able to carry out the synthesis of heat-shock proteins. Interestingly, transcription of σ^{70} (the *rpoD*

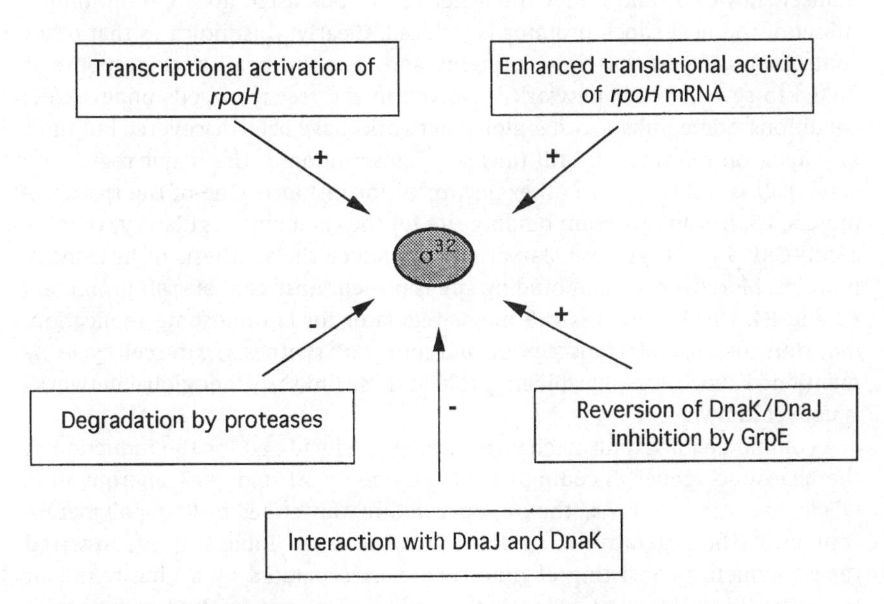

Figure 8.2 Scheme of the heat-shock response. The alternative sigma factor responsible for the transcription of heat-shock genes, σ^{32}, is shown as a grey oval in the centre of the diagram. Different processes that cause activation or repression of the concentration or activity of σ^{32} are boxed, and their activating or repressing effect on σ^{32} is indicated by + or –, respectively.

gene) is also under control of σ^{32}. This allows the cell to recover from heat-induced inactivation of σ^{70} and to return to normal growth after the heat-shock response.

The question about the nature of the signals and sensors that trigger induction of the heat-shock response cannot yet be completely answered. It is very likely, however, that several signal transducing pathways exist. Feedback loops involving heat-shock proteins may in part be responsible for a derepression of σ^{32} activity. The pool of DnaK or DnaJ in the cell has thus been proposed as a *cellular thermometer*. On the other hand, temperature-dependent changes in the secondary structure of the *rpoH* mRNA are proposed as a sensor triggering enhanced translational efficiency of σ^{32} (see above). Clearly, the occurrence of denatured proteins shifting the steady-state concentration of heat-shock proteins (chaperones and proteases) through binding has an important signal character. A very different proposal has been made, suggesting that small ribosomal subunits are directly involved as sensors for heat- as well as cold-shock. In this respect it should be noted that induction of heat-shock proteins can also be triggered by stress factors other than heat. A long list of damaging agents or conditions is known to activate the synthesis of heat-shock proteins. This list includes stress induced by ethanol, osmotic shock, pH shock, heavy metals, DNA damaging agents or oxidative stress. Heat-shock is the most efficient inducer, however, and under the other conditions listed above, often only a subset of the heat-shock proteins is induced. Clearly, this indicates that other controls must exist for these proteins and that the heat-shock response is linked to several other networks involved in the rescue of cells under stress conditions. Some links to other global networks have been discovered but their physiological relevance is still unclear. Transcription of the major regulatory gene *rpoH* is linked to two other networks, for instance. One of the *rpoH* promoters, P5, has an upstream binding site for the catabolite regulatory complex cAMP–CRP. In fact, glucose starvation can induce the synthesis of heat-shock proteins. Moreover, a DnaA binding site is present upstream of *rpoH* promoters P3 and P4. DnaA, which is the major regulator for chromosome replication, may thus control *rpoH* transcription and couple σ^{32} synthesis to the cell cycle. As mentioned above, the physiological role of these links to other global networks is not yet known.

A completely different mechanism has been elucidated for the induction of the heat-shock genes encoding the chaperone *groESL* and *dnaK* operons in *B. subtilis*. In contrast to *E. coli*, the two operons are transcribed by RNA polymerase containing the vegetative *B. subtilis* σ factor (σ^{43}, see Table 2.4). An **inverted repeat** sequence consisting of nine base pairs separated by a nine base pair spacer has been shown to be located directly downstream of the core promoter (Fig. 8.3). This inverted repeat can be found in many bacterial species where it precedes the *dnaK* and *groESL* operons, and apparently presents one of the most highly conserved sequences in eubacteria. The inverted repeat has been termed **CIRCE** (controlling inverted repeat of chaperone expression). The CIRCE

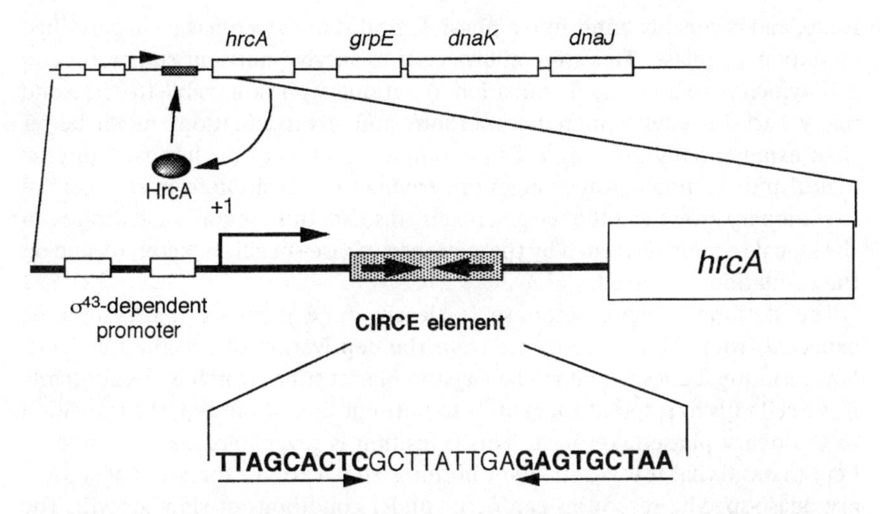

Figure 8.3 Role of the CIRCE element in the expression of the *B. subtilis dnaK* operon. The structure of the *dnaK* operon is schematically presented. The lower part of the figure gives an enlarged view. Transcription is started from the σ^{43}-dependent vegetative *B. subtilis* promoter characterized by two small open boxes. The transcription start site is indicated by an arrow. The genes *hrcA*, *grpE*, *dnaK* and *dnaJ* are represented by large open boxes. The product of the *hrcA* gene acts as a repressor (HrcA, grey oval structure). It binds to an inverted sequence element designated CIRCE (controlling inverted repeat of chaperone expression). The sequence of the CIRCE element with two convergent arrows indicating the repeat sequences (bold type) is shown underneath.

sequence acts as a negative *cis* element in the expression of heat-shock genes, which encode the molecular chaperone machines GroESL and DnaK. Recent findings suggest that the CIRCE element acts as target sequence for binding of a repressor protein. Hence, it appears to function as an operator site. The repressor appears to be the **HrcA** protein (heat regulation at CIRCE), whose activity is modulated by the GroESL chaperone machine. HrcA is the first protein encoded in the *dnaK* operon.

The CIRCE element is also found in bacterial *groESL* operons which, in addition, are under the control of a σ^{32}-dependent promoter. In one such case it has been shown that the CIRCE element does not seem to be involved in heat-shock induction but rather mediates cell cycle control.

8.3 Stationary phase control

Under conditions of limiting nutrients some Gram-positive bacteria, e.g. *B. subtilis*, react by forming spores. *E. coli* and related enteric bacteria take a different route, and reversibly enter into a physiological state which has been described as stationary phase. This state allows cells to survive nutrient starvation and cells which have undergone transition to stationary phase are able to withstand many harmful environmental conditions and stress situations much better than exponentially growing cells. A complete summary of the many physiological and morphological changes underlying the transition from exponential to stationary growth cannot be given here. Instead, this section concentrates on the global regulon governed by the stationary phase-specific σ factor, σ^s, and on the regulation of σ^s itself.

The stationary phase response is clearly more complex than might be expected from changes resulting from the deprivation of specific nutrients. Furthermore, it does not depend on a particular nutrient which is missing. Only if the cell fails to respond adequately to nutrient limitations will the transition to stationary phase take place. This transition is reversible, and it should be kept in mind that it is *not* an 'all or nothing' step. Partial expression of stationary phase-specific responses can occur under conditions of slow growth. The unifying intracellular signal(s) that coordinately induces the transition from exponential to stationary phase has not yet been identified. The regulator responsible for the expression of many genes involved in the stationary phase transition is known, however. It is a special σ factor, σ^s, the product the **rpoS gene**, which controls stationary phase transcription (see Section 2.3.1). Sometimes σ^s is also designated as 'master regulator' of stationary phase control. The expression of the stationary phase-specific σ factor is itself a very complex network, and it is linked to many other global regulators and control circuits.

RpoS or σ^s is a 37.9-kDa protein (330 amino acids in length) with strong homology to the σ^{70} family. It controls a large number of genes. More than 50 genes are known presently whose expression is affected by σ^s. In most cases σ^s acts as a positive effector; however, there are some genes which are affected in a negative way. Many of these genes are not directly required for growth at stationary phase, and often they function to increase the chances of survival under extreme growth conditions. In other words, σ^s is not only responsible for transcription of genes involved in the stationary phase response but it also plays an important role during exponential growth in the expression of genes required under stress conditions, such as osmotic upshift or acid stress, for instance. Table 8.3 summarizes the σ^s-dependent genes.

Consistent with the very close structural similarity of σ^s to σ^{70}, the promoter sequences recognized by the different RNA polymerase holoenzymes $E\sigma^s$ and $E\sigma^{70}$ are very similar. In fact, the promoter consensus hexamer sequences are almost identical. The –10 regions are shared between the two types of promoters. In contrast, the –35 hexamer sequences appear to be slightly different. Discrimination between $E\sigma^s$ and $E\sigma^{70}$, therefore, is very likely to involve the –35 promoter region. A different view has been obtained from *in vitro* studies. Exchanging all possible combinations of upstream and downstream elements from promoters known to exhibit differential sensitivity to σ^s and σ^{70} has indi-

Table 8.3 σ^s -dependent genes

Gene	Function	Additional modulator
aidB	DNA methylation damage repair	Lrp
aldB	Aldehyde dehydrogenase	CRP, FIS
appA	Acid polyphosphatase	AppY
appY	Acid polyphosphatase, regulator	
bolA	Cell division factor	
cbpA	Curved DNA-binding protein, chaperone	H-NS
cdbAB	Cytochrome bd oxidase	AppY
cfa	Cyclopropane fatty acid synthesis	
csgBA	Curli fimbriae	H-NS
csgCDEF	Curli fimbriae	
csiA-F	Six-carbon starvation gene	CRP
dps	DNA-binding protein	
ficA	Control of cell division	
fimA *	Fimbrial protein	
frd	Fumarate reductase	
ftsQAZ	Cell division	
galETK	Galactose operon	CRP
glgA	Glycogen synthase	
glgS	Glycogen synthesis	CRP
glnQ *	Glutamine transport	FIS
glpD	Glycerol-3-phosphate dehydrogenase	
gor	Glutathione oxidoreductase	
himA	Integration host factor (IHFα)	ppGpp, IHF
hmp	Haemoprotein	ppGpp, IHF
htrE	Pili construction protein	IHF
hyaA-F	Hydrogenase I	AppY
katE, katG	Catalase	
lacZ	Lactose operon	CRP
ldcC	Lysine decarboxylase	
mglA *	Methyl-galactoside transport	FIS
mutH, mutS *	Methyl-directed mismatch repair	
osmB	Outer membrane protein	
osmE	Lipoprotein	
osmY	Periplasmic protein	Lrp, CRP
otsA, otsB	Trehalose-6-phosphate synthase	H-NS
poxB	Pyruvate oxidase	
proP, proU	Glycine betaine and proline transport	FIS, H-NS
rob	Right oriC binding protein	
sdhA *	Succinate dehydrogenase	FIS
topA	Topoisomerase I	
treA	Trehalase	
wrbA	Trp repressor	
xthA	Exonuclease III	
xylP *	Xylose transport	FIS

* Negatively affected by σ^s. Data is taken from Loewen *et al.* (1994).

cated that the discriminating signal resides within the downstream element. Moreover, it is undisputed that sequences outside the promoter core regions also contribute to the discrimination for Eσ^s and Eσ^{70} RNA polymerase holoenzymes. DNA curvature can be found upstream of many σ^s-dependent promoters, and the local DNA topology may also play an important role in the discrimination of the two different holoenzymes. The observation that there is less clear homology to the –35 consensus sequence at σ^s-dependent promoters indicates that additional regulatory factors almost certainly participate in the transcription of σ^s-dependent promoters. In fact, additional transcription factors, such as cAMP–CRP, Lrp and the DNA structuring proteins H-NS, FIS and IHF, have been shown to participate in the transcription of many σ^s-controlled genes. With the known exception of the *fic* promoter (encoding a gene involved in folate synthesis), which is exclusively recognized by Eσ^s, σ^s-controlled promoters can also be recognized by Eσ^{70}. The conditions for optimal binding differ, however. On the other hand, there are many σ^{70}-controlled promoters which are not transcribed by σ^s. Stringent regulated promoters belong to this group, for instance (see Section 8.5). Another particular important aspect in the selection of promoters is the relative abundance of σ^s and σ^{70} in the cell. Under stationary growth conditions the level of σ^s increases from almost undetectable levels to about one third the amount of σ^{70}. It should also be recalled that the existence of specially modified RNA polymerases which appear to accumulate during stationary phase has been described (see Section 2.5). Often, transcription of genes expressed preferentially under stationary phase conditions can be initiated at more than one promoter. The additional promoters may be under the control of other regulatory networks, independent of σ^s, which explains the link in the expression of many σ^s-controlled genes to multiple regulatory circuits. Consistent with such a notion is the apparent observation that the global regulator H-NS specifically inhibits the recognition of promoters by the Eσ^{70} RNA polymerase holoenzyme, thus supporting transcription by Eσ^s.

As mentioned above, the level of σ^s is not constant during different growth phases. The σ^s concentration is highest at early stationary phase. It should be noted, however, that a low level of σ^s is also present during exponential growth, consistent with the observation that some σ^s-dependent genes are also induced at exponential growth, and not only during stationary phase. As discussed below, the cellular concentration of σ^s is subject to regulation at several levels, namely transcription, translation and protein stability. Moreover, an increase in the amount of σ^s is not only induced at the transition to stationary phase but also during exponential growth; for example, at osmotic upshift or under various stress conditions. It can be observed that the level of σ^s increases under all starvation conditions, and σ^s concentration is apparently linked to the growth rate. In rich medium there is clear evidence for transcriptional regulation of the *rpoS* gene. The *rpoS* gene is the second gene in an operon together with ***nlpD***, which encodes a lipoprotein involved in cell wall synthesis. Transcription of the *nlpD* gene starts from two promoters which are not growth phase-sensitive. The two promoters are likely to be responsible for a low level expression of *rpoS* at exponential growth. The major *rpoS* promoter is located within the reading frame of the *nlpD* gene. This promoter (controlled by Eσ^{70}) exhibits growth phase-dependence. In line with this, the low molecular weight effector **guanosine tetraphosphate (ppGpp)**, known to be a central growth rate effector (see Section 8.6), has a major regulatory role in the activation of σ^s synthesis. It is not entirely clear, however, whether ppGpp alters the promoter efficiency or the elongation of *rpoS* transcription. In addition, other low molecular weight compounds are effective in modulating the synthesis of σ^s. **Homoserine lactone** or polyphosphates, for instance, stimulate *rpoS* transcription. **UDP–glucose**, in contrast, inhibits σ^s synthesis at the post-transcriptional level. There is additional evidence that transcription from the *rpoS* promoter is under negative regulation by cAMP–CRP. No evidence for a transcriptional feedback at the *rpoS* promoter involving σ^s has been found, however.

The level of σ^s is not only determined by transcriptional regulation. Major controls occur post-transcriptionally, and involve modulation of the efficiency of translation of the *rpoS* mRNA as well as changes in the protein stability. Translation of the *rpoS* mRNA is stimulated fivefold during the late exponential phase or osmotic upshift. Regulation must occur at the translational level because mRNA levels are not affected but the rate of σ^s synthesis increases rapidly after osmotic upshift. Several models have been suggested to explain this translational regulation. Changes in the secondary structure of the *rpoS* mRNA and the participation of several protein factors are the major elements of these models. A plausible model is based on the predicted secondary structure of the *rpoS* mRNA, which shows stable base pairing at the translational initiator region. This stable structure precludes translation. Under inducing conditions the stable secondary structure might be changed, allowing initiation of translation. The switch in RNA secondary structure is thought to be brought about by interacting proteins. Suitable candidates are **H-NS** and **HF-I**. Both proteins have been shown to be involved in the efficiency of *rpoS* mRNA translation. H-NS has already been described as a DNA-binding protein and transcriptional regulator (see Section 7.5.3). HF-I, the product of the ***hfq* gene**, is an RNA-binding protein and is known to act as host factor for the replication of phage Qβ RNA in addition to its function as a positive control factor for the synthesis of σ^s. It is assumed that binding of HF-I to the *rpoS* mRNA suffices to disrupt the stable secondary structure, thus allowing translation to become more efficient. (A similar mechanism for the stimulation of Qβ RNA replication has been proposed. The phage RNA has a highly structured 3′ domain which inhibits minus (–) strand replication. Binding of HF-I to the structured 3′ domain is thought to cause a conformational change of the phage RNA, allowing replication to occur.)

The inducing signal for the translational control of RpoS synthesis is not known. It is clear, however, that the reaction is rapid. Since H-NS also affects *rpoS* translation it might be involved in the same mechanism. H-NS negatively affects σ^s synthesis and *hns* mutants show increased *rpoS* translation as well as reduced σ^s turnover. It has been reported that H-NS and HF-I are able to interact with each other. Possibly H-NS might therefore interfere with the activity of HF-I by direct protein–protein interaction.

H-NS is not only involved in controlling the level of σ^s. It also plays a direct role in the expression of a large number of genes under osmotic stress and at stationary phase. The expression of many of these genes is completely independent of the action of σ^s, however. For instance, H-NS, but not σ^s, is an important regulator for the transcription of the *proU* operon, which encodes a transport system for the osmoprotectants glycine betaine and proline under osmotic shock.

A further interesting control mechanism which affects the synthesis of σ^s has been discovered which involves a small untranslated RNA. This RNA molecule, **DsrA**, plays a positive role in RpoS expression, especially at low temperatures. It has been shown that DsrA antagonizes the repressing effect of H-NS (see Section 7.5.3). DsrA stimulates σ^s synthesis post-transcriptionally, presumably by increasing the efficiency of *rpoS* translation or through stabilization of the *rpoS* mRNA. No obvious homology exists between the DsrA RNA and the *rpoS* mRNA, and the mechanism of translational activation is not known.

There is another post-transcriptional mechanism in regulating the level of σ^s. In exponentially growing cells σ^s is unstable, and the half-life is very short ($t_{1/2}$ is less than 2 minutes), whereas at the onset of starvation the stability increases fivefold. After hyperosmotic shock σ^s is practically stable, with a half-life exceeding 45 minutes. Protein turnover is therefore an important aspect in modulating the cellular level of σ^s. Although the degrading protease is known, the pathway that leads to σ^s turnover is not yet clear. The enzyme responsible for σ^s turnover is **ClpXP**, a cytoplasmic ATP-dependent protease. A protein has recently independently been identified by several groups which negatively modulates the activity of σ^s. The protein is termed **RssB** (regulator of σ^s), but has also been designated as SprE or MviA. RssB is not itself a protease but it may be necessary to present σ^s to the degrading enzyme ClpXP. Interestingly, RssB has a structure reminiscent of the two component response regulators (see Section 7.6.1). Its N-terminal 110 amino acids exhibit strong similarity with the typical structure of receiver domains. The way in which RssB interacts with σ^s, and how this putative complex is transferred to the protease is not known. It is noteworthy, however, that cleavage of σ^s by ClpXP occurs within a similar position of the conserved amino acid domain structures as in the case of σ^{32}, the heat-shock-specific σ factor. Degradation of σ^{32} is mediated by the protease FtsH, and the accessory system composed of the chaperones DnaK, DnaJ and GrpE is involved in presentation of σ^{32} to proteolysis (see Section 8.2). Thus the DnaK chaperone machine and RssB probably serve similar functions in modulating the activity and half-life of the different σ factors σ^s and σ^{32}. The main regulatory features controlling the concentration and activity of the stationary phase regulator σ^s are outlined in Fig. 8.4.

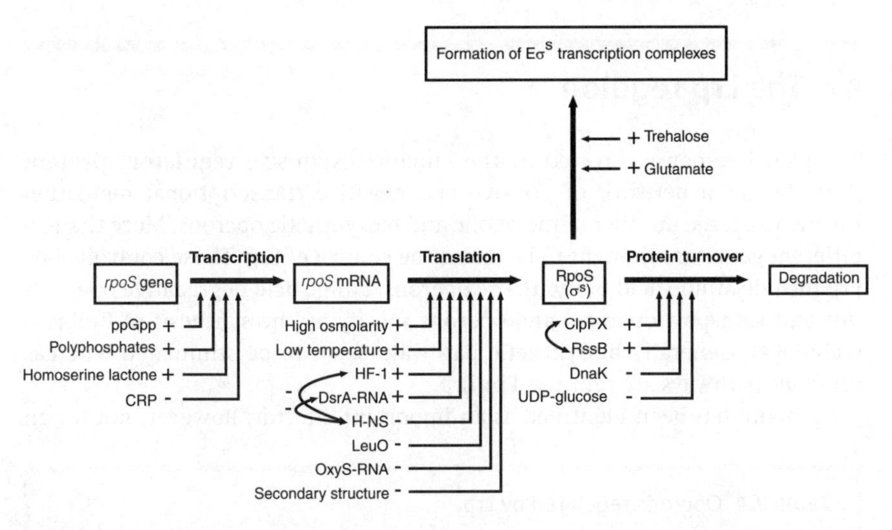

Figure 8.4 Regulatory circuits controlling the concentration and activity of σ^s. Control of the concentration and activity of σ^s is exerted at the stages of transcription, translation, protein turnover or during the formation of σ^s-controlled transcription complexes as indicated by the bold arrows. Components or conditions marked with a + or a – indicate a positive or negative effect on the respective process. Bent arrows with double arrowheads indicate interactions or influences between regulators.

Most stationary phase-inducible promoters contain various binding sites for transcription factors related to other regulatory circuits. For instance, upstream of the *dps* gene binding sites for IHF and the regulator OxyR can be found. The *dps* gene encodes a small 19-kDa protein, **Dps**, which is implicated in chromosome organization, and which has several regulatory and protective functions against oxidative DNA damage. Transcription of the stationary phase-activated *osmY* gene, encoding a periplasmic protein, is affected by Lrp, IHF and cAMP–CRP. Other examples include the regulators H-NS, which affects transcription of the *csg* operons involved in the production of curli fimbriae, or FIS and cAMP–CRP, which coregulate the expression of the aldehyde dehydrogenase AldB. Links of RpoS-dependent expression to other regulatory networks are not only maintained through the involvement of the various transcription factors in σ^s-dependent transcription as outlined above. In addition, the transcription of several regulators occurs in a σ^s-dependent manner. Examples are the regulators AppY, a member of the AraC family of transcription factors which controls operons that are stimulated under anaerobic conditions, or the subunits *hip* and *himA* constituting the regulator IHF. The protein Dps, already mentioned above, which functions in the protection of DNA and has a global effect on gene expression, is also under the control of σ^s.

8.4 The Lrp regulon

The global response directed by the **leucine-responsive regulatory protein** (**Lrp**) defines a network of positive and negative transcriptional regulation involving a large number of metabolic and biosynthetic operons. More than 70 different genes are thought to be under the control of Lrp. Those controlled by Lrp include amino acid biosynthetic operons, amino acid degradative operons, nutrient transport operons, and operons specifying the synthesis of fimbriae (Table 8.4). Generally biosynthetic pathways seem to be stimulated whereas catabolic pathways are repressed by Lrp.

L-leucine has been identified as an important effector; however, not for all

Table 8.4 Operons regulated by Lrp

Operon	Function	Regulation by Lrp
fimB	Regulation of fimbriae	+
gcvTHP	Glycine cleavage	+
gltBDF	Glutamate synthase	+
glyA	Serine–glycine interconversion	–
ilvIH	Isoleucine, valine, leucine biosynthesis	+
kbl-tdh	Threonine cleavage	–
lacZYA	Lactose utilization	+
leuABCD	Leucine biosynthesis	+
livJ	Isoleucine, valine, leucine uptake	–
livKHMG	Isoleucine, valine, leucine uptake	–
lrp	Lrp regulator	–
lysU	Lysyl-tRNA synthetase	–
malEFG	Maltose utilization	+
malT	Maltose utilization	+
ompC	Outer membrane porin	–
ompF	Outer membrane porin	+
oppABCDF	Oligopeptide uptake	–
papAB	Pyelonephritis-associated pili	+
papI	Regulation of pyelonephritis-associated pili	+
pntAP	Pyridine-nucleotide transhydrogenase	+
sdaA	Serine deamination	–
serA	Serine biosynthesis	+

+ Activation; –, repression.

promoters under Lrp control (see below). Another feature of the Lrp regulon is exceptional. There is no unifying stimulus known underlying the Lrp regulon. Since Lrp apparently decreases the synthesis of proteins preferentially needed during growth in rich medium, and induces the synthesis of biosynthetic proteins primarily needed in poor medium, it has been suggested that the Lrp regulon adapts the bacterial metabolism to the special situations of 'feast' or 'famine'.

The regulator Lrp can act as an **activator** or **repressor** of transcription. Lrp is a rather unique transcription factor and shares no similarity with other transcription factor families, such as LysR or the two component regulatory proteins (see Section 7.21 and 7.6). A limited homology (about 25%) has been identified with the AsnC protein, which regulates the synthesis of AsnA, the asparagine synthetase A. Lrp is a small basic protein (pI = 9.2) of 163 amino acids with a molecular weight of 18.8 kDa. The active form is a dimer. There are about 3000 copies of Lrp per cell, but this number is not constant. The Lrp concentration is maximal at log-phase and is clearly dependent on the quality of the growth medium. In minimal medium there is a three to fourfold higher concentration of Lrp compared to rich medium. The level of Lrp depends particularly on the external abundance of amino acids, and also on the type of carbon source. As discussed below, the synthesis of Lrp seems to be inversely regulated to the cell growth rate and is stimulated by the growth rate effector molecule ppGpp (see Section 8.6).

Lrp has a modular organization, which is typical for transcription factors, and three different functional domains can be distinguished. A helix-turn-helix-like motif responsible for DNA-binding has been identified within the N-terminal one-third of the molecule. The central part of the molecule contains a sequence region involved in activation, while the C-terminal one-third has been demonstrated to function in the leucine-response. Lrp binds specifically to the upstream DNA of sensitive promoters. A common recognition sequence is somewhat difficult to define because a number of variations occur at different Lrp sites. Nevertheless, an approximate consensus sequence **AGAATTTTATTCT** has been derived from a large number of DNA sites known to be the target for Lrp interaction. As a probable consequence of the sequence variations of the different Lrp binding sites the different promoters exhibit largely different affinities for the regulator. Thus Lrp comprises a large range of regulation. In addition to the degenerated consensus sequence specific binding of Lrp requires a bent DNA conformation. As can be anticipated from helix-turn-helix DNA-binding proteins, interaction of Lrp with DNA occurs predominantly with the major groove. Circular permutation studies (see Box 6.3) have shown that binding of Lrp further increases the DNA-bending. Occupation of a single DNA site creates a 52° angle, while a bend of 135° was measured when Lrp was bound to two adjacent sites. Promoters under Lrp control generally have several Lrp operator sequences, and binding of the regulator occurs in a highly cooperative manner. Based on the results of DNaseI footprints, which reveal a phased pattern of enhanced and protected DNA regions, it was suggested that Lrp–DNA interaction may cause wrapping of the DNA around multiple Lrp molecules. Lrp might therefore constitute an architectural element that facilitates the assembly of a nucleoprotein complex. This view is consistent with the observation that Lrp, depending on different promoters, interacts with other transcription factors, including the DNA-bending or architectural proteins IHF, H-NS or the stationary phase-specific σ factor σ^s.

As mentioned above, leucine can affect Lrp at several but not all promoters. In some cases the effect of Lrp is dramatically modulated by leucine, while in others the presence of leucine has little or no effect. Compared to other transcription factors, where the cofactor is essential for DNA-binding and/or regulation, e.g. cyclic AMP in the cAMP–CRP complex (see Section 7.3.2), leucine plays a differential role as effector in the Lrp regulon. Clearly, the presence or absence of leucine does not enable or abolish the binding of Lrp to the target sites in an 'all or nothing' manner. It modulates the activity of Lrp rather than turning it on or off. Examples where leucine enhances, antagonizes or has no effect on Lrp-dependent transcription can be found for promoters under positive as well as under negative control. For example, the *ilvIH* operon, encoding the enzyme for branched-chain amino acid synthesis, is activated by Lrp. The presence of leucine reduces this activation. In contrast, leucine enhances the activating effect of Lrp in the case of *fimB* or *fimE* transcription. The two genes are involved in promoter switching during pili synthesis. No effect of leucine is observed, for instance, during Lrp-dependent transcription activation of the *papAB* operon which encodes the genes important for P pili synthesis. Similar combinations of leucine effects are found at promoters which are repressed by Lrp. The effect of leucine is very likely to be mediated by a concentration-dependent modulation of the binding affinity of Lrp to the promoter upstream DNA sites. Thus, leucine has little effect on promoters with high affinity for Lrp, while promoters with low affinity for Lrp are particularly sensitive to leucine.

The amount of Lrp in the cell is carefully regulated. The level of Lrp depends on the external abundance of amino acids and also on the type of the carbon source available. In rich medium the amount of Lrp is lowest. Regulation occurs at the transcriptional level. Transcription of the *lrp* gene, for instance, is subject to autoregulation. This regulation is independent of the presence of leucine. There are binding sites for Lrp of relatively low affinity upstream of the *lrp* promoter between positions –32 and –80, consistent with the relative high concentration of Lrp molecules in the cell. Lrp binding immediately upstream of the *lrp* promoter probably affects RNA polymerase interaction with the promoter inhibiting transcription. Additional Lrp molecules bound more distal to the promoter help to stabilize the nucleoprotein complex. If the Lrp sites are moved much further upstream the regulatory effect gets lost, indicating that no 'action-at-a-distance' mechanism or DNA looping applies for the autoregulation at the *lrp* promoter.

Transcription of the *lrp* promoter is additionally repressed at high growth rate in rich media, even in the absence of functional Lrp protein. Transcription of *lrp* is thus clearly linked to the cell growth rate. Promoters that are inversely regulated with the growth rate are often designated as **gearbox promoters** (see Fig. 8.12 and Section 8.6 below). They are characterized by a CAA sequence motif at the –35 region and a CGG motif at the –10 hexamer. Although the *lrp* promoter has a perfect –35 consensus element for σ^{70}-dependent promoters, which is quite unusual for gearbox promoters, some limited homology of the *lrp* promoter with the gearbox-type promoters can be discerned. One of the major effectors mediating growth rate control is ppGpp. Cellular ppGpp levels generally show a linear inverse correlation to the growth rate (see Section 8.6). It can be shown that *lrp* expression directly correlates with the cellular ppGpp concentration. Thus, it can be concluded that the mechanism determining the Lrp level in the first place is the inverse coupling to the growth rate. This coupling may be mediated by the concentration of ppGpp. The growth rate itself is in turn dependent on the composition of the growth media.

Until now a small number of the Lrp-dependent regulatory regions have been studied in detail. Knowledge stems mainly from the analysis of the upstream regions of the *ilvIH* operon, the *gltBDF* operon or 'switch regions' controlling the sites for **phase variation** during the synthesis of fimbriae. The *ilvIH* operon encodes acetohydroxy acid synthase III, one of the enzymes involved in the synthesis of branched-chain amino acids. The upstream region of the *ilvIH* operon contains six binding sites for Lrp. Two high-affinity sites are clustered at positions –250 and –219, while the remaining four sites are located at positions –137, –103, –74 and –54 relative to the transcription start site. Five of the six sites are important for Lrp-dependent activation of the *ilvIH* promoter. Deletion of the Lrp site at position –54 does not alter the activating effect. Binding to the five upstream sites occurs cooperatively, and activation (about threefold) is a direct effect of Lrp binding. Binding of Lrp to the *ilvIH* promoter regulatory region is necessary and sufficient for activation. The presence of leucine reduces transcription four to sevenfold. Only the Lrp bound to the promoter proximal binding site is supposed to be responsible for a direct contact with RNA polymerase. The structure of the *ilvIH* operon is schematically presented in Fig. 8.5.

Interestingly, the regulatory protein H-NS, known to affect the DNA conformation (see Section 7.5.3), interferes with Lrp-dependent activation of the *ilvIH* operon at stationary phase or under stress conditions. Linking the activity of the two regulators may involve a change in the bending of the upstream regulatory DNA region, possibly through heterologous interactions of the two proteins.

The *gltBDF* operon encodes the large and the small subunit of glutamate synthase GltB and GltD, as well as for a regulatory protein GltF. Three binding sites for Lrp have been identified some distance upstream of the transcription start site. They centre at positions –152, –215 and –246. All three sites are occupied by Lrp in a cooperative way, and the upstream DNA is considerably bent

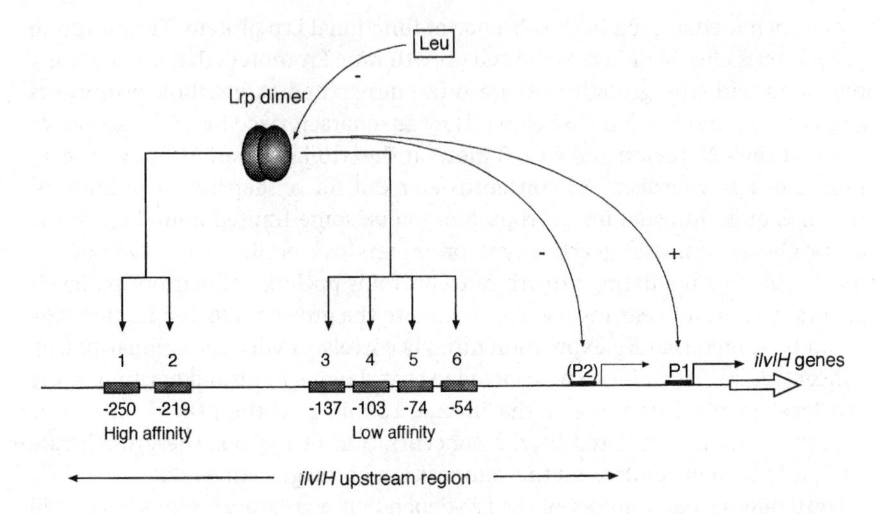

Figure 8.5 Lrp regulation of the *ilvH* operon. The upstream region of the *ilvH* operon is shown as a representative model for regulation by Lrp. Transcription is controlled from two tandem promoters P1 and P2. Two clusters of Lrp binding sites are shown as stippled boxes with numbers indicating their nucleotide positions relative to the transcription start of promoter P1. The Lrp binding sites are numbered 1 to 5. Lrp dimers, illustrated by two oval grey spheres, bind to the two upstream Lrp sites with high affinity. Binding to the downstream sites 3–6 occurs with lower affinity. Binding to the Lrp sites occurs cooperatively, with the exception of binding site 6. Occupation of Lrp sites has a differential effect on the activity of promoters P1 and P2. Promoter P1 is activated through Lrp (+), while promoter P2 is repressed (–). Binding of leucine to Lrp reduces the effect of Lrp strongly.

upon Lrp binding. All three sites must be occupied for efficient activation, and Lrp–DNA complexes must be in phase with the start site of transcription.

Analysis of one of the pili encoding operons, *papBA*, and the *papI* gene in *E. coli* reveals a particularly interesting aspect of Lrp-dependent regulation. Expression of receptor-specific pilus–adhesin complexes, which aid adherence to eukaryotic host cells, is regulated by a mechanism in which alternation between transcriptionally active ('phase on') and inactive ('phase off') situations occurs. Apart from several other regulatory proteins (cAMP–CRP, PapI and PapB) Lrp plays a major role in this regulation. In this case, Lrp-dependent regulation is controlled by the methylation state of two GATC sequences in the upstream regulatory region of the divergently orientated *papBA* and *papI* promoters (see Section 6.1). Methylation occurs by deoxyadenosine methylase (Dam) following replication. Activation of both promoters is dependent on the correct state of methylation. One GATC site must be methylated, the other unmethylated. Changing the methylation state requires replication. Methylation of one site is inhibited by Lrp binding. This is the phase off situation. In the presence of the 8.8-kD regulatory protein PapI, the product of the *papI* gene,

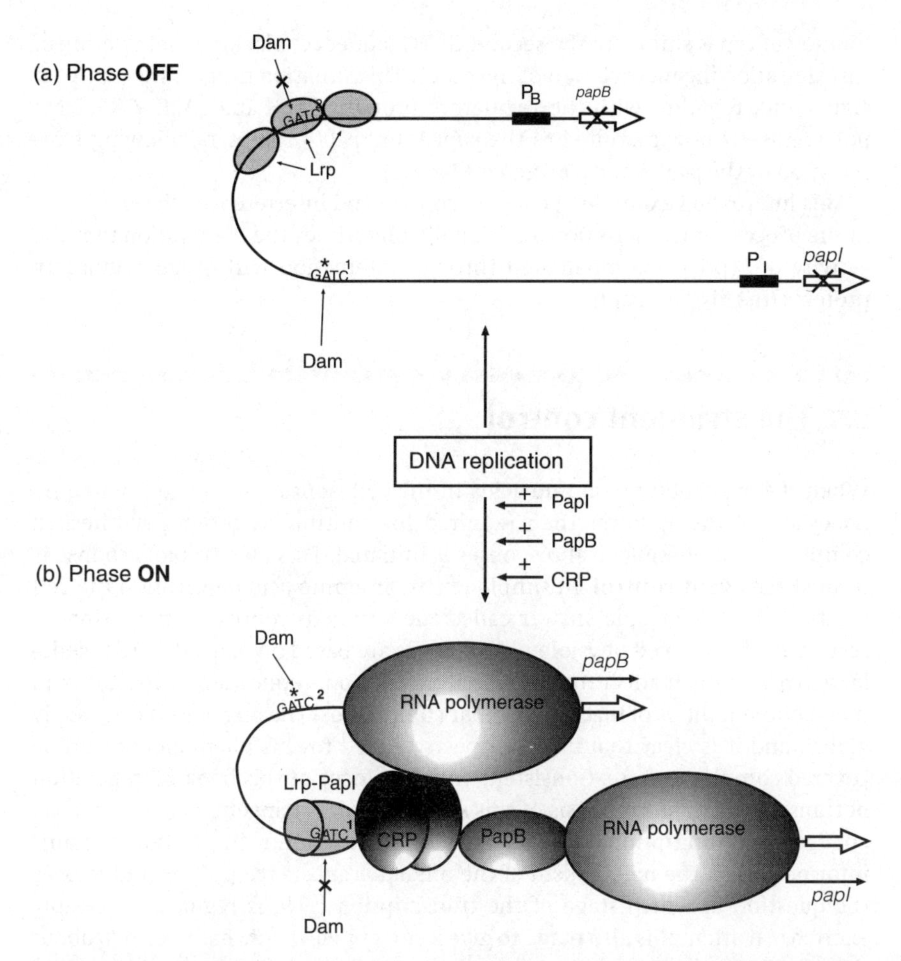

Figure 8.6 Scheme for the regulation of the *E. coli pap* operon. The regulatory region for phase variation between the *papB* and *papI* genes is shown as a thin line with two Dam methylation sequences $GATC^1$ and $GATC^2$. The *papB* and *papI* promoters are shown as small dark boxes. (P_B and P_I respectively). The $GATC^2$ sequence lies within a region of three Lrp binding sites (shaded ovals). The $GATC^1$ sequence overlaps with a binding site for the Lrp–PapI complex. During phase variation fimbrial expression switches reversibly between ON and OFF states. Phase variation of the *E. coli pap* operon is sensitive to methylation of the two sequences $GATC^1$ and $GATC^2$. The two transcription situations, Phase OFF (inactive) and Phase ON (active), are shown ((a) and (b), respectively). DNA replication is required for the transition between the two states. Methylation occurs through the activity of deoxyadenosine methylase (Dam). In the OFF state bound Lrp proteins inhibit methylation of $GATC^2$. In the presence of the positive regulators (+) PapI, PapB and CRP RNA polymerase makes favourable contact with the P_B and the P_I promoters. Lrp and PapI form a complex which, upon binding to $GATC^1$, inhibits methylation. Cooperative protein–protein contacts support transcription initiation at both promoters P_B and P_I. The figure is arranged according to Low *et al.* (1996).

binding of Lrp is shifted to the second GATC sequence, blocking methylation at this site, allowing methylation of the other. This situation triggers the phase on signal and, together with the regulatory proteins PapB and cAMP–CRP, RNA polymerase is now recruited to the *papBA* and *papI* promoters, allowing transcription of the *papBA* and *papI* genes (Fig. 8.6).

This interesting example of coupled control and interference of methylation in the process of transcription additionally underlines the observation that the activity of Lrp can be modulated through interaction with other regulatory proteins (in this case PapI).

8.5 The stringent control

When the availability of amino acids in the cell, sensed by the ability to aminoacylate tRNAs, is lower than required for continuous protein synthesis a complex set of physiological responses is initiated. This pleiotropic response is termed **stringent control**. In simple terms, at amino acid deprivation the cell reacts with a pleiotropic answer called the stringent control or the stringent response. The induced physiological changes are based on many different cellular activities which affect transcription, translation, replication, transport, and metabolic stabilities of macromolecular compounds. The response is thus really global, and it is clear that not all aspects related to this phenomenon can be covered completely here. Consistent with the focus of this book on regulation of transcription, the emphasis of this section on the stringent control is put on changes in transcriptional activity. In addition, the following section contains information on the metabolism of the mediator substance ppGpp, and tackles the question at which stage of the transcription cycle is regulation possibly exerted. An attempt is also made to give a current view on what is known about the mechanism(s) of the stringent response. A word of caution should be given here in advance with respect to the latter item. Despite almost four decades of intensive research there is less than complete understanding of the principal mechanistic steps of this significant bacterial control system. The situation is further complicated by the fact that a large number of contradictory reports has accumulated in the literature over the years. This has given rise in turn to many disparate views on this important subject. The controversial situation has certainly to do with the very complex nature of the stringent response. Moreover, since the stringent response affects a large number of other networks, studies in which physiological changes are induced or mutants are employed almost inevitably cause perturbations in one or the other network as well. Some of the conflicts mentioned may be explained by this complexity. An attempt is made in the following section to give an objective presentation, but since several findings presented below are based on work in my own group I apologize that I may not have completely succeeded in so doing.

8.5.1 Synthesis of the effector nucleotides (p)ppGpp

As outlined in the introduction, the stringent response is provoked following amino acid deprivation in the cell. The response coincides with the rapid accumulation of the small effector nucleotide **guanosine 3′,5′-bis(diphosphate)**, commonly abbreviated as **ppGpp**. A second compound, the corresponding 5′ triphosphate, **pppGpp**, is usually synthesized together with ppGpp. It has been shown in several studies that the two compounds do not differ obviously in their physiological function, and they will collectively be denoted **(p)ppGpp** for sake of clarity. The signal for induction of (p)ppGpp synthesis is not simply the level of *free* amino acids in the cell. Rather the level of charged (aminoacylated) tRNAs required for protein synthesis provokes the accumulation of the effector molecule. This is why the stringent response can also be induced by defective aminoacyl tRNA synthetases. Note that the levels of *all* amino acids or a specific amino acid need not drop down to induce the stringent response. Each single amino acid which is used for protein synthesis will suffice to cause the signal if the concentration is too low to warrant efficient aminoacylation of the cognate tRNA. The following mechanism has been established when bacteria are deprived of amino acids in one way or the other. The lack of one or several amino acids means that the respective tRNAs, specific for the missing amino acid, can no longer be aminoacylated (charged) at their 3′ acceptor ends. It follows that the ratio of charged to uncharged tRNAs decreases. Because of limitations in the availability of charged tRNAs, translating ribosomes will stall at those codons for which the cognate tRNA is missing ('idling' ribosomes). In such a situation it can happen that uncharged tRNA binds in a codon-dependent way to the free ribosomal acceptor site which contains the 'hungry codon'. This in turn triggers a ribosome-associated protein, **RelA**, which starts to synthesize (p)ppGpp. Substrates for this reaction are ATP and GTP or GDP. During the synthesis the transfer of a pyrophosphoryl group of the β, γ-phosphate from ATP to the ribose 3′ hydroxyl group of GDP or GTP is catalysed. The same K_m values for GTP or GDP for this reaction (about 0.5 mM) indicate that the preferred substrate in the cell is probably GTP since GDP concentrations are generally lower. Hence, the resulting product is mainly pppGpp, which is readily converted to ppGpp by a phosphohydrolase in a subsequent step (Fig. 8.7).

Kinetic measurements of the time of appearance of ppGpp and pppGpp in the cell after induction confirm the above assumption. The conversion of pppGpp to ppGpp is rapid, and the molar ratio of the two effectors in the cell can vary widely. ppGpp is not only synthesized under conditions of amino acid starvation. There is a basal level of ppGpp in normally growing cells. The steady-state concentration of ppGpp at exponential growth is around 10–30 μM. Under conditions of amino acid limitation the effector nucleotides accumulate very fast. Synthesis rates increase rapidly and reach a maximum within several seconds. After 15 minutes, a new steady-state level about 20-fold higher than the

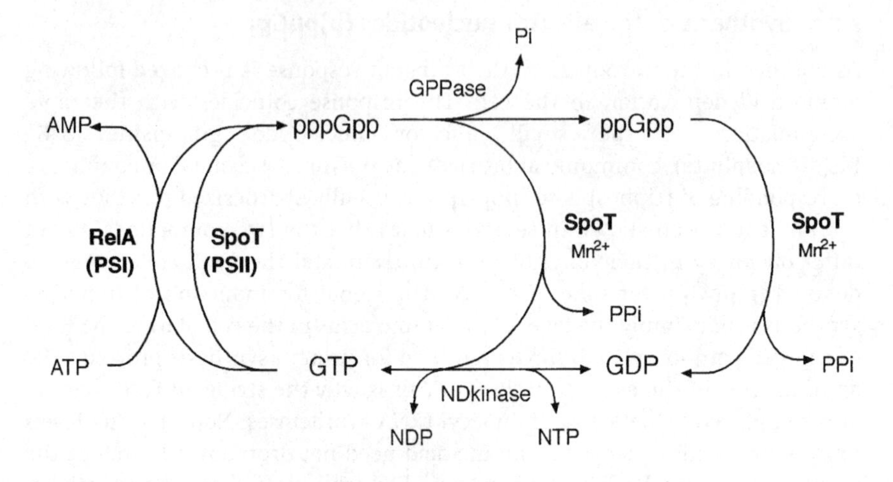

Figure 8.7 Scheme for the synthesis and degradation of (p)ppGpp. Two enzymes, RelA (PSI) and SpoT (PSII), are responsible for the synthesis of (p)ppGpp which is synthesized by pyrophosphate transfer from ATP to the 3′ OH group of GTP. The pentaphosphate is converted rapidly to the tetraphosphate by GPPase. A second activity of SpoT is responsible for the degradation of both pppGpp and ppGpp with release of pyrophosphate (PPi), yielding either GTP or GDP. GDP is recycled to GTP by the action of NDkinase. The scheme is adopted from Cashel *et al.* (1996).

basal level at exponential growth, is reached. During the stringent response the level of ppGpp is almost completely regulated by *de novo* synthesis. This can be inferred from the high turnover number caused by the rapid degradation of ppGpp to GDP (ppGpp has a half-life of roughly 10 seconds).

The ribosome-associated protein RelA, which is responsible for the rapid ppGpp synthesis, was formerly known as the **stringent factor**. Today it is often termed **ppGpp synthetase I (PSI)**. It is a protein of 84 kDa (743 amino acids in length) encoded by the *relA* gene. There is some uncertainty as to how many ribosomes in the cell contain bound RelA protein. Numbers determined for the ratio of ribosomes that contain associated RelA to ribosomes without RelA vary greatly and range from 1:1 to 1:200.

It has been known for a long time that strains lacking the *relA* gene are not completely ppGpp deficient. Consistent with this finding, a second activity for the synthesis of ppGpp has been identified. This activity is mediated by the **ppGpp synthetase II (PSII)** or **SpoT**, the product of the *spoT* gene. The SpoT protein has a molecular weight of 79 kDa (around 700 amino acids) and was first characterized as the enzyme responsible for the rapid degradation of ppGpp. During this reaction 3′ pyrophosphate is removed. The reaction requires Mn^{2+} and is very fast, as was seen above. It should be noted, however, that SpoT is a real bifunctional enzyme, and the synthetic capacity is not just the enzymatic reversal of the hydrolytic reaction. Apparently SpoT has separate activities for the synthesis and degradation of ppGpp, which means it has both a 3′-pyrophosphohydrolase and a 3′ pyrophosphotransferase activity for (p)ppGpp. Both reactions have a different cofactor requirement, for instance. The two synthetases RelA and SpoT share considerable amino acid sequence homology (31% identical amino acids) with the exception of the roughly 50 N-terminal amino acid positions. In contrast to RelA, the SpoT protein is not ribosome-associated but cytosolic. Mutants depleted in both genes *relA* and ***spoT*** have been isolated. These mutants are free of detectable levels of ppGpp and show a $ppGpp^0$ phenotype. SpoT is mainly responsible for maintaining the basal levels of ppGpp in the cell and senses differences in the substrate pools for energy production. Its activity is thus related to the growth rate (see Section 8.6). It plays no major role in the induction of the stringent response. The *spoT* gene is encoded together with two other genes in the *spo* operon. Interestingly, one of the cotranscribed genes is *rpoZ*, the structural gene for the ω factor of RNA polymerase. As outlined in Chapter 2 the actual function of the ω subunit is not known. Based on *in vitro* transcription analysis ω has been proposed to be a mediator for the stringent control, but this view has been disputed according to studies performed with *rpoZ* mutants.

8.5.2 Effects of increased ppGpp levels

The effector nucleotide ppGpp has also been termed an **alarmon** which indicates the **hormone-like** function in situations in which the survival of cells is threatened. No metabolic use or benefit from possible phosphate transfer has so far been established for ppGpp or the pentaphosphate derivative pppGpp. This is surprising since, because of the rapid turnover of (p)ppGpp, the number of these high energy phosphate molecules synthesized and degraded during the period of starvation is enormous. The turnover number is comparable to the number of peptide bonds formed in an exponentially growing cell (around 150 000 $cell^{-1}s^{-1}$).

Cells that contain mutations in *relA* show a **relaxed** phenotype. That is, they are unable to induce the stringent response. The most pronounced effect induced by the stringent response, and the first one that was characterized as an immediate consequence of amino acid starvation, is the instantaneous decline in **stable RNA synthesis**. This reaction is not seen in *relA* mutants, which has led to the term relaxed. Inhibition of stable RNA synthesis as a consequence of the stringent control is certainly the most striking response provoked by amino acid starvation. As mentioned earlier, however, the reactions of the stringent response are by far more pleiotropic. The most important effects are listed below:

1. The synthesis of the components of the translation apparatus is inhibited (rRNAs, ribosomal proteins, translation factors IF-3, EF-Tu, EF-Ts, EF-G, and tRNAs).
2. The synthesis of the RNA polymerase subunits is inhibited; the stress-specific σ factors σ^s and σ^{32} are activated.
3. The synthesis and metabolism of amino acids are activated.
4. Proteolysis is activated.
5. Transport is inhibited; an exception is the uptake of branched amino acids, which is activated.
6. Carbohydrate metabolism is activated.
7. Phospholipid metabolism is inhibited.
8. Cell wall synthesis is inhibited.
9. DNA synthesis is inhibited.
10. The fidelity of translation is reduced.
11. The mutation frequency is enhanced.
12. Penicillin tolerance and peptidoglycan synthesis are changed.

There appears to be a common logic that can be discerned behind the list of these pleiotropic changes in cell metabolism. Almost all the changes induced are suitable for overcoming amino acid limitations. Moreover, the responses are suitable for adjustment of the capacity of the protein synthesizing machinery to the altered level of the cellular substrate pool. Clearly, not all the changes observed are *direct* consequences of the stringent response. Inspection of the distribution of the total proteins synthesized before and after induction of amino acid limitation by two-dimensional protein gels reveals that approximately 50% of all proteins synthesized are changed in their expression level. Of course it is very difficult to distinguish *direct* from *indirect* effects. In many cases the changes observed may be secondary consequences of superordinated primary events. This is the typical complicated situation when several networks affect each other. A distinction between 'cause' and 'effect' is sometimes very difficult or even impossible.

The strong inhibitory effect on transcription of stable RNAs has hardly been repeated convincingly in quantitative terms employing *in vitro* experiments with isolated templates and purified RNA polymerase. Many different parameters, e.g. the salt concentration or the topology of the template, have been recognized to be important, and have consequently been optimized. Particular attention has been paid to the relative concentration of substrates and reactants. Nevertheless, only a moderate inhibition was found under almost any *in vitro* conditions, which never matched the strong inhibition known to occur *in vivo*. This has led to the consideration that one or more important factor(s) might be missing in the *in vitro* studies. There have been several attempts to isolate such factors in the past. Several proteins that accumulate rapidly under conditions of amino acid starvation, such as the **stringent starvation protein** **(Ssp)**, for instance, have been considered as likely candidates to mediate or support the stringent response. It transpires, however, that none of the supposed candidates have obvious effects on the efficiency of the *in vitro* systems. This could mean that either there is no such factor or that the *in vitro* systems employed are much too simple to reflect the complex events underlying the mechanism of the stringent response. One important aspect that is completely unknown, for instance, concerns the amount (and recycling) of active RNA polymerase under conditions of elevated ppGpp levels. Furthermore, if one considers the highly complex scenario that leads to the accumulation of ppGpp in a RelA-dependent way, involving active aminoacyl synthetases, translating ribosomes, ATP, GTP and a delicately balanced ratio of charged to uncharged tRNAs, the latter assumption may well be correct. It could well be that the *in vitro* systems employed so far are just too simple and the combined effects of other indirectly related cellular components and substrates are crucial to exert the full stringent control during transcription.

8.5.3 Where does ppGpp act?

One of the major questions in determining how the stringent control works was to define the possible target(s) of (p)ppGpp interaction. Early genetic evidence indicated that RNA polymerase is probably directly involved in mediating the ppGpp effect during stringent control. This conclusion was derived from the independent analysis of several RNA polymerase mutants with altered response to elevated ppGpp concentrations. Mutations in the *rpoB* gene, encoding the RNA polymerase β subunit, often conferred a **relaxed phenotype**, which means that stable RNA synthesis was not as severely reduced as in the wild-type under conditions that normally induce the stringent response. Several such mutants in the *rpoB* gene have been characterized in more detail. Mutations normally clustered in the centre of the molecule close to the sites where mutations conferring rifampicin resistance were located (see Fig. 2.8). Although strains carrying the *rpoB* mutations were able to relieve the toxic effect of high levels of ppGpp *in vivo*, a specific resistance towards the effector could not always be established with purified polymerases from such strains. This might not be too surprising since the *in vivo* effect of the RNA polymerase mutants on the inhibition of stable RNA synthesis was often only about twofold compared to wild-type strains. In other cases it was shown that the cellular level of ppGpp in the mutant did not reach the wild-type level after induction of the stringent response.

Since transcription initiation complexes of stringent-controlled promoters are considered to be intrinsically unstable, RNA polymerase mutants were scanned for the same effect. Such mutants, weakening the interaction with promoters, have been found and again, the mutations mapped in the β subunit. When transcription of a collection of stringently regulated promoters was tested with the mutant RNA polymerases, they behaved as if the stringent response had been induced. Interestingly, the mutant polymerases also increased expression from promoters under positive stringent control. The effects were explained by a different capacity to melt the stringent promoters. Melting in turn was considered to be dependent on the discriminator sequences flanking the promoters under stringent control (GC-rich for negative and AT-rich for positive regulated promoters; see below). The suboptimal spacing of the –10 and –35 promoter elements often associated with stringent-controlled promoters and/or the local superhelicity of the template are also considered to be important determinants for the altered melting behaviour assumed to be characteristic of stringent promoters. Although differences in the melting behaviour of the promoters being studied have not been explicitly demonstrated, the results underline the role of RNA polymerase in mediating the stringent response.

A different set of RNA polymerase mutants has recently been characterized. These mutants were isolated as phenotypic suppressors of ppGpp^0 strains, which are *relA*$^-$ and *spoT*$^-$. Such strains are unable to grow on amino acid-free minimal medium, showing multiple amino acid auxotrophy. The mutants functionally mimicked the action of ppGpp for the expression of both negative and positive regulated genes in the absence of the effector. Several of the mutations identified mapped in the conserved region 3.1 of the *rpoD* gene encoding the σ^{70} subunit (see Section 2.2.4). This does not mean that ppGpp binds to the σ^{70} subunit. Region 3 has been shown, however, to be involved in the association of σ^{70} with the RNA polymerase core enzyme. It was concluded, therefore, that ppGpp may alter transcription initiation by a changed stability of the holoenzyme caused by a reduction in the affinity between σ^{70} and the core polymerase. In fact, the mutant σ^{70} subunits were shown to have a different association profile with the core polymerase when cellular extracts from wild-type and mutant cells were analysed by glycerol gradients. The study does not prove, but adds evidence to the conclusion, that RNA polymerase is the direct target of ppGpp binding.

Several attempts have been made to identify the putative target for ppGpp interaction employing spectroscopic and biochemical methods. Supporting indications for a direct ppGpp–RNA polymerase interaction have come from measurements of the changes in circular dichroism. In addition, chemically reactive derivatives of ppGpp have been synthesized and used for cross-linking experiments. Although several RNA polymerase subunits have been identified as targets for the reactive chemical compound from these cross-linking studies, the results have not been conclusive, since in none of the reported cases was genuine ppGpp used as the affinity label. Rather, modified chemical analogues, such as azido-ppGp, whose specificity is probably significantly different from ppGpp, were employed. Cross-linking studies with structurally unaltered ppGpp and high energy ultraviolet (UV) irradiation have failed to demonstrate direct RNA polymerase contacts so far or, at least, have yielded inconclusive results.

Better indications for a direct interaction between ppGpp and RNA polymerase stem from studies employing ppGpp labelled with a fluorescent probe at the terminal phosphates. The fluorescent derivative was bound to RNA polymerase and distance measurements using fluorescence resonance energy transfer (FRET) indicated that the derivative was located about 27 Å away from the rifampicin-binding domain of RNA polymerase. This study excluded the σ subunit as a binding target. Although competition experiments and *in vitro* inhibition studies indicated that the fluorescent ppGpp derivative reacted almost identically to unmodified ppGpp, and therefore appeared to mimic the specificity of the original effector, minor doubts remain as to whether the reaction conditions actually represent the *in vivo* situation of ppGpp-mediated transcriptional repression.

In summary, the evidence that RNA polymerase is in fact the target for ppGpp interaction is compelling. Unequivocal proof of a direct RNA polymerase–ppGpp binding and determination of where exactly the site(s) of interaction is (are) located, is still lacking, however. In particular, the participation of auxiliary factors has not yet been ruled out completely. The elusive transduction pathway from synthesis to the site of action of ppGpp therefore still awaits further experimental clarification.

8.5.4 The major target of stringent regulation: stable RNA synthesis

Inhibition of the synthesis of stable RNAs (rRNAs and tRNAs) is by far the most pronounced effect of the stringent control, which follows immediately after amino acid starvation. This stringent relationship originally led to the term stringent control. The synthesis of stable RNAs makes up more than 90% of cellular transcription (see Table 1.1). These molecules do not require translation but are essential for the translation apparatus and the translation process. On the other hand, induction of the stringent response does not cause a strong increase in degradation of the fraction of stable RNAs. It is clear, therefore, that the inhibition must be exerted at the level of transcription. Transcription initiation was envisaged to be the major step in the mechanism of inhibition of stable RNA synthesis during the stringent response. This has led to the comparison of promoter sequences of genes known to be under negative stringent regulation. From such comparisons, it has been concluded that a conserved CG-rich sequence between the −10 region of stringently controlled promoters and the transcription start site probably functions as a **discriminator**. This idea has stimulated a number of detailed investigations to find out whether or not specific promoter sequences are responsible for the stringent control. In fact, most transcription units under negative stringent control in *E. coli* contain the so-called **discriminator motif GCGC** immediately downstream of the −10 region. Mutations in the GCGC discriminator sequence of the promoter upstream of the *tufB* operon which, among other genes, encodes the translation factor EF-Tu, clearly indicated that every position within the GCGC sequence was required to confer stringent control *in vitro*. Transcription of rRNA operons is initiated from two promoters P1 and P2 in tandem orientation (see Section 8.7). From a comparison of *E. coli* rRNA promoters it is known that all P1, but not P2, promoters contain the perfect GCGC discriminator motif. This observation is consistent with the notion that stringent regulation is mainly exerted at the rRNA P1 promoters. Recent *in vivo* studies have shown, however, that *E. coli* rRNA P2 promoters are also under stringent regulation, albeit not to the same extent (see below). When a mutant P2 promoter with a perfect GCGC discriminator motif was analysed *in vitro* and *in vivo*, the mutation conferred stringent (as well as growth rate) control to the variant P2 promoter in the same way as was known for the P1 promoter. In contrast, the same GCGC sequence, when linked to the synthetic P*tac* promoter, does not convert this promoter to either stringent or growth rate regulation. The results indicate that the GCGC sequence motif is a *necessary* but not *sufficient* element in providing stringent sensitivity (and also growth rate control, see Section 8.6 below). To determine whether additional promoter elements downstream or upstream of the GCGC discriminator motif might be required for stringent and growth rate control, the upstream and downstream sequence portions relative to the GCGC motif of the regulated rRNA P1 promoter and the unregulated P*tac* promoter were exchanged in all possible combinations. Analysis of these hybrid promoters revealed that none of the promoter sequence elements on its own was able to restore the original regulatory behaviour of their parent promoters. These results imply that the signal for stringent sensitivity (and growth rate control; see below) is not restricted to a particular sequence motif within the promoter region (Fig. 8.8). The GCGC element is certainly a necessary component but obviously the complete promoter context determines whether or not negative stringent control is exerted at this promoter.

Several attempts have been made to compare the kinetics of transcription initiation in the presence and absence of ppGpp. The results of such studies are controversial. The comparability of the results probably suffers a lot from the fact that the choice of the *in vitro* reaction conditions employed, which most importantly also includes the activity of the RNA polymerase used, differed greatly in each case. In an attempt to study growth rate regulation the kinetics of initiation complex formation and dissociation have been analysed in a heterologous system employing *E. coli* RNA polymerase and the *rrnB* P1 and P2 promoters from *B. subtilis*. The two promoters have been shown previously to be functional in *E. coli*. Interestingly, in *B. subtilis* P2 promoters appear to be the major target for regulation. This means that they exhibit differential control. In contrast to the *E. coli* rRNA promoters only the downstream *B. subtilis* promoter P2 is sensitive to ppGpp. Analysis of the kinetics of formation and dissociation of RNA polymerase promoter complexes was performed. The existence of a three-step initiation mechanism with three distinct intermediate polymerase–promoter complexes was deduced from the kinetic measurements. According to the study the initiation pathway is thought to involve the formation of

Promoter	-35 -10 Discriminator	Stringent sensitive	Growth rate sensitive
P1:	AAATTTCCTC**TTGTCA**GGCCGGAATAACTCCC**TATAAT**GCGCCACCAC	+++	+++
P2:	AAATAAATGC**TTGACT**CTGTAGCGGGAAGGCG**TATTAT**GCACACCCCG	+	+
P2F:	AAATAAATGC**TTGACT**CTGTAGCGGGAAGGCG**TATTAT**GC**G**CACCCCG	+++	+++
PtacW:	AAATGAGCTG**TTGACA**ATTAATCATCGGCTCG**TATAAT**GTGTGGAATT	-	-
PtacM:	AAATGAGCTG**TTGACA**ATTAATCATCGGCTCG**TATAAT**G**C**G**C**GGAATT	-	-
P1TM:	AAATTTCCTC**TTGTCA**GGCCGGAATAACTCCC**TATAAT**GCGCGGAATT (P1 ← → PtacM)	-	+
TMP1:	AAATGAGCTG**TTGACA**ATTAATCATCGGCTCG**TATAAT**GCGCCACCAC (PtacM ← → P1)	-	-

Figure 8.8 Effects of the discriminator sequence on stringent and growth rate sensitivity of rRNA promoters. A series of promoter sequences with differences in the discriminator region (highlighted in grey) and changes in the upstream and downstream core promoter elements is shown. The –10 and –35 elements are underlined and presented in bold type. The sequences include the two *rrnB* promoters P1 with a natural perfect discriminator sequence GCGC. The ribosomal promoter P2 differs at one position from the ideal discriminator consensus. P2F represents a variant P2 promoter which has an A to G base change conferring a perfect GCGC discriminator sequence. PtacW represents the wild-type tac promoter which does not have a consensus discriminator. PtacM is identical to PtacW except for a perfect GCGC discriminator sequence. Promoter P1TM is a hybrid with the sequence upstream of the discriminator derived from promoter P1 and the sequence downstream from the discriminator derived from PtacW. The promoter TMP1 is the reverse construct with the PtacW sequence upstream of the discriminator and the P1 sequence downstream thereof. The response to stringent control or growth rate sensitivity is indicated in the adjacent columns: +++ corresponds to strong sensitivity, + weak sensitivity, – not sensitive. The classification reflects results obtained both *in vitro* and *in vivo*.

primary heparin-sensitive closed complexes which isomerize to initial complex intermediates and finally to heparin-resistant open complexes. The presence of ppGpp did not affect any one step of this pathway between the regulated (*rrnB* P2 in the case of *B. subtilis*) and the unregulated (*rrnB* P1) promoters by more than a factor of two. It was concluded therefore that ppGpp might not be a regulator of the *early* steps of rRNA transcription initiation. It should be noted that ppGpp-sensitive promoters in *B. subtilis* do not share the same discriminator motif known to be characteristic for stringent and growth rate regulated *E. coli* promoters. This indicates that the RNA polymerase may play a direct and specific role in the ppGpp-mediated regulation. Alternatively, these results might be taken as an indication that there are species-specific differences in the mechanism of the stringent control.

In a different study three *E. coli* promoters were analysed in a homologous *in vitro* system. The three promoters had been previously analysed *in vivo* and shown to exhibit differential ppGpp sensitivity (see Fig. 8.8). Measurements of the rates of RNA polymerase–promoter complex formation and dissociation allowed the derivation of a kinetic scheme consistent with the function of ppGpp to trigger a different initiation pathway at sensitive promoters. The kinetic scheme was based on the surprising result that in the presence of ppGpp the initial binary closed complexes were stabilized at sensitive promoters. Open complex formation, however, was impeded and the ternary complexes formed subsequently (in the presence of NTPs) appeared to be structurally altered. The results are consistent with a kinetic mechanism in which ppGpp-altered RNA polymerases are preferentially bound to sensitive promoters, where they enter a different initiation pathway. This phenomenon has been designated '**trapping**'. The results indicate that discrimination between stringent and non-stringent transcription regulation occurs at an early step of the initiation pathway. Effective inhibition is caused at later steps of the transcription cycle, involving promoter clearance and transcription elongation (Fig. 8.9).

The derived model is compatible (and actually extends) a formerly postulated **RNA polymerase partition model**, which was developed to explain the differential regulation of stringently and non-stringently controlled promoters. According to the RNA polymerase partition model two interconvertible populations of RNA polymerase exist in the cell. The two types of polymerase have different affinities for stringent and non-stringent promoters; that is, stable RNA promoters and mRNA promoters. The interconversion between polymerases with different activities in the transcription of mRNA genes and stable RNA genes is mediated by ppGpp (probably through binding to the β subunit). The partition model fits into the branched kinetic model described above. Only those RNA polymerases that have been modified by ppGpp enter the alternative kinetic pathway at sensitive promoters, where they are trapped in the form of a closed complex.

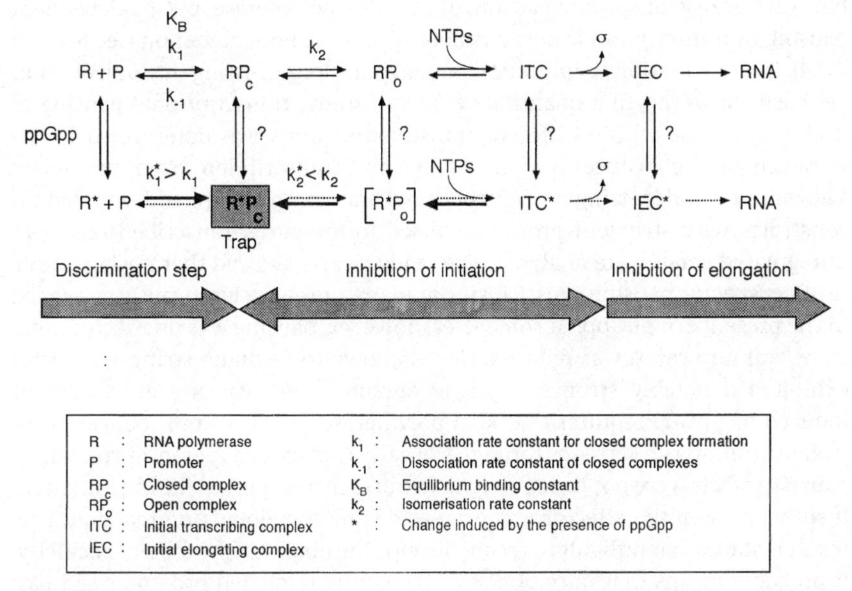

Figure 8.9 Kinetic model for ppGpp-dependent inhibition at stringently and non-stringently controlled promoters. The upper pathway represents the simplified mechanism of transcription initiation in the absence of ppGpp as shown in Figs 3.1 and 3.7, respectively. The lower pathway represents transcription initiation in the presence of ppGpp. In the presence of ppGpp a partition of polymerases to the lower pathway is observed. Changes in the kinetic constants or the complexes mediated by the presence of ppGpp are indicated by an asterisk. In the first step, designated the discrimination step, the ppGpp modified reaction path leads to a stabilization of the binary closed complex (R^*P_C) which sequesters ppGpp modified RNA polymerase (R^*) and thus functions as a trap for productive transcription initiation complex formation. The concentration of the open complex (R^*P_O) in the presence of ppGpp is diminished compared to the situation in the absence of ppGpp. A major effect of the known ppGpp-dependent inhibition is also due to later steps of the transcription initiation pathway, namely abortive cycling and promoter escape. Finally, ppGpp-dependent inhibition is additionally observed during elongation after σ has left the transcribing complex. ? indicates that it is not known whether modification of kinetic steps at intermediate complexes by ppGpp is possible or reversible. The figure is adapted from Heinemann and Wagner (1997).

8.5.5 ppGpp affects the rate of RNA chain elongation

From many studies in the past it is clear that the effect of ppGpp on transcription is not restricted to the initiation stage. Early observations demonstrated that *in vitro* transcription of the *E. coli rrnB* operon could be specifically inhibited by the ppGpp-dependent reduction of the transcription elongation rate. This effect was not the result of a generally lowered step time during NTP addition but was because of specific pausing of the RNA polymerase. ppGpp-dependent pausing of transcription is now a generally accepted phenomenon (see Section 4.3.4). Its contribution to inhibition under conditions of stringent control is not yet clear, however. In a quantitative *in vitro* study, transcriptional pausing of RNA polymerase within different transcription units was determined in the presence of high concentrations of ppGpp. Transcription from promoters known to be under stringent control or not was compared. In addition, hybrid constructs with stringent promoters fused to the early transcribed region of unregulated genes were analysed. The study clearly showed that ppGpp is able to affect specific pausing sites. Pausing at many sites was significantly enhanced in the presence of ppGpp. At some sites, however, pausing was unaffected, and, in several rare cases, was reduced. Genes known to be under stringent control exhibited a notably stronger pausing enhancement. Pausing enhancement induced by ppGpp requires that RNA polymerase initiates from stringent controlled promoters *and* passes through the early transcribed region. Surprisingly, pausing effects were not dependent on the presence of ppGpp during initiation. It sufficed when the effector was present during the elongation reaction. This result is somewhat difficult to reconcile with the observed promoter specificity. It probably means that initiation at a stringently regulated promoter and passage of the RNA polymerase through the early transcribed region results in a structural transition that is more accessible or sensitive to the action of ppGpp. The results are consistent with earlier studies on transcription elongation of stable RNA genes in the presence of ppGpp. Strong ppGpp-dependent pausing sites in a sequence region downstream of the *E. coli* rRNA promoter P1, and close to promoter P2, led to the proposal of a **turnstile attenuation** mechanism which regulates rRNA transcription because of RNA polymerase queuing and **promoter occlusion**. As exemplified in Fig. 8.10, a small pause of several seconds close to the promoter may initiate quite a dramatic reduction in the overall synthesis of transcription units initiated at high frequency. How far these findings reflect the *in vivo* situation is yet to be deduced from future studies.

Experiments in which the transcription elongation rates at elevated ppGpp levels were assessed *in vivo* have been performed, but a direct comparison between the *in vitro* and *in vivo* data sets is difficult. In the *in vivo* studies a synthetic inducible hybrid promoter consisting partly of the strong T7A1 promoter fused to a fragment from the *lac* promoter containing two operator sites was used. The appearance of transcripts (the *infB* and *lacZ* mRNAs as well as truncated stable RNA products) with a specific length was quantified by hybridization. These studies clearly showed a significantly reduced transcription rate in the presence of high concentrations of ppGpp for the mRNAs tested. The measured elongation rates for the mRNA transcripts was reduced to about half the rate before the stringent control was provoked. In contrast, transcription elongation rates for the hybrid rRNA construct were not reduced. Moreover, the study revealed that the elongation rate of the rRNA construct was about twice

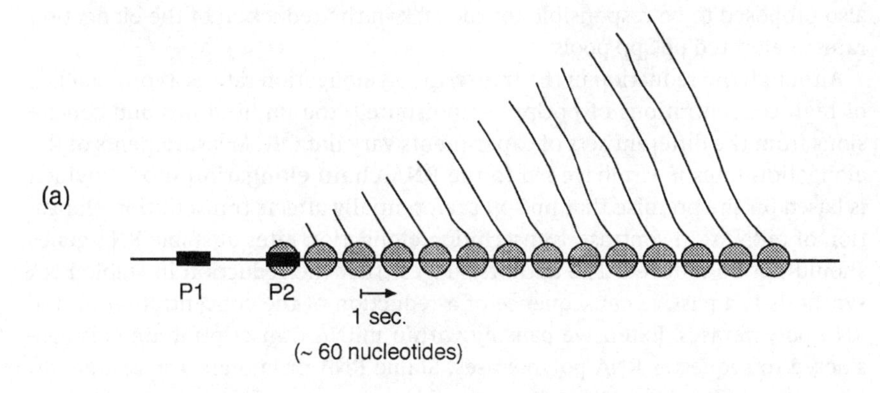

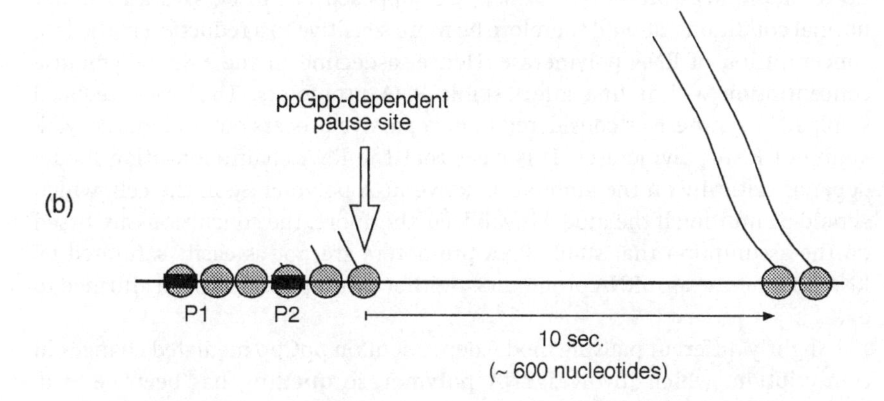

Figure 8.10 ppGpp-dependent promoter occlusion because of RNA polymerase pausing. The transcription situation of an rRNA operon with the two tandem promoters is schematically depicted. (a) The strong rRNA promoters load RNA polymerase at a very high rate of about one enzyme per second. No higher initiation frequency is possible as the rate of elongation (about 60 nucleotides per second; see Section 4.2) is limiting, and one RNA polymerase molecule is almost in contact with the next. (b) If a pause is induced by ppGpp in the proximity of promoter P2, as can be observed *in vitro*, stalling RNA polymerases will block the upstream promoters by a turnstile mechanism. A pause of only 10 seconds will cause a significant effect on overall transcription as it creates a gap of about 600 nucleotides without transcribing polymerase. The figure is adapted from Kingston *et al.* (1981).

the level of the mRNA rate in the absence of ppGpp. The presence of a particular sequence region, a minimal functional *nut* site, known to be necessary for antitermination (see Section 5.4) was responsible for the high transcription elongation rate. According to the above studies the antitermination sequence is also proposed to be responsible for the lack in the reduction of the elongation rates at elevated ppGpp pools.

Although the reduction in the transcription elongation rate as a consequence of high concentrations of ppGpp is undisputed, the implications and conclusions from the different sets of experiments vary notably. Measurements of the elongation rates *in vivo* have led to the **RNA chain elongation model**, which is based on the premise that ppGpp preferentially affects transcription elongation of mRNAs. In contrast, transcription elongation rates of stable RNA genes should not be reduced. The authors suggest that the reduction in stable RNA synthesis is a passive consequence of a reduction of the concentration of free RNA polymerases. Extensive pausing within mRNA transcription units is considered to sequester RNA polymerases. Stable RNA promoters are assumed to have a low affinity for RNA polymerase, in contrast to mRNA promoters which are considered to be saturated by RNA polymerase under normal growth conditions. Stable RNA promoters, which are supposed not to be saturated under normal conditions, should therefore be more sensitive to a reduction in the free concentration of RNA polymerase. Hence, a decline in the RNA polymerase concentration will at first affect stable RNA promoters. Therefore, reduced stable RNA synthesis is considered to be a passive process due to limited availability of RNA polymerases. It is clear that the RNA chain elongation model depends critically on the amount of active RNA polymerase in the cell, which should be limiting if the model is valid. Furthermore, the conclusions are based on the assumption that stable RNA promoters are not as easily saturated by RNA polymerase as mRNA promoters. Neither assumption has been affirmed to date.

A slightly different pausing model dependent on ppGpp mediated changes in transcription, which involves **RNA polymerase queuing** has been derived. The queuing model is based on theoretical calculations and takes into account that a linear inverse correlation between the growth rate and the cellular concentration of ppGpp can consistently be observed under almost all conditions. This model has been put forward mainly to explain ppGpp-dependent growth rate control and more details are presented in Section 8.6.

8.5.6 Positive stringent control

One of the interesting aspects of the stringent control is the fact that transcription regulation mediated by the same signal (ppGpp) inhibits a large number of transcription units while at the same time transcription of a special set of genes is activated. To the group of operons activated under conditions of amino acid starvation belong many operons encoding enzymes for the biosynthesis of amino acids but also genes such as *rpoS* which encode the stationary phase-specific σ factor. As indicated above, this positive regulation makes physiological sense with respect to the re-establishment of the desired levels of amino acids. The promoter regions of several positively regulated amino acid biosynthetic genes were compared. In a similar way to which the promoters control negatively regulated genes, promoters directing the positively regulated operons share some common characteristics. The –10 region and the flanking promoter downstream sequences of positively regulated promoters have proven to be important for transcription activation at elevated ppGpp concentrations. Among the positively regulated operons the *Salmonella typhimurium his* operon has been studied in more detail. The –10 region of the *his* promoter has the sequence TAGGTT instead of the σ^{70} consensus TATAAT. In contrast to the GC-rich discriminator found at negatively ppGpp regulated promoters the comparable discriminator for positively regulated genes is A-rich. In the case of the *S. typhimurium his* promoter the sequence immediately downstream of the –10 region and upstream of the transcription start site is **AAAAGGT**. The significance of the two promoter elements in conferring positive stringent regulation has been demonstrated by a series of mutations within these regions. Base changes in the two elements clearly affect the positive control exerted by ppGpp *in vitro* and *in vivo*. Positive control can no longer be seen when the –10 region is converted to the σ^{70} consensus sequence. Promoter strength is increased but transcription is no longer dependent on the presence of ppGpp. Interestingly, fusion of the *his* promoter to the discriminator sequence GCGC, characteristic for negatively regulated genes, causes neither repression nor activation by ppGpp, underlining the notion that the discriminator is not a sufficient requirement for both positive and negative stringent control.

Many effects of enhanced gene expression in the presence of ppGpp are probably indirect. It has been documented in several cases of positively regulated genes that the mRNA stability increases significantly after induction of the stringent response. The decay of mRNAs, for instance, is certainly dependent on the efficiency of translation, which itself is affected by ppGpp. Notable examples where positive control is mainly affected by mRNA stability are the stringent starvation protein (Ssp), which is in fact identical to the product of the *ompA* gene. In addition, part of the effect on the ppGpp-dependent increase in *rpoS* expression is certainly due to post-transcriptional effects. The fact that *rpoS* expression is positively linked to enhanced levels of ppGpp is a clear indication that several effects observed under stringent control may not be directly due to the presence of ppGpp but might be mediated indirectly by σ^{s} induction.

8.6 Growth rate regulation

Growth rate regulation can be distinguished physiologically from stringent control, although both control systems share many common properties and may in fact be mechanistically very similar. Growth rate regulation describes the phenomenon that the rate of stable RNA synthesis in cells growing at steady-state is roughly proportional to the square of the growth rate. In contrast, the synthesis rates of mRNAs or the protein synthesis rates remain approximately constant. Hence, the synthesis of stable RNAs is said to be under growth rate control (Fig. 8.11). Whether or not transcription of a given gene is growth rate-dependent is defined by the promoter structure. Comparison of many promoters and results from mutational studies have revealed that the determinants for growth rate control and stringent control are very similar, if not identical. Sequences responsible for growth rate regulation have been restricted to the promoter core regions, including the discriminator element. The upstream sequences (UAS or UP; see Section 6.2.1) important for enhancing promoter efficiency are not responsible for growth rate regulation. In addition,

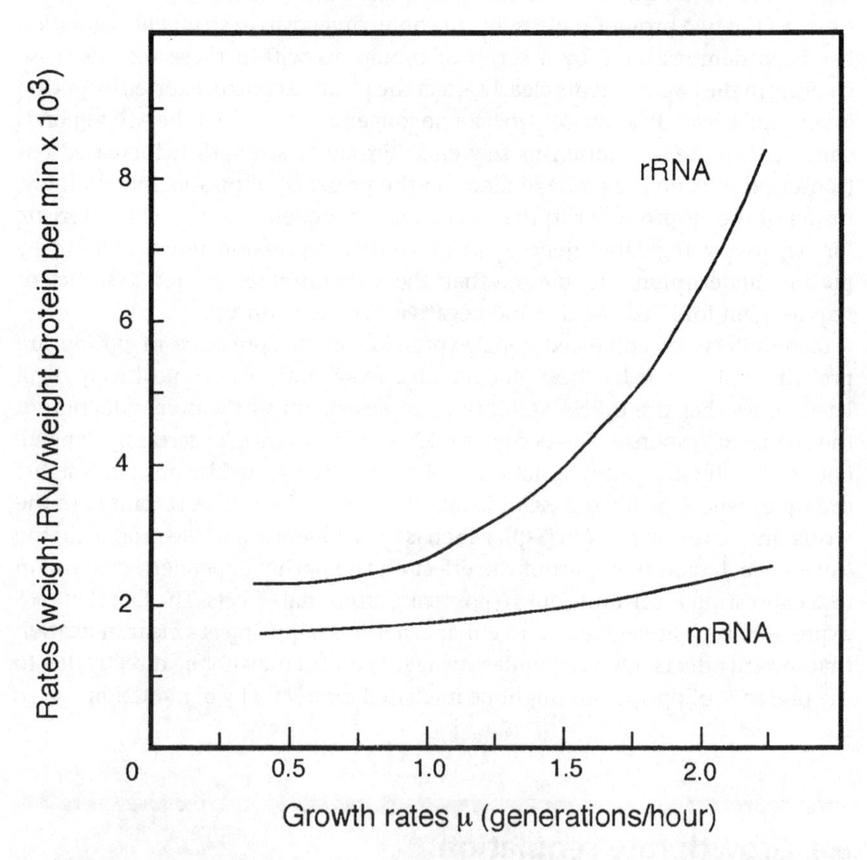

Figure 8.11 rRNA synthesis rates are growth rate-dependent. The rates of average mRNA synthesis and rRNA synthesis are plotted as a function of the growth rate μ (generations/hour). While the mRNA rates do not show a significant dependence on the growth rates, such a dependence is apparent for the synthesis rates of rRNAs. The rRNA synthesis rates are roughly proportional to the square of the growth rate. This correlation is expressed as the growth rate law.

characteristic regions where transcription factors generally interact are not necessary for this kind of control. There are also some promoters which respond inversely to an increase in the growth rate. These promoters have been termed **gearbox promoters** (see Section 8.4). Examples of gearbox promoters are found upstream of the *ftsQAZ* operon encoding genes important for cell division, e.g. the *bolA1* promoter, encoding a cell shape regulator protein, or the promoter directing transcription of the *mcb* gene required for immunity to the antibiotic microcin B17. Gearbox promoters are characterized by specific sequences upstream of the transcription start point as well as by their –10 region. They generally have an AT-rich UP region. Their –35 consensus element is **CTGCAA** rather than the TTGATA of standard σ^{70} promoters. The –35 region is separated between 14 and 16 base pairs from the –10 region and has the consensus sequence **CGGCAAGT** rather than TATAAT (Fig. 8.12). Gearbox promoters provide inverse growth rate dependence. This means that genes controlled by gearbox promoters show increased expression when the growth rate decreases. These promoters are generally induced during stationary phase but their inverse regulatory behaviour is also observed during exponential growth.

Apart from a characteristic promoter structure, growth rate regulation requires a mechanism that links the activity of the promoters to the physiological state of the cells. Hence, from the early studies of bacterial physiology such a link was sought. Since these early studies the concentration of the effector ppGpp had been considered to be responsible for growth rate regulation. This conclusion was based on the fact that a linear inverse correlation of the ppGpp concentration and the growth rate can be demonstrated in nearly all conditions (reported exceptions are conditions of pyrimidine starvation). The basal level of ppGpp varies between 10 and 80 μM depending on the growth rate and on the particular strain. This concentration is predominantly maintained through the activity of SpoT. The much higher concentrations (around 1 mM)

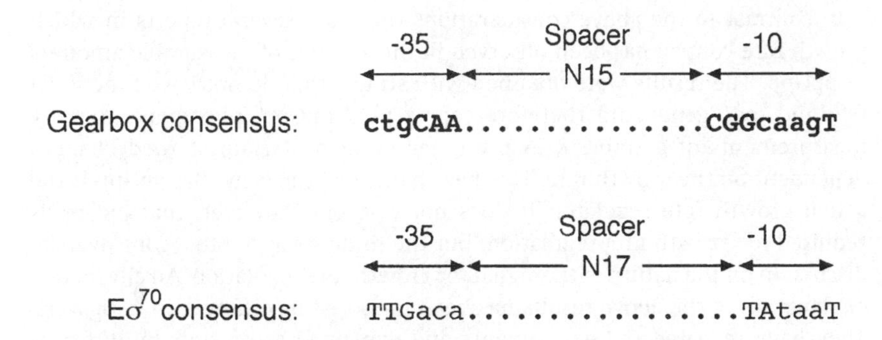

Figure 8.12 Consensus structure of gearbox promoters. The consensus sequence of typical gearbox promoters is compared with the σ^{70} consensus structure. Capital letters indicate more invariant positions. N stands for a non-defined sequence. The gearbox consensus sequence is given according to Aldea *et al.* (1990).

found during amino acid starvation require the activity of RelA. The relationship between the basal ppGpp level and the growth rate does not only hold true when the growth rate is affected by variations in the growth medium. The same correlation has been observed when strains with different mutations in the *spoT* gene were analysed. A collection of different mutations affecting the activity of SpoT, which is responsible for the basal ppGpp level in the cell, was tested for differences in their growth rates. The different *spoT* mutations resulted in different growth rates when cells were grown in the same minimal medium. Measurements of the basal ppGpp concentrations of these mutants revealed the same linear inverse relationship between growth rate and ppGpp level as in wild-type cells grown in different media.

A largely constant correlation to the level of ppGpp has also been found for the rate of stable RNA synthesis which defines growth rate control. The rate of stable RNA synthesis is commonly determined as the ratio of the rate of stable RNA synthesis (r_s) to the rate of total RNA synthesis (r_t), briefly $\mathbf{r_s/r_t}$. The ratio r_s/r_t shows a fixed relation to the levels of ppGpp both in relaxed and in stringent strains. At very low ppGpp levels nearly all RNA transcription is devoted to stable RNA synthesis and the ratio approaches 1 (in ppGpp^0 strains the r_s/r_t ratio is 0.5 because of an increase in mRNA synthesis rather than a decrease in stable RNA synthesis). At high levels of ppGpp r_s/r_t reaches a minimum level of roughly 0.25, with most transcripts originating from promoters that are not under ppGpp control. Together, these observations have led to the proposal that ppGpp directly determines changes in the growth rate and affects stable RNA synthesis. In addition, the fact that mutations disrupting or changing the stringent control almost always affect growth rate control in the same manner has led to the conclusion that both mechanisms are very similar and probably share the same determinants. Many experiments support but do not prove the assumption that ppGpp is actually the signal for growth rate regulation (see Fig. 8.8).

In contrast to the above considerations there are several reports in which growth rate control has been observed in the absence of measurable amounts of ppGpp. The results were obtained with strains that do not have functional *relA* and *spoT* genes and therefore represent a ppGpp^0 phenotype. Activity measurements of a stable RNA promoter in such a strain showed changes dependent on the growth rate. This has challenged the view that ppGpp is the actual growth rate regulator. It does not rule out, however, that ppGpp is required for growth rate regulation, but the finding has led to a controversial discussion on the nature of the signal for growth rate regulation. An alternative explanation of the above results has been proposed by different investigators. They have repeated the experiments and explained the results by different activities of the RNA polymerase under the conditions employed. However, the controversy is not yet resolved. If the initial interpretation of the results reported is correct, it means that there must at least be a second ppGpp-independent mechanism for growth rate regulation. In addition, the analysis of promoter mutants which are sensitive to stringent control but apparently do not show growth rate regulation has been interpreted as an indication that the mechanisms of stringent control and growth rate regulation are not identical. If one considers, however, that quite high concentrations of ppGpp are required to induce the stringent response as opposed to the small changes observed at different growth rates, the mutant promoters that exhibit apparent differential regulation may have lost some of the sensitivity required to exhibit both types of control.

Metabolic growth rate control has also been proposed to be the consequence of subsaturation of the principal macromolecular synthesis reactions such as transcription and translation with substrates and catalytic components. In this proposal, growth rate-related changes in gene expression are explained as a consequence of medium-induced changes in the kinetics of RNA and protein *chain elongation* owing to the amount of free ribosomes and free RNA polymerases.

Changes in the growth rate definitively affect the rate of stable RNA synthesis. An important finding is the observation that the effect is exerted at the level of both transcription initiation and elongation. As already outlined in Section 8.5.5, the transcription elongation rate depends on the ppGpp concentration. RNA polymerase pausing is most likely to be critical step affecting the elongation rate. Several investigations have shown that the RNA chain elongation rate depends on the growth rate. Different models related to this finding have already been discussed above (see Section 8.5.5). A modified model of transcriptional regulation in response to ppGpp-dependent pausing was derived from theoretical calculations (**queuing model**). The authors start from the premise that inhibition of stable RNA synthesis occurs mainly at slow growth and high concentrations of ppGpp. mRNA synthesis in turn is inhibited mainly during fast growth when ppGpp concentrations are low. The proposed pausing is assumed to occur specifically in the early transcribed regions of mRNAs in response to low ppGpp levels. This would lead to RNA polymerase queuing and cause **promoter occlusion**. RNA polymerases are subsequently diverted to stable RNA promoters. Stable RNA transcription can thus be stimulated at high growth rates and low ppGpp levels. This model deviates from the **chain elongation model** outlined in Section 8.5.5 in several respects. Whereas the queuing model rests on the frequent occurrence of ppGpp-dependent pauses not too far downstream of the promoter to enable promoter occlusion, the chain elongation model assumes many pauses distributed throughout all of the mRNA genes. Hence, regulation is not mediated by promoter occlusion but through sequestering RNA polymerase because the average transcription rate is slowed down and more RNA polymerases are associated with the DNA segments encoding mRNA genes. The models differ further in the assumption, predicted only by the authors of the queuing model, that the intrinsic promoter strength can be changed by ppGpp and high levels of ppGpp inhibit the initiation step at stable RNA promoters.

An often cited hypothesis to explain growth rate regulation without the participation of ppGpp is known as the **ribosome feedback control** mechanism. This proposal is based on the experimental finding that increasing the gene dosage of intact rRNA operons does not cause a comparable increase in the synthesis of rRNAs. The cells react as if excess concentration of ribosomes represses further rRNA synthesis. No such repressing activity was observed when the gene dosage was increased by truncated rRNA operons or operons encoding non-functional rRNAs. Under these conditions transcription from rRNA promoters increase roughly in proportion to the gene copies. It was at first proposed that free functional ribosomes may act as feedback regulators for stable RNA synthesis. It was later shown that, by limiting the amount of the translation initiation factor IF2, free ribosomes actually stimulated rRNA synthesis. Therefore it was concluded that not free but translating ribosomes may provide the feedback signal. Despite several attempts to verify the repressing activity of ribosomes no conclusive proof has been obtained for this conjecture. In addition, no other compound of the translation machinery has been demonstrated to function as a feedback regulator. On the other hand, experiments have been performed that measure the ppGpp content of strains which carry additional rRNA operons. It was found that the ppGpp level was increased and the r_s/r_t ratio was correspondingly lowered. Based on these findings the feedback control was considered to be only one facet of the pleiotropic actions mediated by ppGpp. The ribosome feedback control mechanism is supported, on the other hand, by a study in which several of the seven *E. coli* rRNA operons were deleted. In such strains the deficiency in rRNA operons was compensated by an increase in rRNA transcription from the remaining operons. The compensation was achieved not only by a higher initiation frequency but also by a higher chain elongation rate. No explanation for the latter effect has been given. Since it was found that the ppGpp level was apparently unchanged in the mutant strain the increase in transcription initiation has been taken as an indication for a feedback control.

One important argument of the ribosome feedback control hypothesis is based on the apparent lack of a specific repressor for stable RNA synthesis. Such a potential repressor has recently been discovered, however. The DNA-binding protein H-NS has been shown to bind to the rRNA promoter region and to antagonize the activating effect of FIS. Moreover, H-NS was shown to cause repression of rRNA transcription at late growth phases and under conditions of reduced growth *in vivo*. H-NS fulfils all the requirements expected for a growth rate regulator of rRNA synthesis. It has binding sites overlapping the rRNA core promoter elements, its activity is linked to the growth rate, and it acts only at *rrn* P1 promoters. H-NS is therefore likely to constitute a (redundant) control system for rRNA transcription, which particularly takes effect under conditions when ppGpp-dependent mechanisms of growth rate control fail to act. There is good reason to conclude that H-NS is partly responsible for the observations related to the feedback phenomenon of stable RNA synthesis.

Recently a different idea has been put forward to explain ppGpp-independent growth rate regulation. It is known for several rRNA promoters that they will only form stable initiation complexes in the presence of the correct initiating NTPs *in vitro*. The cellular NTP concentrations for initiation at *rrn* P1 promoters (ATP and GTP) are known to increase with increasing growth rates, however. This increase correlates with the rRNA P1 promoter activities. Based on these notions it was assumed that the concentration of NTPs is sensed by the instability of the initiation complexes before NTPs are used in catalysis as substrates of transcription. The putative NTP-dependent regulation of the rRNA synthesis rates has been taken to suggest a model for the homeostatic regulation for the synthesis of ribosomes. It is not clear, however, whether the *in vivo* NTP concentrations are really limiting.

8.7 Lessons from a manifold regulated system—the synthesis of ribosomal RNA

In this final section an attempt is made to summarize the wealth of known facts on the regulation of rRNA transcription in bacteria. Ribosomal RNAs are central molecules for all living organisms. They are not only structural constituents but also important catalytic components of ribosomes, the cellular machinery for translation. In bacteria rRNA synthesis determines the capacity of the protein synthesizing machinery. Since the rate of protein synthesis is in turn directly linked to cell growth the importance of the control of rRNA transcription for cell growth is immediately evident.

The regulation of rRNA synthesis shows outstanding complexity, involving many different mechanisms. rRNA transcription is linked to various global regulatory networks, some of which have been described in the preceding chapters. The participation of many different regulatory mechanisms and the high complexity of the control systems involved certainly underlines the importance of rRNA molecules. The study of the control mechanisms of rRNA synthesis furthermore provides a nice example indicating that important cellular constituents are generally carefully regulated, very often in redundant fashion. The many established links to different control circuits further support the central role that rRNAs and ribosomes play in the growth and general metabolism of living cells. The sections below explain this central role of bacterial rRNAs in more detail. The introductory remarks are followed by a paragraph on the regulatory implications which are based on the structure of the rRNA operons and their promoters as well as consequences related to the arrangement of the different rRNA operons on the bacterial chromosome. The important function of sequences upstream of the transcription start sites, and the effects of *trans*-acting factors in both the activation and repression of rRNA synthesis are then described. Aspects of the preceding sections which are fundamental to the synthesis of rRNAs, namely stringent control and growth rate regulation, will briefly be recalled. Finally, regulatory mechanisms acting at the level of transcription elongation, such as RNA polymerase pausing and antitermination, are presented, and the coupling to post-transcriptional control mechanisms will be summarized.

This final section aims to serve another function. The complexity of the regulation of rRNA synthesis is an impressive example underlining the concerted action of the different mechanisms of transcription regulation which have been the subject of preceding chapters. This section should also remind the reader of the introductory statements that the transcription process in bacteria cannot be regarded as an isolated process, but that transcription is intimately linked to other macromolecular synthesis reactions and metabolic pathways that occur simultaneously in the cell. Last but not least, because of the large number of different regulatory aspects involved in the control of rRNA synthesis, this section deliberately acts as a practical example, and repeats and deepens some of the information presented in the sections above.

8.7.1 The importance of rRNA synthesis for ribosome formation and cell growth

In their natural habitat bacteria encounter a constantly changing environment. Situations which are optimal for growth and conditions of starvation or environmental stresses, like osmotic shifts, temperature changes, pH shifts, or competition by other organisms change rapidly. The life cycle of bacteria has thus been described as changing between '*feast*' and '*famine*'. Bacteria are exceptionally well adapted to such changes. They have evolved many protective mechanisms, some of which (e.g. heat-shock, stringent control or SOS response) have been summarized in the preceding chapters. In addition, they are able to adapt their growth rate immediately in response to the rapidly changing conditions.

Bacterial growth is directly linked to the capacity of the cells for protein biosynthesis. The organelles responsible for carrying out protein synthesis are ribosomes. Changing the growth rate means that the protein synthesis rate must also change. There is, however, a natural upper limit at which a single ribosome can catalyse polypeptide synthesis. This does not normally exceed 20 amino acids per second. Hence, an increase in total protein synthesis can only be achieved by an increase in the number of ribosomes. In fact, the number of ribosomes in rapidly growing cells increases in proportion to the cell mass and in a single cell, under optimal nutritional conditions, more than 50 000 ribosomes are found, accounting for about half of the cell mass. On the other hand, the synthesis of ribosomes is a very expensive matter for the cell in terms of energy and substrate consumption. Each ribosome is composed of three different RNA molecules transcribed as precursor molecules from a long transcription unit. In addition, more than 52 different proteins have to be synthesized in a coordinated way. Consequently, a large amount of metabolic energy and cellular substrates are consumed for ribosome biosynthesis. It is not surprising therefore that exponentially growing cells devote most of their metabolic energy to the synthesis of ribosomes. More than 50% of all transcripts generated under such conditions are rRNAs. Within the regulatory framework of ribosome biogenesis rRNAs are the key molecules that determine the rate of ribosome formation. The synthesis of ribosomal proteins (r-proteins) is a subordinated process. The rate of many, if not all, r-proteins is controlled by a **translational feedback** mechanism and is linked to the available amount of rRNA. Many r-proteins bind directly to rRNA but also to their own mRNA. When these proteins are in excess, binding to defined sequences of their corresponding mRNA will coordinately repress translation of those r-proteins which are encoded on the same mRNA. The S10 operon has been presented as an example in Section 5.3, and Fig. 8.13 summarizes the principle.

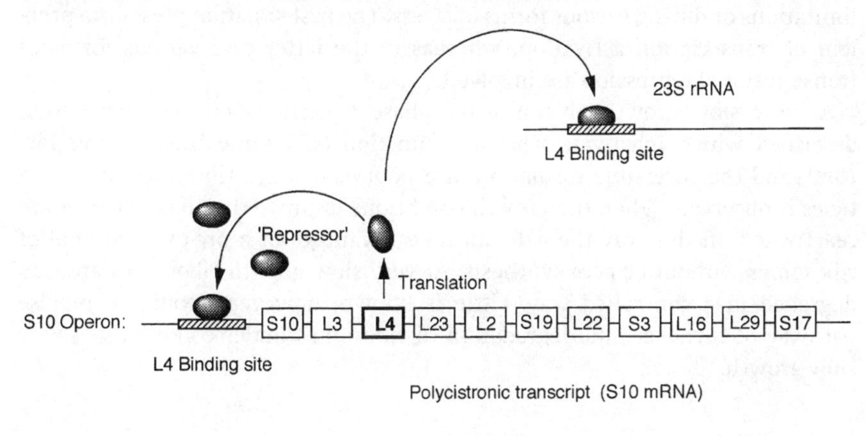

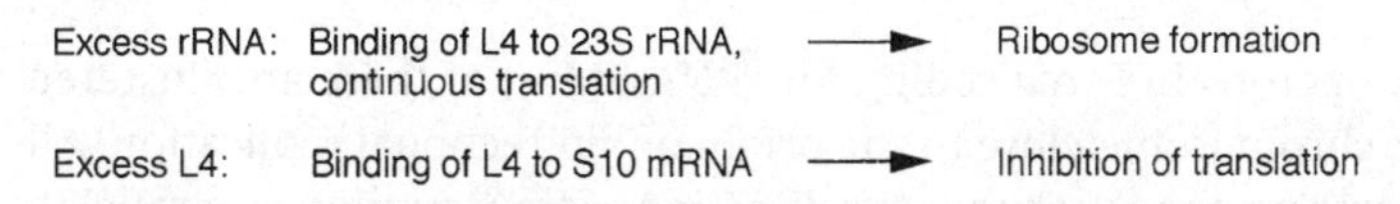

Figure 8.13 Translational feedback regulation of the S10 operon. The S10 operon, encoding 11 genes for proteins of the large and the small ribosomal subunits, is schematically presented. L4, encoding the translational repressor, is marked in bold. The L4 protein, shown as grey spheres, binds directly to the target 23S rRNA at a specific binding site indicated by a shaded box. A similar binding site for L4 is present in the leader region of the S10 operon mRNA transcript. Binding of L4 to this site inhibits translation of the downstream genes. If 23S rRNA is limiting L4 will bind to the leader sequence of the S10 operon mRNA and inhibit translation of the complete operon. At excess rRNA L4 will bind preferentially to the 23S rRNA and no inhibition of S10 mRNA translation occurs. Translation of ribosomal proteins thus depends on the availability of rRNA and occurs only if sufficient rRNA is synthesized.

There are some other mechanisms known that couple the synthesis of r-proteins to the available amount of rRNA. These include transcriptional repression or induced degradation of r-protein mRNAs. As a result of several such control systems, it follows that the rate of rRNA synthesis determines the total rate of ribosome formation.

When cells enter increasingly unfavourable external conditions they cannot afford the high costs of excessive ribosome synthesis. Ribosomal RNA synthesis is turned down and the cells slow down growth according to the environmental conditions. This response is carefully regulated over a wide range. Ribosomal RNA synthesis can either be shut off instantaneously or smoothly adapted according to the cellular demands in a balanced way. It follows that the regulation of r-RNA synthesis is not only the major determinant for the rate of ribosome formation during exponential growth. The rate of rRNA synthesis is also responsible for the downregulation or shut off in ribosome synthesis under conditions where growth rates have to be reduced in response to nutritional limitations or during various forms of stress. The first situation presents a problem of transcription activation, whereas in the latter case various forms of transcriptional repression are involved.

At very slow growth or stationary phase conditions proteins have been described which inactivate ribosome function (**ribosome inactivating factors**), and the reversible formation of a pool of inactive ribosomal 100S particles is observed. When the growth conditions improve the 100S particles are reactivated. In this way the cell can take advantage of a pre-existing pool of ribosomes without *de novo* synthesis. At very slow growth ribosomes are also degraded in a controlled way. Little is known, however, about the precise pathway of turnover and degradation of ribosomes under conditions of very slow growth.

8.7.2 Implications from the structure and arrangement of rRNA operons

There are *seven* operons in *E. coli* coding for rRNAs. Most of these are clustered on the circular chromosome close to the origin of bidirectional replication. All rRNA operons are arranged such that the direction of transcription is parallel to the direction of replication. It is believed that, because of this striking arrangement, interference between replication and transcription at high frequency is minimized (see Section 4.5). The individual *E. coli* rRNA operons are named *rrnA*, *rrnB*, *rrnC*, *rrnD*, *rrnE*, *rrnG* and *rrnH*. Their location on the *E. coli* chromosome and their approximate map position is given in Fig. 8.14. During rapid growth several rounds of replication may be simultaneously initiated on the same circular chromosome. As a consequence the circular DNA is partly polyploid. Hence, the number of rRNA operons is notably higher than seven under such conditions. Those operons which are particularly close to the origin of replication exist at a higher gene dosage compared to the distal operons. If

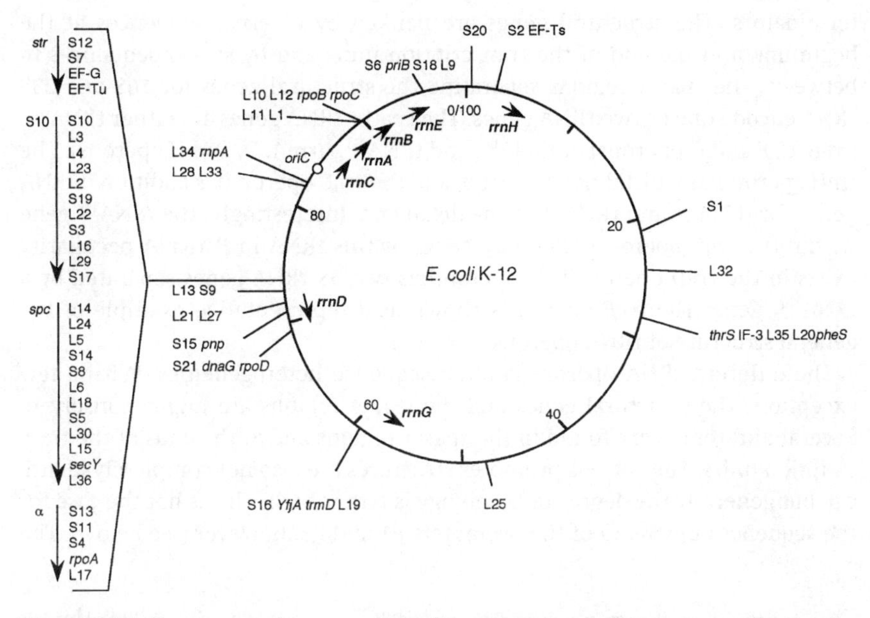

Figure 8.14 Chromosomal arrangement of the seven *E. coli* rRNA operons. The *E. coli* chromosome is represented as a circle divided into 100 minutes, indicated by numbers on the inside of the circle. The positions of ribosomal protein genes are marked on the outside and labelled correspondingly. The location of the seven rRNA genes (*rrnA*, *rrnB*, *rrnC*, *rrnD*, *rrnE*, *rrnG* and *rrnH*) is indicated by arrows which point in the direction of transcription. OriC denotes the origin of replication, from which replication starts bidirectionally. Note that the direction of transcription of all rRNA operons is parallel to the direction of replication.

not limited by special feedback mechanisms, such an increase in the gene dosage may potentially contribute to the strong increase in rRNA synthesis under rapid growth. Several attempts have been made to reduce the number of rRNA transcription units, and to study the effects of a reduced rRNA gene dosage on cell growth and the transcription efficiency of the remaining operons. Surprisingly, removal of up to five of the seven *E. coli* rRNA transcription units does not strongly impair maximal growth rates but all seven rRNA operons are required for rapid adaptation if the medium conditions change. Two immediate questions are related to such studies. First, do individual rRNA transcription units encode specific information necessary for specific growth conditions, and second, are the different rRNA operons regulated differently? These questions can in part be answered by sequence comparison of the seven *E. coli* rRNA operons. Their general structure is fairly similar. All operons encode the genes for 16S, 23S and 5S rRNA in this order. Transcription is started from two promoters, P1 and P2, arranged in tandem and separated by approximately 120 base pairs. Generally transcription is terminated by two strong Rho-independent tandem terminators. The structural genes are flanked by external sequences at the beginning and the end of the transcription units and by spacer sequences in between. The spacer regions separating the structural genes for 16S and 23S rRNA encode one or two tRNA genes. The spacer tRNA genes are either $tRNA^{Glu}{}_2$ (*rrnB, C, E* and *G* operons) or $tRNA^{Ile}{}_1$ and $tRNA^{Ala}{}_{1B}$ (*rrnA, D*, and *H* operons). The *rrnH* operon has a distal $tRNA^{Asp}$ gene, and the *rrnC* operon has additional tRNA genes for $tRNA^{Asp}$ and $tRNA^{Trp}$ at the distal end. Interestingly, the $tRNA^{Trp}$ gene within the *rrnC* operon is the only gene for this tRNA in *E. coli*. A peculiarity exists in the *rrnD* operon. This operon has two 5S rRNA genes separated by a $tRNA^{Thr}{}_1$ gene. The *rrnB* operon is shown as a representative example of the general structure of rRNA operons.

The different rRNA operons contain sequence heterogeneities. With a few exceptions the structural genes and the leader regions are highly conserved. Several differences are found in the spacer regions and at the ends of the transcription units. The various promoter structures are also not completely identical, but generally the degree of homology is rather high. This is not the case for the sequences upstream of the promoters P1 and P2, however (see below). The

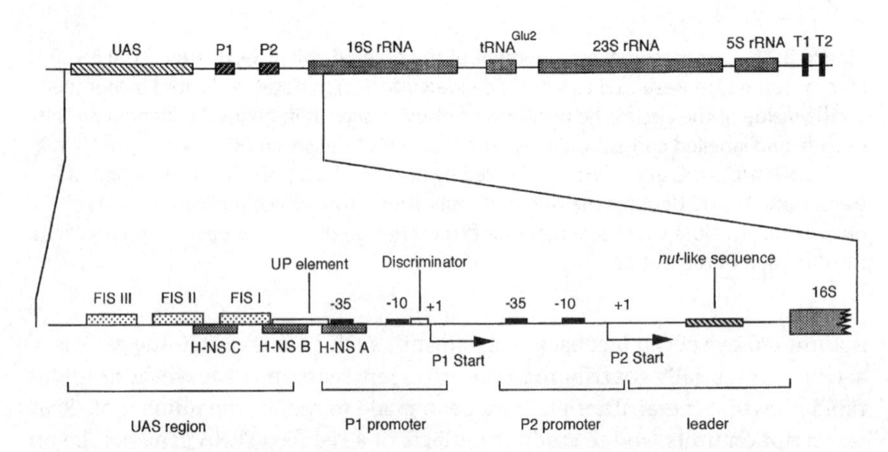

Figure 8.15 Structure of the *rrnB* operon. The schematic arrangement of the *rrnB* transcription unit is presented. The upper part shows the complete operon with the upstream activation sequence (UAS), the tandem promoters (P1 and P2), the structural genes for 16S rRNA, the spacer $tRNA^{Glu2}$, the 23S rRNA, the 5S rRNA and the two tandem Rho-independent terminators (T1 and T2). In the lower part the upstream promoter region is presented enlarged. The UAS region, two promoters P1 and P2 and the leader regions are indicated by brackets. Upstream of P1 three binding sites for the transcription activator FIS (FIS I, FIS II and FIS III), indicated as stippled boxes, and three binding sites for the repressor H-NS (H-NS A, H-NS B and H-NS C), indicated as grey boxes, are located. The UP element, adjacent to the –35 region of promoter P1, is marked. The stringent control discriminator sequence, immediately upstream of the transcription start site of promoter P1, is shown as a small open box. Within the leader region the location of the *nut*-like sequence is marked by a shaded box.

existing heterogeneities at these regulatory regions may give rise to differential expression. Selective expression of special rRNA operons in response to different growth stages has been observed in some parasites. One example is *Plasmodium berghei* which causes malaria. This parasite expresses different rRNAs when it lives in the mosquito and in the mammalian host. The different rRNAs are probably related to stage-specific ribosome populations. Such a differential expression makes sense when the rRNA genes expressed at different conditions are not identical. This is in fact the case for *P. berghei* which certainly presents an extreme case. No comparable differential expression of rRNA operons can be expected in *E. coli.* Minor sequence changes, however, do occur in *E. coli* and may provide a potential advantage for a better adaptation to different growth conditions or to environmental changes. In fact, there is evidence that the rates of transcription are not identical for all seven *E. coli* rRNA operons under all growth conditions. For instance, a lower expression of the *rrnH* operon has been noted; this might be related to the lower gene dosage because of the greater distance to the replication origin. A different inhibitory response in the expression of the *rrnB* and *rrnD* operons has also been reported to occur under conditions of the stringent control. Moreover, under steady state slow growth conditions, transcription from *rrnB* P1 is repressed more strongly than compared to transcription from *rrnD* P1. Growth rate regulation may therefore affect different operons in a different way (see below). Differences in stringent or growth rate control should be related to the promoter sequences. What is known about the promoters P1 and P2 from the different rRNA operons? The first striking notion is certainly that none of the ribosomal promoters shows a perfect match to the consensus sequence of σ^{70} promoters, although they are the strongest promoters of the cell. Note that only a few rRNA promoters among an excess of about 2000 mRNA promoters are responsible for more than 50% of all the transcripts made during exponential growth. The most significant difference is certainly the suboptimal distance of 16 rather than 17 base pairs between the –35 and –10 regions found for all rRNA promoters. In addition, the –35 region for all promoters P1 and P2 deviates at least at one position from the σ^{70} consensus. The same is true for the –10 region of P2 promoters. Only the –10 regions of P1 promoters have a perfect match with the σ^{70} consensus sequence. Mutational studies have shown that changing the sequence of *rrnB* P1 promoter to make it closer to the σ^{70} consensus increases promoter strength *in vitro* and *in vivo*. The sensitivity for regulation of the mutant promoters is lost, however. Apparently the rRNA promoter core sequences have not evolved for maximal efficiency but instead appear to be optimized for complex regulation. All rRNA P1 promoters, but not the P2 promoters, have a perfect GCGC discriminator sequence necessary for stringent and growth rate regulation (see Sections 8.5 and 8.6). This is in line with the early observation that all the major regulatory mechanisms act predominantly at promoters P1. For a long time it was considered therefore that regulation mainly occurs at P1 while P2 promoters are rather insensitive to regulation, and are only responsible for a basal level of rRNA transcription. This notion has been challenged by more recent *in vivo* studies and analyses of the strength of isolated P2 promoters (see below). The relatively low activity and response to regulation of the downstream P2 promoters can in part be explained by promoter occlusion from RNA polymerases initiated at high frequency at P1. On the other hand, a low promoter escape rate at P2 will cause queuing of RNA polymerases into the P1 start site. It will thus affect P1-dependent transcription by a mechanism described as **turnstile attenuation** (see Fig. 8.10). The physiological significance of such models are not yet clear, however. The invariant existence of tandem promoters in all rRNA operons is not precisely understood today, and is a challenge for future research.

Although rRNA promoters are certainly among the strongest promoters in the cell it was found that their activity is generally only modest or even rather weak when assayed *in vitro*. There are two main reasons for this observation. First, sequences upstream of the core promoters and *trans*-acting factors are required for full activity in the case of P1 promoters. This point is discussed in more detail below. The other important issue concerns the dependence of rRNA promoters on the template topology. It is known that the activity of rRNA promoters is strongly supercoil-dependent. Transcription of rRNA genes requires a superhelical template for optimal efficiency, therefore. It was seen in Section 6.3 that negative supercoils can facilitate promoter melting. The melting and isomerization of closed to open initiation complexes might be rate-limiting for rRNA promoters. This notion is supported by the observation that the open complexes of rRNA promoters are unusually labile and have a strong tendency for the backward reaction. The GC-rich discriminator sequence immediately downstream of the –10 region of P1 has been suggested as one putative explanation of this property. The stored energy of negative supercoils potentially facilitates base pair opening and might thus stabilize the open complexes. Another reason for the supercoil-dependence of the rRNA promoter activity can be attributed to the suboptimal spacer length, with only 16 base pairs. The optimal angular arrangement of the –35 and –10 core promoter elements relative to the RNA polymerase may require a defined twist promoted by negative supercoiling. The effect of the superhelicity on rRNA transcription is probably not restricted to the initiation steps. There is evidence that supercoiling is also effective during elongation. Measurements of RNA polymerase pausing during *in vitro* transcription of the early region of the *rrnB* operon, employing templates with different superhelical densities, have been performed. Such studies have indicated that the transcription elongation rate is optimal only at a defined degree of negative superhelical density. Local changes in the superhelical density of the template DNA may therefore provide an important mechanism for the fine-tuning of the transcription elongation rate (see Section 8.7.6).

Some further characteristics of rRNA promoters are noteworthy. First, as already mentioned above, the open complexes between rRNA promoters and

RNA polymerase are exceptionally unstable. Furthermore, they are sensitive to salt. They rapidly dissociate in the presence of intermediate monovalent salt concentrations or competitors such as random DNA or the polyanionic compound heparin. To obtain stable complexes *in vitro* the first two initiating NTP substrates must be added. Thus at least *in vitro* only the ternary complexes formed at rRNA promoters appear to be stable. NTP hydrolysis or phosphodiester bond formation does not seem to be required for stabilization, at least not for the *rrnB* P2 promoter.

Interestingly, the initiating NTPs for all P1 promoters are purines. P1-directed transcripts start with ATP or, in the case of the *rrnD* operon, with GTP. In contrast, P2 promoters require the less frequently used initiating pyrimidine nucleotide CTP. Here, the *rrnG* operon is an exception, where P2 starts with GTP. The requirement for high concentrations of ATP or GTP for stable initiation complex formation at P1 promoters might be of interest since it has been proposed that the cellular NTP levels could play a role in growth rate regulation (see Section 8.6 and below).

The spacing between the –10 region and the transcription start site of rRNA promoters is also exceptionally long. This distance is nine base pairs for the P1 promoters. P2 promoters have an additional peculiarity. Their initiation is not restricted to a defined position but occurs at similar frequency from several neighbouring sites within a cluster of cytosines.

Under certain *in vitro* transcription conditions (limiting substrate NTPs) an aberrant transcription initiation at the *rrnB* P1 promoter can be observed. In this case RNA polymerase slips back during initiation, and transcription is initiated efficiently from the –3 position. The process is termed **primer slippage** (see Section 3.4.2). The reaction is indicative of a second tight RNA-binding site in the RNA polymerase complex, consistent with present models of the transcription elongation complex (see Chapter 4). The relevance of this reaction is unknown, however, and transcripts initiated at –3 have not been detected *in vivo*.

A sequence comparison of P1 promoters has revealed that intermingled with the σ^{70} consensus sequence, determinants can be discerned for the heat-shock-specific σ factor, σ^{32}. Transcription initiation with the RNA polymerase $E\sigma^{32}$ holoenzyme has been shown to be possible *in vitro*. It has been suggested, therefore, that under conditions of heat-shock $E\sigma^{32}$ might be involved in rRNA transcription. However, this has not been shown *in vivo* to date, and the significance of the observation is unclear.

8.7.3 Upstream activating sequences

The reason for the high *in vivo* activity of rRNA promoters is clearly related to sequences upstream of the core promoter elements. The DNA sequences between positions –150 and –50 have been shown to be indispensable for the high activity of P1 promoters. The corresponding sequences have consequently been designated as **UAS regions** (see Section 6.2.1). Sequences flanking the *rrnB* P2 promoter have also been proposed to have an effect on the promoter strength, but no comparable systematic investigations on potential upstream sequences of P2 promoters and their possible regulatory effects have been performed. Results pertinent to the rRNA P1 UAS only are therefore discussed. The UAS regions for the seven rRNA P1 promoters are not conserved. The only striking similarity between the different UAS regions is the presence of regularly spaced tracts of AT sequences. The role of such AT clusters in **DNA-bending** has been discussed in Chapter 6. In line with theoretical calculations the DNA of all rRNA UAS regions has clearly been characterized as being bent. Detailed studies in the case of the *rrnB* UAS have shown that the centre of the bend is located approximately at position –90. For many prokaryotic promoters it is known that bending contributes to promoter strength. Replacement of the *rrnA* P1 UAS with an unrelated strongly curved DNA fragment from the kinetoplast *Crithidia fasciculata* restores *in vivo* promoter activity to a large degree. There is good evidence for a second curvature within the DNA of rRNA promoters. The centre of this bend can be localized within the promoter core. The direction of this curvature is not the same as the upstream curvature (see Fig. 6.4). The activating effects of the UAS region are consistent with the assumption of a stabilization of the binary RNA polymerase–promoter complex by extended contacts. A higher affinity for RNA polymerase can be concluded from an increase in K_B (see Section 3.6) which is dependent on the presence of the UAS region, as well as from extended protections observed in DNaseI footprinting experiments. Later steps in the initiation cycle may also be affected, however. This has been concluded from a study employing *rrnD* promoters where RNA polymerase–promoter complex formation (K_B) *and* isomerization (k_2) is involved in UAS-dependent activation.

The major effects on the promoter activity in the absence of *trans*-acting factors, can be attributed to a region outside the strong UAS curvature. A region directly flanking the –35 element and extending up to position –60 dramatically activates P1 promoter activity. This sequence element has been termed the **UP element**. UP elements are not restricted to rRNA promoters but have been described to occur at other transcription systems as well (see Section 6.2.1). Interestingly, the UP elements are also AT-rich but apparently do not bend DNA notably. Base changes introduced at positions –47 to –58 of the *rrnB* P1 UP element do not change the electrophoretic mobility of corresponding promoter fragments. UP elements act as separate modules and can be exchanged functionally. They are able, for instance, to activate hybrid promoters that do not normally have an UP element. The UP element is specifically recognized by the C-terminal domain of the RNA polymerase α subunit (αCTD). Hence, in addition to the recognition by the σ subunit rRNA promoters make contacts to the α subunit by direct binding to the UP element. Therefore, the UP element defines an *extended* promoter and has been designated as a *third* promoter recognition element. Two segments within the αCTD, comprising amino acid residues

265–269 and 296–299, have been identified which are responsible for the interaction with UP elements. These segments are located in an exposed region of the protein secondary structure and form part of an α-helix and an unstructured loop. This region of the protein has no similarity to any other known DNA-binding motif. The same region has also been identified as contact site for several transcription factors (see Sections 2.2.1 and 7.2.1). It must be concluded therefore that activation by the αCTD can either occur through direct contacts to the UP element or to transcription factors. Both possibilities are feasible because the αCTD is linked to the rest of the molecule by a rather flexible linker of approximately 10 amino acid residues. The flexibility necessary for the different αCTD contacts is further underlined by the fact that the UP element can be displaced by one helical turn without significant loss of the activating function.

8.7.4 Effects of transcription factors

The UAS regions of all seven rRNA P1 promoters have been shown to contain binding sites for *trans*-acting proteins. Three binding sites for the transcription factor FIS are localized upstream of P1 (see Fig. 8.15). They centre around positions –70 (site I), –100 (site II) and –140 (site III). In addition to the degenerated recognition motif for FIS (see Section 7.5.2) which is present in all binding sites, the curvature of the UAS region is important for FIS binding.

Detailed analyses have been performed to study the effect of FIS on transcription of the *rrnB* operon. From these studies it is known that binding of FIS activates P1-directed transcription significantly. From the different promoter activities determined in *fis*$^+$ and *fis*$^-$ strains a 10-fold activation *in vivo* has been concluded. Activation is related to the concentration of FIS during the growth cycle. The level of FIS is related to the growth phase and increases sharply at the onset of exponential growth. Clearly, FIS is responsible for the peak in rRNA synthesis that can be observed at exponential growth. Therefore, FIS is one component involved in the growth phase-dependent regulation of rRNA transcription. On the other hand, cell growth of *fis*$^-$ strains is not significantly retarded, and transcription of rRNA in such cells appears to be reduced only by a factor of 2.5. However, cells unable to express functional FIS molecules show significantly reduced rRNA synthesis rates at nutritional upshifts, and they do not reach the very high growth rates of wild-type cells in rich medium. Other mechanisms in addition to activation through FIS must be involved in growth phase-dependent regulation.

Although three FIS sites are present upstream of all rRNA operons, FIS binding to the three sites is not cooperative. The binding constants for all three sites are similar (in the nanomolar range), and occupation with FIS dimers occurs in a sequential way. Not all three binding sites contribute to activation to the same extent. FIS binding site I is responsible for the major activating effect. FIS-dependent activation probably involves direct interaction between FIS and RNA polymerase. Activation is dependent upon the localization of bound FIS dimers relative to the core promoter. Full activation requires that FIS molecules and RNA polymerase are bound to the same face of the DNA helix. The angular orientation is critical. Displacing an FIS binding site by a complete helical turn (approximately 11 base pairs) does not abolish activation, whereas activation is completely lost by displacement of half a helical turn (five base pairs). FIS is known to require curved DNA for binding, and is also capable of bending the target DNA. FIS-dependent activation cannot simply be explained, however, by a change in the upstream DNA curvature. This has been concluded from the analysis of mutant FIS proteins, which are still able to bind to DNA and also to bend DNA, but which have lost the activating function. Analysis of the initiation complex formation at the *rrnB* P1 in the presence of FIS is consistent with an increase in the binding affinity (K_B) as a major contribution to the increase in promoter strength. In a similar study with the corresponding *rrnD* P1 promoter it was found that later steps during initiation, namely the conversion from the closed to the open promoter complex, are also affected. The activating function of FIS has also been explained with the formation of small DNA microloops which brings RNA polymerase and the activator into close spatial proximity. In addition, such microloops may facilitate melting and promoter escape by **torsional transmission** (see Section 7.5.2).

The mechanism of FIS-dependent activation is probably explained by a direct interaction between FIS and RNA polymerase. It has been proposed that FIS interacts with the αCTD; however, studies with mutant RNA polymerase have shown that αCTD is not required for FIS-dependent activation. Other experiments point to the σ subunit as the direct contact point for FIS, but again, direct proof is lacking. A final answer as to where FIS–RNA polymerase contacts occur can therefore not yet be given.

A second regulatory protein has been shown to interact specifically with the upstream regions of rRNA P1 promoters. The protein has been identified as H-NS, a small neutral protein known to regulate the transcription of many genes and to affect the structure of DNA (see Section 7.5.3). Binding sites for H-NS have been mapped within the *rrnB* P1 promoter upstream region. Three such binding sites have been characterized by high resolution footprinting techniques and their centres have been localized at positions –25 (site A), –50 (site B), and –80 (site C). Sites B and C thus clearly overlap the FIS binding sites I and II. In addition, site A overlaps the P1 promoter core sequence (Fig. 8.16). Most strikingly, despite the overlap in binding sites, interaction of FIS and H-NS is not mutually exclusive. As revealed by footprint analyses and binding competition experiments, FIS and H-NS occupy different faces of the DNA helix. H-NS does not recognize a specific primary sequence but requires curved DNA for binding. Moreover, it is known that H-NS binding changes the curvature of its target DNA. In the case of the *rrnB* P1 upstream region the two proteins H-NS and FIS obviously bind to two separate centres of local curvature. *In vitro* studies with the two proteins varied over a wide range of concentrations have shown that only very high amounts of one of the two proteins will replace the respective other one from the P1 UAS region. Hence, the two proteins do not compete through sterical hindrance of overlapping protein radii. Rather, competition appears to be mediated by alternative DNA conformations (see Fig. 8.16).

Recent studies have shown that inhibition of P1 promoter activity by H-NS does not involve displacement of the RNA polymerase from the promoter. In addition, open complex formation is not significantly affected either. Inhibition is most likely to be the result of a conformational alteration of the initiation complex triggered by an H-NS-dependent deformation of the DNA structure, such that it is rendered inadequate for productive transcription.

What is the physiological function of the simultaneous binding of FIS and H-NS to the upstream rRNA P1 promoters? First, binding of the two proteins is not restricted to the *rrnB* P1 upstream sequence but occurs to the UAS regions of all seven rRNA operons. Second, H-NS has been shown to affect selectively transcription from P1 promoters *in vitro* and *in vivo*. It acts as a repressor, specifically antagonizing FIS-dependent activation. *In vivo* studies with *fis* and *hns* mutants have shown that during the cell cycle H-NS is involved in the rapid shut off that is known to occur at the end of the exponential phase. Regulation of rRNA transcription through the interplay of the two antagonistic transcription factors FIS and H-NS correlates well with the known levels of the two proteins during the cell cycle. The rapid accumulation of FIS during early logarithmic growth promotes a rapid increase in stable RNA transcription. H-NS, in contrast, reaches a maximal cellular concentration at the end of the

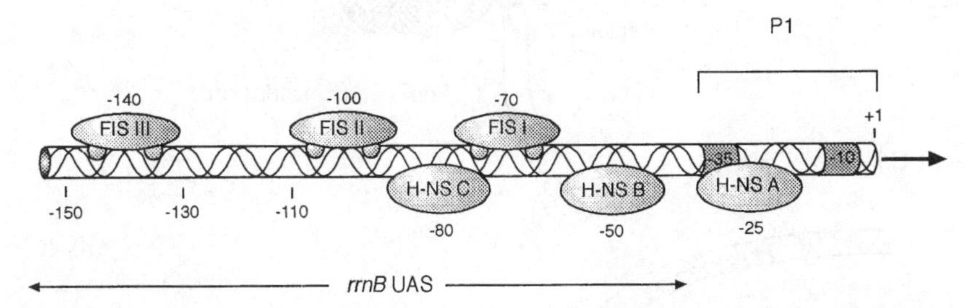

Figure 8.16 Steric arrangements of H-NS and FIS binding sites at the *rrnB* P1 UAS. The *rrnB* P1 promoter upstream region is depicted as a cylinder with the helical path of the DNA schematically indicated. The –35 and –10 sequences of promoter P1 are shown as grey bands. Three FIS molecules occupy binding sites FIS I, FIS II and FIS III at the indicated positions (numbers correspond to nucleotide positions relative to the P1 start). Binding of FIS involves helix-turn-helix motifs which fit into two adjacent major grooves of the DNA (indicated by small circles). From the helical projection it is apparent that all three FIS molecules bind roughly to the same face of the DNA helix. On the other hand, the three H-NS molecules, bound with their centres at positions –25 (H-NS A), –50 (H-NS B) and –80 (H-NS C), occupy the opposite side of the helical DNA. This arrangement explains that binding of the different transcription factors is not mutually exclusive although FIS and H-NS binding sites are partly overlapping.

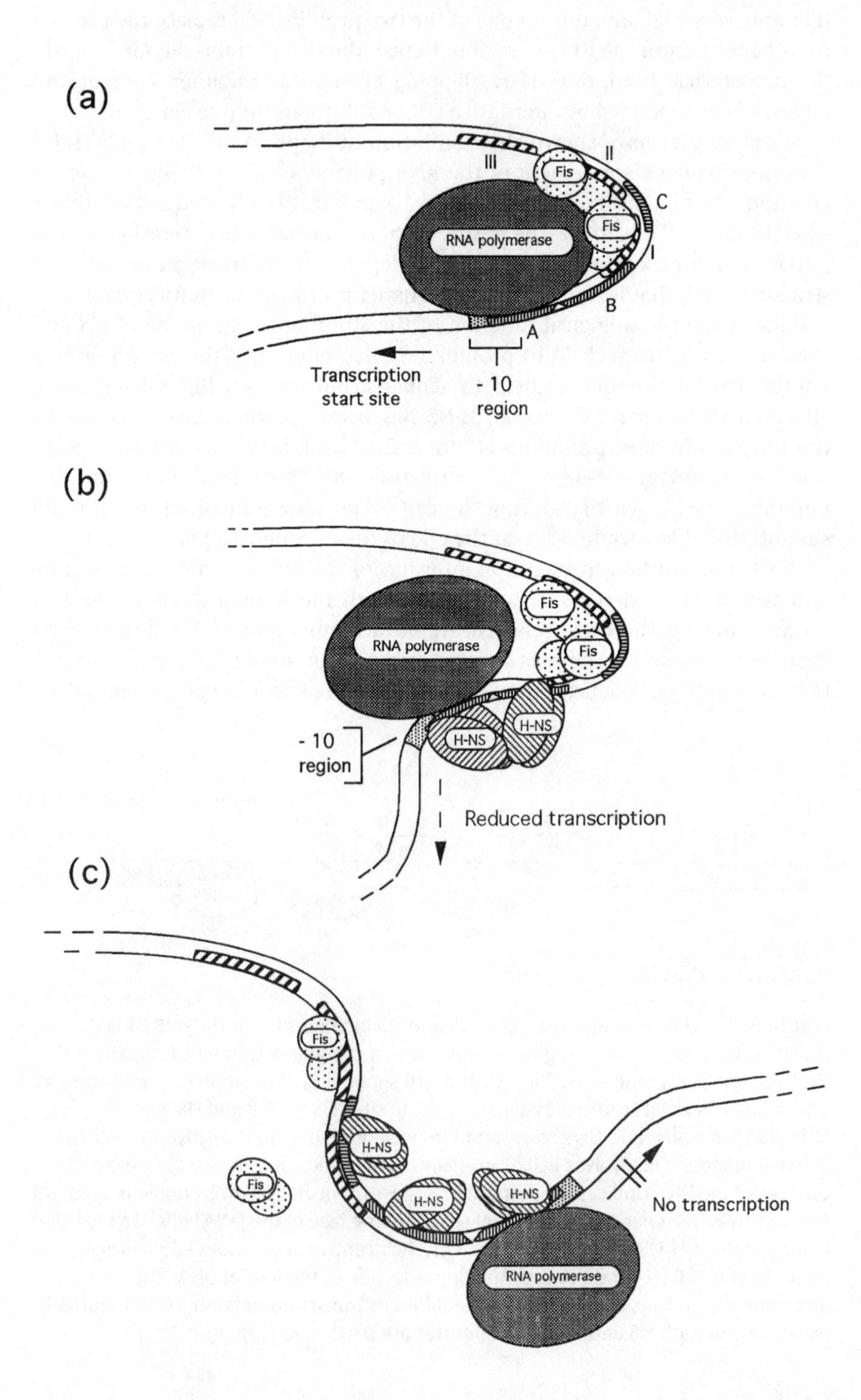
(a)
III
Fis
II
C
Fis
RNA polymerase
I
B
A
Transcription
start site
- 10
region
(b)
Fis
RNA polymerase
Fis
H-NS
H-NS
- 10
region
Reduced transcription
(c)
Fis
H-NS
Fis
H-NS
H-NS
No transcription
RNA polymerase

exponential phase when rRNA synthesis rates are known to decrease sharply. The two proteins thus serve opposing functions. FIS is responsible for the growth phase-dependent activation of rRNA synthesis, allowing fast growth and rapid changes to nutritional upshifts. In contrast, H-NS slows down rRNA transcription under limiting conditions. A mechanistic model of the FIS–H-NS antagonism during rRNA transcription involving the conformational distortions brought about by the two regulatory proteins is presented in Fig. 8.17.

In summary, H-NS is an important element responsible for the rapid adaptation of rRNA synthesis to accommodate reduced growth requirements when cells leave the most active metabolic state and enter stationary phase. It has been shown that H-NS-dependent downregulation of rRNA synthesis at very low growth rates does not involve the effector nucleotide ppGpp. H-NS may thus be a candidate for an alternative growth rate regulator independent of the activity of ppGpp. Furthermore, the shut-down in rRNA synthesis under certain stress conditions may be accomplished through H-NS, which is known to affect the synthesis of many genes under various forms of stress.

The glycolytic enzyme **Fda** (**fructose-1,6-diphosphate aldolase**) has been suggested to function as a growth phase-dependent *trans* regulator. This was concluded from a temperature-sensitive mutant in the *fda* gene (*ts8*). The mutation causes significant reduction in rRNA synthesis with only a small effect on protein synthesis at the restrictive temperature. The effect is indirect, however, and is almost certainly linked to the ppGpp-dependent growth rate regulation.

8.7.5 Stringent and growth rate regulation

The two regulatory phenomena stringent control and growth rate regulation have been discussed in detail in Sections 8.5 and 8.6. Since both mechanisms are of outstanding importance for the regulation of rRNA synthesis a brief summary of what is pertinent to the control of rRNA transcription is repeated here.

Figure 8.17 Model of the FIS–H-NS conformational antagonism. (a) The rRNA promoter upstream region is shown as a curved structure with RNA polymerase bound to the P1 promoter at the inside of the curvature. The transcription start is marked by an arrow. The curved DNA structure is fixed to the polymerase by two FIS dimers (shown as stippled spheres) bound to FIS site I and II (shown as shaded areas). Transcription is activated by the favourable DNA conformation and the increased complex stability due to FIS–RNA polymerase contacts. Three H-NS binding sites (A, B and C) are not occupied. (b) Occupation of H-NS sites A and B by dimeric repressors (shaded ovals) change the DNA conformation. The FIS–RNA polymerase interactions are disrupted and no further activation of P1 transcription is observed. (c) High occupation of H-NS sites destroys the activation conformation of the UAS DNA completely. The resulting structure is inadequate for transcription and repression of the P1 promoter is observed. Because of strong conformational distortion of the DNA, FIS molecules may be displaced from their binding sites. The figure is taken from Tippner *et al.* (1994).

Stringent control and growth rate regulation have been known for a long time to be the major control systems related to the synthesis of rRNAs. Both types of control are considered to be hallmarks of rRNA transcription regulation. The two control systems can be distinguished physiologically. Stringent control is induced at amino acid deprivation. It involves a RelA-dependent increase of the effector nucleotide ppGpp, which is triggered by the codon-dependent binding of uncharged tRNA to the ribosomal A-site. ppGpp accumulation causes an immediate shut-down in rRNA transcription by mechanisms not yet precisely resolved.

In contrast, growth rate regulation can be observed in cells growing at steady-state. The phenomenon of growth rate regulation is characterized by the coupling of the rRNA synthesis rates to cell growth rates, such that rRNA transcription increases to the square of the cell growth rates (see Fig. 8.11).

Both types of control have been found to be accomplished by the *rrn* P1 promoters. Recent studies have shown, however, that the isolated P2 promoters are also sensitive to stringent control, although not to the same extent as P1 promoters. *In vivo* measurements of the relative promoter efficiencies have demonstrated that most of the rRNAs synthesized at stationary phase is due to rRNA P2 promoter activity. The same is true when growth rates are reduced or in strains that do not express a functional FIS protein. In such strains, even at exponential growth, the activity of P2 exceeds that of P1 promoters by a factor of more than 2.5. The reverse ratio is observed in the presence of FIS. This explains that growth rates of *fis*⁻ strains are almost normal and, except for conditions of very fast growth or nutritional upshift, the amount of rRNA synthesized is only slightly reduced. Together these results show that promoter P2 is well suited to take over regulatory functions when the activity of P1 is repressed. In addition to a higher P2:P1 promoter activity ratio at slow growth rates or during stationary phase different operons are not affected to the same extent. Reduction of the P1 promoter activity is about threefold higher for the *rrnB* operon than the *rrnD* operon. This differential regulation might be explained by differences in the repression mediated by H-NS. It has been shown that H-NS binds with different affinities to the UAS regions of the *rrnB* and *rrnD* operons.

A GC-rich sequence motif between the −10 promoter hexamer and the transcription start site has been identified to be a *necessary* but not *sufficient* element for both stringent control and growth rate regulation. This element is termed the **discriminator**. The discriminator motif of P2 promoters is incomplete, consistent with a generally reduced sensitivity for stringent and growth rate regulation of P2 promoters. Experiments with hybrid promoters from stringently and non-stringently controlled genes have revealed that, beside the discriminator sequence, the complete promoter context appears to be responsible for stringent as well as for growth rate control (see Fig. 8.8). Regulatory elements such as the UAS regions or the UP element of rRNA promoters are not involved in stringent or growth rate regulation. However, the mechanism(s) underlying stringent or growth rate sensitivity have been correlated to effects brought about by negatively supercoiled templates. The susceptibility to melting affected by negative supercoiling might of course be affected by the GCGC discriminator. Several different mechanisms as to how ppGpp might affect rRNA transcription have been suggested. In almost all the mechanisms proposed, ppGpp-modified RNA polymerase directly participates in regulation. Reversible modification of RNA polymerase by ppGpp should render the enzyme incapable of initiating effectively from stringent-sensitive promoters. The idea of a reduced affinity of ppGpp modified polymerases has been put forward as the **RNA polymerase partition model** (see Section 8.5.4). Other models assume a reduced capacity of RNA polymerase for open complex formation in the presence of ppGpp. A **trapping model** has been derived from kinetic analysis of the formation and dissociation of initiation complexes with stringent and non-stringent promoters in the presence of ppGpp. According to this model RNA polymerases undergo an alternative initiation pathway at stringently regulated promoters in the presence of ppGpp. The formation of closed complexes is enhanced and the isomerization rates to the open complex are reduced. Thus RNA polymerases are trapped in stabilized closed complexes. The open complexes formed in the presence of ppGpp are structurally different compared to complexes formed in the absence of ppGpp. The capacity of the ternary complexes to form productive transcripts is reduced. The kinetic mondel explains that discrimination between stringent and non-stringent transcription occurs at an early step of initiation, while the actual inhibition occurs at later stages and during promoter clearance (see Section 8.5.4).

Although ppGpp plays a role in both stringent and growth rate regulation, the latter control has been described to occur in strains devoid of ppGpp. More than one mechanism may, therefore, account for the adaptation of rRNA synthesis to the growth rate. A ribosome feedback mechanism was discussed in Section 8.6. In this model an unknown repressor is proposed to prevent excess rRNA synthesis over the cellular demand. Recent results imply that the regulatory protein H-NS may act as a ppGpp-independent growth rate regulator (see Section 8.7.4). Another hypothesis for the ppGpp-independent growth rate regulation of rRNA transcription has recently been made which starts from the premise that initiation complexes between RNA polymerase and rRNA promoters are notoriously labile. The hypothesis is based on the observation that binary initiation complexes between rRNA promoters and RNA polymerase are generally very short-lived *in vitro* and require initiating NTPs for stable open complex formation. It is thought that efficient initiation *in vivo* also requires high ATP and GTP concentrations (the initiating nucleotides for P1 transcription starts). Decreasing NTP concentrations mimic the effect of ppGpp and further destabilize rRNA P1 promoter complexes. If the assumption is made that the cellular ATP and GTP concentrations, which increase or decrease with increasing or decreasing growth rates, become limiting, then a mechanism for growth rate-dependent transcription is feasible (see Section 8.6).

As specified in more detail in Section 8.5 and 8.6, inhibition of transcription under stringent control or during growth rate regulation is accomplished at the initiation *and* elongation steps of transcription. While the major mechanisms currently discussed for the repression of rRNA transcription during *initiation* have been summarized above, some aspects of ppGpp-dependent inhibition during *elongation* follow here. Generally, ppGpp-dependent effects on the transcription elongation rate involve RNA polymerase pausing (see Sections 4.3.4 and 8.5.5). There is evidence from both *in vitro* and *in vivo* studies that transcription elongation of rRNA genes is affected by ppGpp, but also by other factors, such as the superhelicity of the template (see above), or special transcription factors (see below). Results of a quantitative analysis of ppGpp-dependent RNA polymerase pausing during transcription through rRNA and mRNA genes *in vitro*, employing rRNA promoters and promoters not under stringent or growth rate control, allow the following conclusions: ppGpp does not create new pauses but has a significant effect on some natural pauses. Most pauses are enhanced, while a few examples of reduced pauses can also be observed. The promoter structure *and* transcription through the region immediately downstream of the promoter (about 30 nucleotides) determine ppGpp-dependent pausing. To affect pausing the presence of ppGpp is only required during elongation. Reduced elongation rates and pausing have often been found to synchronize transcription and translation (see Section 4.3). Since rRNAs are not translated, pausing during transcription of rRNA genes must serve a different purpose. It is supposed that fine tuning the transcription elongation rate is important for structure formation and assembly of ribosomes (see below). Furthermore, the data is consistent with a model that has been proposed to explain the ppGpp-dependent shut-down of rRNA transcription by promoter occlusion (see Fig. 8.10).

8.7.6 Pausing, attenuation and antitermination

Pausing and attenuation during rRNA transcription also has a functional significance in the absence of ppGpp. Sequences similar to the *nut* sites in the early transcribed regions of phage λ are known to exist in the leader regions upstream of the 16S rRNA genes as well as in the spacer regions between the 16S and 23S RNA genes. In phage λ the *nut* sequences, together with several Nus proteins, are essential for an antitermination mechanism which regulates the expression of delayed-early and late phage genes by readthrough of termination signals (see Section 5.4). The *nut*-like elements within rRNA operons were initially also considered to suppress premature termination during transcription of the long and untranslated rRNA operons (see Section 5.4.4). *nut*-like sequences within the rRNA operons consist of the conserved boxA sequence $^{5'}$-**UGCUCUUUAACA**-$^{3'}$ which is homologous to the phage λ boxA sequence (Fig. 8.18).

The boxB is represented by a conserved stem-loop structure with no primary

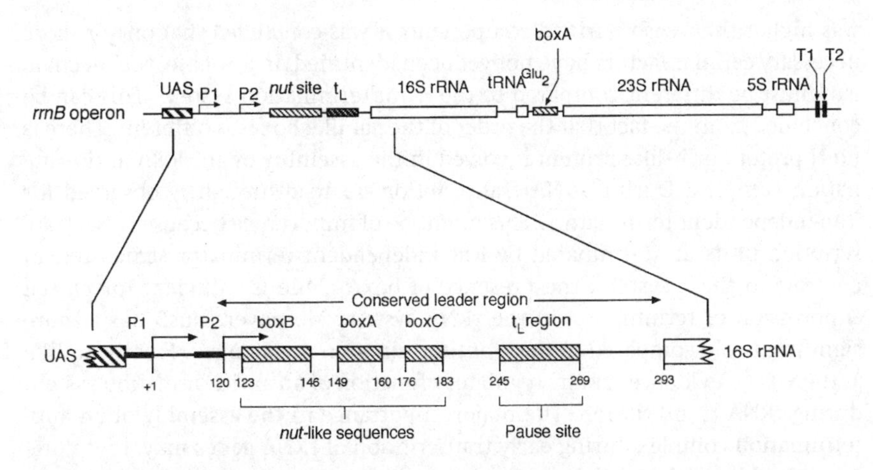

Figure 8.18 Location and arrangement of the *nut*-like sequences within the *rrnB* leader region. The *rrnB* operon is shown with the structural genes as open bars, the tandem promoters P1 and P2 as arrows, the two terminators as black bars and the UAS region as a shaded box. Between the P2 promoter and the structural gene for 16S rRNA two shaded areas mark the *nut*-like sequence region and a structure denoted t_L. A second boxA sequence is located between the structural genes for the spacer tRNA and the 23S rRNA. The lower part of the figure presents an enlarged projection of the leader region. The components boxB, boxA and boxC of the *nut*-like sequence region are indicated as shaded boxes and their exact location with respect to the P1 transcription start is marked. In addition, the t_L region, known as a transcriptional pausing site, is shown as a shaded box and the location is marked by nucleotide numbers.

sequence conservation. The order of the two *nut*-like elements in rRNA operons is different as in the phage λ. The rRNA *nut*-like sequences are thought to form a similar complex with the transcribing RNA polymerase and the Nus factors NusA, NusB, NusE and NusG as the elongation control particle characterized in the case of λ transcription (see Fig. 5.12). Of course, there is no N protein present in uninfected *E. coli* cells, nor has any analogue for the N protein been characterized to function in an *rrn* antitermination complex. Of particular interest is the fact that one of the Nus factors, NusE, known to be a constituent of the antitermination complex, is identical to the small subunit r-protein S10. Another r-protein, namely S1, has also been shown to be able to interact with the rRNA boxA sequence. S1–boxA interaction inhibits the complex formation of the NusB–NusE heterodimer with the boxA sequence. The relevance of this observation is unclear but it might indicate that the structure that is formed during transcription through the *nut*-like sequence is much more complex than presently thought.

The *rrn* antitermination system has been shown to be able to suppress Rho-dependent transcription termination signals when introduced downstream of the *nut*-like elements. Because the readthrough efficiency with crude extracts was higher than with purified components it was concluded that one or more necessary cellular factors have not yet been identified. In any case, the mechanism must be different compared to the λ antitermination system. This can be concluded from the fact that the order of the *nut*-like boxes is different. There is no N protein or N-like protein involved in the assembly of an rRNA antitermination complex. During rRNA transcription no readthrough is observed for Rho-independent terminators. This might be of importance because rRNA transcription units are terminated by Rho-independent terminator structures. In contrast to the λ system, the presence of boxA alone is sufficient for partial suppression of termination in the rRNA system. Moreover, NusB has a more significant role compared to the λ antitermination mechanism. Finally, unlike λ, there is no evidence for any regulatory function of an antitermination system during rRNA transcription. The major importance in the assembly of an antitermination complex during early transcription of rRNA genes may, therefore, involve additional functions quite different from simply suppressing termination signals. This conjecture is supported by a number of conclusive observations. First, binding of r-proteins and assembly of the small ribosomal subunit starts during rRNA transcription, before the end of the structural gene is reached. One important property of the Nus factors NusA and NusG is related to transcriptional pauses. Adaptation of the proper transcription rate may be of much higher importance than suppression of premature termination. During normal rRNA transcription there is no necessity to suppress premature transcription termination. Even in the absence of a functional antitermination system there is no convincing evidence for premature termination. All the effects that have been observed with natural rRNA transcription units can be explained by post-transcriptional processes. On the other hand, evidence for the involvement of the *nut*-like boxA sequence in determining the transcription elongation rate has been obtained *in vivo* (see Section 8.5.5). The presence of a consensus boxA sequence within a truncated rRNA or an mRNA transcript causes an increase in the RNA chain elongation rate, and largely abolishes the effect of ppGpp on the reduction of the transcription rate. Hence, the presence of the *nut*-like elements within rRNA operons and the putative constitution of antitermination complexes involving the *E. coli* Nus factors, may primarily serve functions related to the synchronization of the transcription rate with the complex events of structure formation, protein assembly and processing of functional ribosomal particles. These latter events all take place simultaneously with transcription, and it is clear that they must be coordinated and tuned in some way.

8.7.7 Coupling of rRNA transcription to post-transcriptional events and ribosome biogenesis

From the preceding paragraph it is clear that the post-transcriptional steps in ribosome biogenesis such as RNA folding, binding of r-proteins and processing of the primary rRNA transcripts are intricately linked to transcription and cannot be regarded as steps subsequent and independent of transcription. Involvement of the conserved *nut*-like sequences within the rRNA leader, and the t_L sequence, which is known as RNA polymerase pausing site, has been demonstrated in a process that is important for the correct assembly and structure formation of the small ribosomal subunit. It is now known that the above sequence elements within the leader of bacterial rRNA operons facilitate the correct folding of the 16S rRNA and support the assembly process. The conserved leader sequences have been proposed, therefore, to function as molecular **RNA chaperones.** The mechanism probably involves transient interaction(s) between conserved sequence elements of the leader (*nut*-like box sequences) with complementary regions in the 5′ domain of the mature 16S rRNA. Such transient contacts between the leader and the 16S RNA have been demonstrated by cross-linking experiments and gel retardation studies. Mutations in both the *nut* region or the t_L structure cause a cold sensitive phenotype. This means that cells carrying such mutations cannot form active ribosomes at low temperature. Obviously, the chaperone-like transient interaction between the conserved leader elements and the mature 16S rRNA is particularly important at low temperatures. Facilitated structure formation through the leader RNA appears to be dispensable when the activation energy barrier necessary for correct particle formation can be overcome by higher growth temperatures. The crucial step may reside in the kinetics of the 16S rRNA secondary structure formation. It has been shown *in vitro* that leader mutations conferring the cold-sensitive phenotype cause aberrant folding kinetics of defined parts within the 5′ domain of the 16S rRNA. Obviously, without the correct leader interactions metastable secondary structures are trapped; these require too high an activation energy to revert or to react in the forward direction. With the trapped secondary structures the correct folding pathway cannot proceed. It can be easily imagined that concomitant with the aberrant rRNA folding the association of ribosomal proteins will be affected. The analysis of 30S particles derived from mutant ribosomal operons did not reveal that individual proteins were missing. However, such particles were altered in their correct three-dimensional structure and overall architecture, as revealed by *in vivo* chemical probing of 30S ribosomes.

Interestingly, during the biogenesis of eukaryotic ribosomes the chaperone-like facilitation of rRNA structure formation, assembly and processing is not brought about by transient interactions of leader RNA transcripts. A separate family of **small nucleolar RNAs (snoRNAs)** is responsible for a similar process. In addition, snoRNAs direct methylations to 2′O-methyl nucleotides and pseudouracil substitutions which occur frequently at many positions within eukaryotic rRNAs. A number of homologous sequence elements can be found between one of the most widespread eukaryotic snoRNAs, namely **U3 RNA**, and bacterial leader RNAs, indicating that similar steps may be performed by these two different RNA molecules. The eukaryotic snoRNAs apparently act in *trans*, and can be used several times, while the bacterial leader RNAs are typical *cis* elements that only affect the transcript to which they are linked. Obviously, the complex mechanisms of ribosome formation contain similar elements conserved during evolution from bacteria to higher organisms.

SUMMARY

Regulation of transcription is generally not restricted to an individual operon which is controlled independently; instead, families of genes are coordinated into networks of variable complexity. Regulation beyond the operon level can be classified in *regulons*, where a common regulator controls several related, but independent operons, or *modulons*, where a common regulator directs the transcription of several operons from different regulons. Often, an operational distinction is made in which regulation is classified according to a common stimulus which affects transcription of many different genes independent of the mechanism of regulation. The ensemble of genes regulated in this way is often described as a *stimulon*.

The *SOS response* represents a typical regulon, where the transcription of different genes involved in the *repair of DNA damage* is coordinately controlled. The *LexA repressor* acts as the common regulator, and is inactivated by *proteolytic cleavage*. The cleavage reaction is induced by the protein *RecA*, which normally functions in DNA recombination, but under conditions of DNA damage is reversibly converted by single-stranded DNA fragments into an efficient *coprotease*. Activated RecA not only inactivates the repressor but can also post-translationally modify the activity of one of the SOS gene products (UmuD) by inducing a similar proteolytic cleavage reaction as in case of LexA. RecA and the repressor LexA are autoregulated, such that the system returns rapidly to the uninduced state when the conditions of DNA damage are overcome.

Cell survival can also be endangered by heat. A protective mechanism comprising a complex regulon, termed the *heat-shock response*, is activated at temperature upshift. Expression of the heat-shock genes is controlled by the specific σ factor σ^{32} (σ^{H}). The heat-shock regulon encodes two types of proteins to cope with heat-damaged proteins, namely proteases and molecular chaperones, which degrade and help to refold denatured proteins, respectively. The transcription of heat-shock genes depends on the *concentration of the* σ^{32} protein. The cellular concentration of σ^{32} is regulated by synthesis through change in activity or altered protein stability. Enhanced synthesis can be the result of *transcription activation* or *enhanced translational activity*. The activity of σ^{32} can be reversibly changed by *binding to the heat-shock proteins* DnaK, DnaJ or GrpE. The protein turnover of σ^{32} is regulated by *proteolytic action* of a heat-shock-specific protease (FtsH). A *second heat-shock regulon* is controlled by the minor σ factor σ^{24}.

This regulon is induced at extreme heat conditions and senses *extracytoplasmic changes*. One of the several promoters controlling the σ^{32} gene is a σ^{24}-dependent promoter. This guarantees transcription of σ^{32} even under conditions of extreme temperature. Transcription of heat-shock genes in *B. subtilis* occurs by a different mechanism. An inverted repeat DNA element downstream of the *dnaK* or *groESL* heat-shock promoters (*CIRCE*) is recognized by a specific repressor (HrcA), whose activity is modulated by the GroESL proteins.

A complex regulon is responsible for the reactions that occur during the transition of Gram-negative bacteria from exponential to *stationary growth*. Most genes preferentially expressed under conditions of *stationary phase* growth or *osmotic shock* are collectively regulated in response to the stationary phase-specific σ factor σ^{38} (σ^{s}). The promoters for σ^{s}-dependent genes or genes expressed under exponential growth (σ^{70}) do not differ greatly. Therefore, σ^{s}-dependent transcription is either controlled by additional regulators or the relative concentration of the different σ factors. The σ^{s} concentration in the cell is regulated by *transcription*, *translation*, *activity* and *protein stability*. Many different regulators cooperate in modulating the concentration of σ^{38} and are therefore linked to the system of stationary phase expression. Transcription of the σ^{s} mRNA is activated by *ppGpp*, *homoserine lactone*, and is inhibited by *UDP–glucose* and *cAMP–CRP*. At the post-transcriptional level regulation may occur because of alterations of the σ^{38} *mRNA structure*, probably through the activity of *H-NS* or the RNA-binding protein *HF-1*. A small untranslated RNA (*DsrA*) antagonizes the repressing effect of H-NS. The stability of σ^{38} is regulated by the *protease ClpXP*, and the activity of σ^{s} is modified by a protein termed *RssB*.

The *regulatory protein Lrp* exerts global control on a large network of metabolic and biosynthetic operons, which is summarized as the *Lrp regulon*. The activity of the transcription factor Lrp is modulated by binding the amino acid L-leucine. Whereas biosynthetic operons are usually activated, catabolic pathways are generally repressed by Lrp. It represses the expression of genes needed during growth in rich medium and activates the transcription of genes preferentially needed in poor media. Lrp, therefore, appears to adapt cellular expression to the quality of the growth medium. The activity of Lrp is often modulated by additional DNA structuring transcription factors, e.g. *IHF* or *H-NS*, indicating the importance of DNA-bending for the regulatory mechanism of Lrp.

A global network regulating all possible biosynthetic activities in the cell is induced when cells are starved of amino acids. The response is summarized by the term *stringent control*. During the stringent control stable RNA transcription is repressed, while transcription of amino acid biosynthetic operons is activated. The control is triggered by the accumulation of the effector compound *ppGpp*, which is synthesized by the ribosome associated *ppGpp synthetase I* (*RelA*) as a consequence of codon-directed binding of uncharged tRNAs to the ribosomal A site. The accumulation of ppGpp not only affects transcription but causes a pleiotropic pattern of metabolic changes. ppGpp probably associates with RNA polymerase, causing a differential switch in transcriptional activity at stringently or non-stringently controlled promoters. Promoters which are under negative stringent regulation are characterized by a GC-rich sequence between the –10 region and the transcription start site called the *discriminator*. The discriminator is a necessary but not sufficient element for stringent regulation. Genes under positive stringent control have an AT-rich discriminator sequence. During the stringent control transcription is affected at the stages of initiation and elongation. The mechanism of inhibition by ppGpp during transcription initiation depends on the respective promoter and can take place at several substeps involving changes in K_B, k_2 and the rate of promoter escape. Regulation at the stage of transcription elongation is brought about by enhancement of RNA polymerase pausing at specific sites. Reactions induced by the stringent control are tightly linked to the regulatory network of the stationary phase-specific σ factor σ^s.

The transcription of a number of genes is regulated such that their rate of transcription is proportional to the square of the growth rate of the cell. These genes are classified as *growth rate regulated* genes. Stable RNAs are the prototypes of growth rate-regulated genes. The same promoter determinants (discriminators) that are responsible for stringent regulation direct growth rate control. The transcription of some genes is under *inverse growth rate control*. Specific promoter structures termed *gearbox promoters* have been identified in these cases. The concentration of the effector molecule ppGpp, which correlates inversely with the growth rate, appears to be responsible for the mechanism of growth rate regulation. The changes in the concentration of the effector ppGpp at steady-state growth are the result of a second ppGpp synthetase activity (*ppGpp synthetase II* or *SpoT*). Growth rate regulation has been suggested to be the result of ppGpp-induced transcriptional pauses which affect the transcription elongation rate. Pausing or queuing RNA polymerases may either cause promoter occlusion or sequester RNA polymerases and thereby initiate a redistribution of polymerases to promoters with higher affinities. Growth rate regulation may also be observed in the absence of ppGpp. Under such conditions regulation may be brought about by the transcription factor H-NS. Alternatively, a ppGpp-independent feedback control mechanism has been proposed, which is considered to link stable RNA expression to cell growth. The responsible regulator has not yet been identified, however. A growth rate-dependent change in the cellular concentration of NTPs required for the formation of stable transcription initiation complexes at sensitive promoters has also been suggested as a possible mechanism for ppGpp-independent growth rate regulation.

The regulation of rRNA transcription represents a vivid example of a complex regulated network. A large number of different control mechanisms act in concert to adjust the synthesis of rRNAs exactly to the cell demands. The complexity of this regulation is not surprising, since the synthesis of rRNAs determines the rate of ribosome formation and thus presents an important parameter for the protein synthesizing capacity and the cell growth of bacteria. In *E. coli* rRNAs are encoded in seven operons which are conserved in structure except for the upstream promoter regions. There is accumulating evidence for operon-specific regulation. All seven transcription units are controlled by tandem promoters P1 and P2, which both contribute to transcription but are regulated differently. P1 promoters are under strong stringent and growth rate control, while P2 promoters respond only marginally to changes in the ppGpp level. rRNA promoters are characterized by a suboptimal spacing between the –10 and –35 promoter core elements, which explains the twist and supercoil dependency of these promoters. The strength of rRNA promoters strongly depends on the curved *upstream activating sequence (UAS)* and a region immediately upstream of the –35 region, the *UP element*. The UAS region contains binding sites for the activating factor *FIS* and the antagonizing repressor *H-NS*. Both transcription factors link rRNA synthesis to the growth phase. The UP element, which provides a direct binding site for the C-terminal domain of the RNA polymerase α subunit (αCTD) contributes to the high efficiency of rRNA promoters. rRNA transcription is further controlled during elongation because of RNA polymerase pausing. Sequence elements closely homologous with the antitermination sequences of lambdoid phages (*nut* sites) suggest the existence of a mechanism during rRNA synthesis which is related to transcription antitermination. The conserved *nut* site sequences within the leader region of all rRNA operons have been shown to participate as molecular chaperones in the correct folding of rRNAs before they are cleaved off by processing. The leader sequences of bacterial rRNA operons thus facilitate structure formation and assembly of the small ribosomal subunits similar to eukaryotic snoRNAs.

References

Aldea, M., Garrido, T., Pla, J. and Vincente, M. (1990) Division genes in *Escherichia coli* are expressed coordinately to cell septum requirements by gearbox promoters. *EMBO Journal* **9**: 3787–94.

Cashel, M., Gentry, D. R., Hernandez, V. J. and Vinella, D. (1996) The stringent response. In: Neidhard, F. C., Curtiss III, R., Ingraham, J. L. *et al.* (eds) Escherichia coli *and* Salmonella *Cellular and Molecular Biology*, Vol. 1. Washington DC: ASM Press, pp. 1458–96.

Heinemann, M. and Wagner, R. (1997) Guanosine 3′,5′-bis(diphosphate) (ppGpp)-dependent inhibition of transcription from stringently controlled *Escherichia coli* promoters can be explained by an altered initiation pathway that traps RNA polymerase. *European Journal of Biochemistry* **247**: 990–99.

Kingston, R. E., Nierman, W. C. and Chamberlin, M. J. (1981) A direct effect of guanosine tetraphosphate on pausing of *E. coli* RNA polymerase during RNA chain elongation. *Journal of Biological Chemistry* **256**: 2787–97.

Loewen, P. C. and Hengge-Aronis, R. (1994) The role of the sigma factor σ^S (KatF) in bacterial global regulation. *Annual Reviews of Microbiology* **48**: 53–80.

Low, D., Braaten, B. and van der Wouden, M. (1996) Fimbriae. In: Neidhard, F. C., Curtiss III, R., Ingraham, J. L. *et al.* (eds) Escherichia coli *and* Salmonella *Cellular and Molecular Biology*, Vol. 1. Washington DC: ASM Press, pp. 146–57.

Tippner, D., Afflerbach, H., Bradaczek, C. and Wagner, R. (1994) Evidence for a regulatory function of the histone-like *Escherichia coli* protein H-NS in ribosomal RNA synthesis. *Molecular Microbiology* **11**: 589–604.

FURTHER READING

Bremer, H. and Ehrenberg, M. (1995) Guanosine tetraphosphate as a global regulator of bacterial RNA synthesis: a model involving RNA polymerase pausing and queuing. *Biochimica et Biophysica Acta* **1262**: 15–36.

Bukau, B. (1993) Regulation of the *Escherichia coli* heat-shock response. *Molecular Microbiology* **9**: 671–80.

Calvo, J. M. and Matthews, R. G. (1994) The leucine-responsive regulatory protein, a global regulator of metabolism in *Escherichia coli*. *Microbiological Reviews* **58**: 466–90.

Condon, C., Squires, C. and Squires, C. l. (1995) Control of rRNA transcription in *Escherichia coli*. *Microbiological Reviews* **59**: 623–45.

Craig, E. A., Gambill, B. D. and Nelson, R. J. (1993) Heat shock proteins: molecular chaperones of protein biogenesis. *Microbiological Reviews* **57**: 402–14.

Gourse, R. L., Gaal, T., Bartlett, M. S., Appleman, J. A. and Ross, W. (1996) rRNA transcription and growth rate-dependent regulation of ribosome synthesis in *Escherichia coli*. *Annual Reviews of Microbiology* **50**: 645–77.

Hengge-Aronis, R. (1993) Survival of hunger and stress: the role of *rpoS* in early stationary phase gene regulation in *E. coli*. *Cell*: **72**: 165–8.

Hengge-Aronis, R. (1996) Back to log phase: σ^s as a global regulator in the osmotic control of gene expression in *Escherichia coli*. *Molecular Microbiology* **21**: 887–93.

Jensen, K. F. and Pedersen, S. (1990) Metabolic growth rate control in *Escherichia coli* may be a consequence of subsaturation of the macromolecular biosynthetic apparatus with substrates and catalytic components. *Microbiological Reviews* **54**: 89–100.

Lamond, A. I. and Travers, A. A. (1985) Stringent control of bacterial transcription. *Cell* **41**: 6–8.

Little, J. W. and Mount, D. W. (1982) The SOS regulatory system of *Escherichia coli*. *Cell* **29**: 11–22.

Loewen, P. C. and Hengge-Aronis, R. (1994) The role of the sigma factor σ^s (KatF) in bacterial global regulation. *Annual Reviews of Microbiology* **48**: 53–80.

Murray, K. D. and Bremer, H. (1996) Control of *spoT*-dependent ppGpp synthesis and degradation in *Escherichia coli*. *Journal of Molecular Biology* **259**: 41–57.

Neidhardt, F. C. and M. A. Savageau (1996). Regulation beyond the operon. In: Neidhard, F. C., Curtiss III, R., Ingraham, J. L. *et al.* (eds) Vol. 1. *Escherichia coli and Salmonella Cellular and Molecular Biology*. Washington, DC: ASM Press, pp. 1310–24.

Newman, E. B. and D'Ari, R. (1992) The leucine-LRP regulon in *E. coli*: a global response in search of a raison d'Étre. *Cell* **68**: 617–19.

Newman, E. B. and Lin, R. T. (1995) Leucine-responsive regulatory protein: a global regulator of gene expression in *E. coli*. *Annual Reviews of Microbiology* **49**: 747–75.

Nomura, M., Gourse, R. and Baughman, G. (1984) Regulation of the synthesis of ribosomes and ribosomal components. *Annual Reviews of Biochemistry* **53**: 75–117.

Reddy, P. S., Raghavan, A. and Chatterji, D. (1995) Evidence for a ppGpp-binding site on *Escherichia coli* RNA polymerase: proximity relationship with the rifampicin-binding domain. *Molecular Microbiology* **15**: 255–65.

Vincente, M., Kushner, S. R., Garrido, T. and Aldea, M. (1991) The role of the 'gearbox' in the transcription of essential genes. *Molecular Microbiology* **5**: 2085–91.

Vinella, D. and D'Ari, R. (1995) Overview of controls in the *Escherichia coli* cell cycle. *BioEssays* **17**: 527–36.

Wagner, R. (1994) The regulation of ribosomal RNA synthesis and bacterial cell growth. *Archives of Microbiology* **161**: 100–9.

Zuber, U. and Schumann, W. (1994) CIRCE, a novel heat shock element involved in regulation of heat shock operon *dnaK*, of *Bacillus subtilis*. *Journal of Bacteriology* **176**: 1359–63.

Glossary

Abortive transcription Synthesis of short RNA transcripts (less than 10 nucleotides) which are released by RNA polymerase during repetitive transcription from the transcription start site. Abortive transcription occurs during initiation while the σ factor of RNA polymerase is still present in the initiation complex. RNA polymerase does not move away from the promoter during abortive transcription.

αCTD C-terminal domain of RNA polymerase α subunit (amino acids 249–329) which has important functions in promoter (UP element) and transcription factor binding. The αCTD is separated by a flexible linker sequence (approximately 10 amino acids from the N-terminal domain; see αNTD.

Amphipathic helix Protein secondary structure consisting of an α helix with alternating polar and unpolar amino acid residues at every seventh position (two helical turns). The surface of the helix is therefore of opposite polarity (charge) at both sides.

Antisense strand Identical in sequence to the template strand of a transcript

Anti-sigma factor Protein that binds to sigma factors and interferes with transcription initiation. The phage T4 anti-sigma factor AsiA, for example, binds to the C-terminal subdomain 4.2 of σ^{70} and inhibits transcription initiation of the host genes.

Antitermination Transcriptional control mechanism to suppress termination. Antitermination involves the concerted activity of Nus factors and RNA sequences termed *nut* sites which form a cooperative complex with RNA polymerase. Antitermination is important for the regulation of phage lambda gene expression but also for the expression of many bacterial genes.

αNTD N-terminal domain of RNA polymerase a subunit (approximately the first 240 amino acids) separated by a flexible linker from the C-terminal domain; see αCTD. The αNTD is required for RNA polymerase subunit assembly.

ApA wedge model Model describing DNA curvature owing to dinucleotide steps ApA which have altered roll and tilt angles compared to normal B-DNA.

AraC Representative of a class of dimeric regulators (AraC-like regulators) that interact with the major groove of DNA through a conserved helix-turn-helix motif. AraC-like regulators are often modulated allosterically by binding of sugar molecules. AraC regulates transcription of the *araBAD* operon via DNA loop formation.

Arrest sites Site where RNA polymerase cannot continue transcription. Arrest sites may be caused by limitations of substrate NTPs or by proteins bound to the template DNA acting as a 'road block'.

Assembly control particle (ACP) Complex consisting of *nut*-like sequences bound to transcribing RNA polymerase through the Nus factors NusA, NusB, NusE and NusG. The ACP which is formed while RNA polymerase transcribes through the ribosomal RNA leader regions has been denoted in analogy to the elongation control particle (ECP) characteristic for phage lambda transcription antitermination. The ACP apparently does not only function to program a termination-resistant RNA polymerase but facilitates ribosomal biogenesis through transient interactions with elements of the preformed ribosomal subunit.

α subunit 35-kDa (*E. coli*) constituent of RNA polymerase. Each RNA polymerase has two α subunits.

Attenuation Transcriptional control mechanism mediated through regulated termination. Attenuation generally involves the formation of mutually exclusive secondary transcript structures which either function as Rho-independent terminators or antiterminators. The best studied examples are the biosynthetic operons for amino acids.

Backtracking Reverse sliding of the RNA polymerase elongation complex involving the reversible RNA–DNA hybrid formation and dislocation of the active centre.

B junction Model describing DNA curvature resulting from the kink derived at the junction between regions of B-DNA and non-B-DNA.

β subunit Second largest subunit of RNA polymerase (*E. coli*: 1342 amino acids, 150.6 kDa).

β′ subunit Largest RNA polymerase subunit (*E. coli*: 1407 amino acids, 155.2 kDa).

cAMP Cyclic adenosine monophosphate. Associates with CRP and enables the transcription factor to bind to its target sequences, regulating a large number of transcription units.

Cap structure Post-transcriptional 5′ end modification of eukaryotic mRNAs. The Cap structure is a 7-methyl guanosine linked via a 5′-5′ phosphotriester to the first nucleotide of the transcript. In addition, the 2′ hydroxyl groups of the first ribose units are methylated.

Chaperone Molecules that assist the correct folding or biogenesis pathway of other protein or RNA molecules without being components of the final structures. Notable members are proteins of the heat-shock family, e.g. DnaK, DnaJ, GrpE.

Chemotaxis Active movement of bacteria according to a gradient in the concentration of chemical substances in the environment.

CIRCE element (controlling inverted repeat of chaperone expression) Regulatory DNA sequence for binding of the HrcA repressor. The CIRCE element acts as operator for the regulation of chaperone gene transcription in *B. subtilis*.

Closed complex First complex formed between RNA polymerase and promoter DNA. The DNA within a closed complex is still entirely double-stranded.

Coding strand See template strand.

Core polymerase RNA polymerase with the subunit composition $\alpha_2\beta\beta'$. The core polymerase is fully active in transcription elongation but is unable to initiate transcription without the additional σ subunit (see holoenzyme).

Corepressor See inducer.

Core promoter Minimal DNA sequence to direct specific transcription initiation. The core promoter is composed of a –10 region, a –35 region and a spacer sequence of 17 ± 1 base pairs.

Coumermycin Antibiotic inhibiting the action of topoisomerase II (gyrase).

CRP Catabolite regulator protein, also sometimes called catabolite activator protein (CAP). CRP, complexed with cAMP, binds to the major groove of a conserved interrupted palindromic sequence and causes positive and negative regulation of a large number of bacterial transcription units.

***dam* methylation** Modification of adenosines within GATC sequences to 6-methyl adenosines by deoxyadenosine methylase (*dam*). Serves to distinguish parental DNA from daughter strands during mismatch repair. Transcription from several promoters that contain GATC sequences is under cell cycle control through *dam* methylation which follows replication.

Daughter strand Newly synthesized DNA strand in a semiconservative replication mechanism.

DEAD box protein Family of proteins with the conserved amino acid sequence Asp-Glu-Ala-Asp, or, in the single-letter, code: DEAD. This sequence element is characteristic for RNA or DNA helicases.

Dead end complexes Arrested ternary elongation complexes which, in contrast to paused complexes, are functionally impaired and incapable of resuming elongation.

Discriminator motif Conserved sequence element (GCGC) between the –10 region and the transcription start site of stringently controlled and growth rate-dependent promoters.

Distamycin Antibiotic that binds into the small groove of AT-rich double-stranded DNA regions where it causes a straightening of curved DNA.

Distributive Distributive enzymes are able to reinitiate at the same position of a macromolecular polymerization reaction where they have previously stopped synthesis and left the polymerizing complex (e.g. DNA polymerases).

DnaJ Heat shock protein (see chaperones).

DnaK Heat shock protein (see chaperones).

DNA Deoxyribonucleic acid.

DSR Sequence region flanking strong (phage) promoters, ranging downstream from positions +1 to + 20.

DsrA Small untranslated RNA involved in regulation. DsrA is capable of antagonizing H-NS-mediated repression

Elongation control particle (ECP) Complex consisting of transcribing RNA polymerase, a *nut* sequence bound by Nus factors (N), NusA, NusB, NusE and NusG. This complex, which carries the tethered transcript, has the characteristics of an antitermination machine programmed to read through terminators at far distance.

Enhancer elements DNA sequences containing regulator protein-binding sites which modulate the activity of promoters. Enhancer elements are not restricted to a position close to the promoter but can affect transcription initiation from distant sites.

Fimbriae Hair-like organelles found on *Escherichia coli* and other enterics conferring adhesive properties towards mammalian host tissues.

FIS (Factor for inversion stimulation) Growth rate-regulated dimeric transcription factor that binds via a helix-turn-helix motif to a degenerated palindromic sequence and bends DNA upon binding. FIS is the major activating factor for the transcription of stable RNA genes. FIS is also involved in DNA recombination and replication.

GalR Dimeric repressor of the *gal ETK* operon (representative of the LacI/GalR family of regulators).

Gearbox promoter Small group of promoters which are regulated inversely with the growth rate. Their consensus sequence deviates from standard *E. coli* promoters.

Gene Functional unit of heredity; part of a DNA molecule that codes for a structural or functional cellular unit.

Genetic code Alphabet of nucleic acid triplets encoding the 20 different amino acids.

GreA/GreB Two proteins of similar function which are associated with elongating RNA polymerase and function as transcript cleavage factors. The two proteins play a role in the escape of arrested transcription complexes and possibly for proofreading.

gRNAs *Guide RNAs;* gRNAs are involved in the process of RNA editing. They are responsible for the insertion and removal of uracil bases from pre-edited mRNA molecules.

Growth rate regulation Increase in transcriptional activity with increasing growth rates. Observed for stable RNA transcription where synthesis rates increase with the square of the growth rates.

GrpE Heat shock protein (see chaperones).

Heat shock proteins (HSP) Family of proteins, conserved from bacteria to mammals, that is transiently expressed at elevated temperature. Heat shock proteins function as molecular chaperones (e.g. DnaK, DnaJ, GrpE) or proteases (Lon, Clp). Their transcription is controlled by specific sigma factors (σ^{32} and σ^{24}).

Helix-turn-helix (HTH) Structural motif of DNA-binding proteins consisting of two α helices linked by a turn. HTH proteins bind to the major groove of DNA.

Heparin Group of glycosaminoglycans with average M_r of 6000–20000 containing N- and O-sulphate groups. Heparin is of polyanionic character and is frequently used as a DNA or RNA competitor, sequestering free nucleic acid-binding proteins.

HF-I RNA-binding protein (host factor for phage Qβ). HF-I affects H-NS-dependent effect on *rpoS* expression, probably by binding to H-NS.

Histone-like proteins (or DNA-structuring proteins) Small abundant bacterial proteins (e.g. HU, IHF, FIS, H-NS) that bind to DNA and help to structure the bacterial nucleoid. Histone-like proteins do not share amino acid sequence homologies with eukaryotic histones, but cause similar compaction of DNA. For many of the histone-like proteins specific functions as transcription factors have been shown.

H-NS Histone-like nucleoid structuring protein, a DNA-binding protein that specifically recognizes curved DNA, not a conserved primary sequence. H-NS binding causes compaction of DNA (see histone-like proteins). In many case it acts, however, as a specific transcription factor (mostly repressor) controlling a broad spectrum of genes (often genes required to overcome cellular stress situations). For the transcription of several genes (e.g. ribosomal RNAs) H-NS functions as an antagonist to the activator FIS.

Holoenzyme RNA polymerase holoenzyme has the subunit structure $\alpha_2\beta\beta'\ \sigma$. The σ subunit provides the ability to initiate transcription. After the initiation cycle the σ subunit is released and RNA polymerase is converted to the core enzyme.

Homoserine lactone α-amino-γ-butyrolactone; Metabolic intermediate signalling stationary phase expression. The structurally related compounds, acylated homoserine lactones, act as extracellular signal in sensing bacterial population densities ('quorum sensing').

Housekeeping genes Standard metabolic genes required during exponential growth. Transcription of housekeeping genes is generally initiated by RNA polymerase associated with the vegetative sigma subunit, σ^{70}.

HU Small heterodimeric protein which binds to and bends DNA. HU shows sequence homology to IHF and in several cases both proteins can substitute each other (see also histone-like proteins).

IHF DNA-binding protein consisting of two subunits (IHF-α and IHF-β). IHF binds to the minor groove of DNA and causes a large bend, similar to a complete U-turn. IHF is involved in the regulation of many transcription units where it often functions as an architectural element, facilitating an adequate nucleoprotein complex for transcription initiation (see also histone-like proteins).

Inchworm model Model describing RNA polymerase translocation as discontinuous movement of front and rear edges of the enzyme along the template DNA.

Inducer Compound that associates with a regulatory protein (usually a repressor), changing the DNA-binding affinity or the mode of action of transcription interference.

Initial transcribing complexes (ITC) Transcription initiation complexes that have already started RNA synthesis (fewer than 10 nucleotides) but not yet released the sigma factor.

Kinetoplast DNA DNA present in large quantities within self-replicating organelles located in an expanded region of mitochondria of members of protozoans, like Trypanosoma, Leishmania and Crithidia. Kinetoplast DNA is characterized by regions of strong DNA curvature.

Leucine zipper Amino acid motif that functions in the dimerization of proteins (mostly transcription factors) by hydrophobic side chains exposed on adjacent sides of two α helices.

LexA See SOS response.

Linking number (L_k) Indicates how many times two topologically closed structures are linked with each other. The linking number is thus a topological constant. It is used to describe the superhelical status of circular DNA according to $L_k = T_w + W_r$ (see writhe and twist).

Lrp (leucine-responsive regulatory protein) Transcription regulator of a complex regulon involving many biosynthetic and metabolic operons. The activity of Lrp is modulated in many cases through leucine.

Messenger RNAs (mRNAs) RNA molecules that encode the sequence for one or more proteins. mRNAs are normally translated by ribosomes.

Modulons (stimulon) Independent operons belonging to different regulons but which are controlled by a common regulator.

Non-template strand Strand opposite to the template strand within double helical DNA. The non-template strand is complementary to the template strand. RNA synthesized during transcription has the same sequence as the non-template strand, except that ribose is substituted for deoxyribose and uracil is substituted for thymine.

Novobiocin Antibiotic inhibiting the action of topoisomerase II (gyrase).

NtrB (NR_{II}) Sensor kinase for the regulation of nitrogen fixation genes. Phosphorylates NtrC.

NtrC (NR_{I}) Transcription activator (response regulator) for the expression of nitrogen-fixation genes. NtrC activates σ^{54}-controlled promoters. The activity of NtrC depends on phosphorylation by NtrB.

Nun protein Specific termination factor encoded by the lambdoid phage HK022

Nus factors (Nus proteins) Derived from N protein utilization substances, this is a group of proteins involved in transcription antitermination, pausing and termination, but also during post-transcriptional processes; e.g. NusA, NusB, NusE, NusG.

***nut* site** RNA sequence containing the conserved sequence elements boxA, boxB and boxC. The *nut* site sequences are known to bind to Nus proteins and RNA polymerase. They are involved in the mechanism of antitermination but also have important post-transcriptional functions, e.g. maturation. *nut*-like sequences in the ribosomal leader RNA are involved in a chaperone-like mechanism facilitating structure formation and assembly of ribosomal subunits.

Open complex Complex between RNA polymerase and promoter DNA in which the transcription start site is no longer base-paired but free to allow base pair recognition of incoming NTPs according to the sequence of the template strand.

Operon Number of prokaryotic genes (usually functionally related) transcribed from a common promoter as a single transcription unit.

Pausing sites Position where the RNA polymerase step time is significantly longer than for an average elongation step. Pausing sites are often, but not exclusively, associated with the formation of stable hairpin structures of the growing transcript.

Persistence length Measure of the tendency of a stiff polymer (e.g. a helical DNA molecule) to continue in the same direction without bending. B-form DNA has an average persistence length of approximately 150 base pairs.

Phase variation The expression of flagella or fimbriae is controlled by a reversible switch between transcription ON and OFF states involving several transcriptional regulators, such as CRP or Lrp, but also GATC sequences.

Plectonemic Coiling of a double-stranded helix in which the two strands are intertwined. They may not be separated without uncoiling.

Polarity This refers to the effects of nonsense mutations (polar mutations) within a translatable mRNA that influence transcription in the downstream region. Because of the coupling of transcription and translation, provoked translational termination induces premature transcription termination of downstream genes (polar transcription). The termination factor Rho is involved in the phenomenon of polarity.

Polysomes Battery of ribosomes translating simultaneously on the same mRNA molecule.

POU-specific proteins Family of cell type-specific eukaryotic transcription factors with a common DNA-binding motif designated according to the representative members Pit-1, Oct-1, Oct-2 and Unc-86.

ppGpp Guanosine tetraphosphate accumulates to high cellular concentration upon amino acid deprivation or reduced growth rates. It is a pleiotropic effector compound involved in the stringent control and during growth rate regulation. The major effect of ppGpp is exerted at the level of transcription. Stable RNA synthesis is repressed at elevated ppGpp concentrations.

Pribnow box Core promoter element for the recognition of RNA polymerase with the consensus sequence TATAAT. Also termed –10 region, it binds to subdomain 2.4 of the σ subunit.

Primer shifting Slippage of the RNA polymerase active centre backward to a template position upstream of the start site after correct initiation and synthesis of the first few nucleotides has occurred.

Processing Cleavage or modification reactions that render the primary products of transcription or translation into functional molecules.

Processive Processive enzymes cannot reinitiate the same polymerization reaction once they have left the reactive complex. The forward reaction to long products is highly favoured. Processive enzymes can only initiate their reaction at a defined position (e.g. a promoter). RNA polymerase is a processive enzyme.

Productive elongation After release of the σ subunit of RNA polymerase following the initiation cycle the transcription complex undergoes a conformational change to an elongating complex. This complex is highly processive. It produces long transcripts and no more abortive products are formed.

Promoter Start site for DNA-dependent RNA transcription; it contains conserved sequences recognized by the holoenzyme of RNA polymerase.

Promoter clearance Step that characterizes the transition of a transcription initiation complex to an elongation complex (see also promoter escape). The RNA polymerase which has released the sigma factor (core enzyme) moves into the downstream direction, leaving the promoter region. As a result of this movement the promoter becomes free for the next round of initiation.

Promoter escape Movement of the RNA polymerase initiation complex in the downstream direction, changing the characteristics of an initiation to an elongation complex. The sigma factor is lost but, in contrast to promoter escape, because of RNA polymerase pausing within the initiation region the promoter is not necessarily free for the next round of initiation.

Promoter melting Disruption of the complementary base pairs (between five and 18) within the transcription start site to allow incorporation of NTP substrates into the growing RNA chain according to base pair complementarity.

Promoter occlusion See turnstile attenuation.

Proofreading Correction or avoidance (error prevention) of errors that might occur because of misincorporation of a nucleotide into the growing transcript. Proofreading is probably achieved by pyrophosphorolysis and/or the transcript cleavage activity of the GreA and GreB proteins.

***qut* site** Antitermination directing sequence at the lambda $P_{R'}$ promoter. The *qut* site acts as a DNA element to which the antitermination protein Q binds and alters RNA polymerase activity, such that it reads through the downstream terminator tR′.

RecA See SOS response.
Regulon Several independent operons controlled by a common regulator.
RelA (ppGpp synthetase I, stringent factor or PSI) Ribosome-associated protein that synthesizes (p)ppGpp from ATP and GTP in response to the cognate binding of uncharged tRNA to the ribosomal A-site.
Replication Synthesis of a new DNA double strand; both complementary parental DNA strands are copied by a semiconservative mechanism.
Response regulator Group of proteins constituting two component regulatory systems. They are usually composed of a receiver domain which contains conserved aspartate residues that are modified through phosphorylation and an output domain which functions as a DNA-binding element.
Rho factor Hexameric protein acting as a transcription termination factor. Rho binds to the nascent transcript, moves along the RNA in 5′ to 3′ direction, approaches the transcription complex and causes the release of the transcript. The reaction requires the hydrolysis of ATP.
Ribozymes Functional RNA molecules that are capable of carrying out enzymatic reactions in the absence of proteins.
Rifampicin Antibiotic inhibiting bacterial transcription initiation by binding to the β subunit.
RNA Ribonucleic acid
RNA polymerase Enzyme that catalyses the DNA-dependent polymerization of ribonucleotide triphosphates to yield RNA; international enzyme nomenclature: EC 2.7.7.6.
Roll The rotation around the long axis of two neighbouring base pairs within a helical structure. The roll is positive (greater than 0) if the two base pairs open up towards the minor groove side.
RpoS (σ^{38}) Alternative sigma factor for genes expressed at stationary phase or osmotic shock.
***rut* site** Rho factor utilization site. Bipartite (*rut* A, *rut* B) binding site on a transcript which is recognized by Rho. *rut* sites have little secondary structure, a high cytosine over guanosine content and often harbour *nut* site sequences boxA and boxB.

Semiconservative Only one strand of DNA is newly synthesized during replication (daughter strand); the complementary strand is the original parental strand.
Sensor kinase Group of specific kinases that function as members of two component regulatory systems. Phosphate transfer occurs from a conserved histidine residue (histidine kinase) which is autophosphorylated depending on a specific stimulus or signal. The phosphate residue is subsequently transmitted from the sensor kinase to an aspartate residue of the corresponding response regulator.
σ subunit Specificity factor for RNA polymerase which directs transcription initiation. It is a constituent of the RNA polymerase holoenzyme, but not present in the core enzyme.
Sliding One-dimensional movement (linear diffusion) of an entirely electrostatic (non-specifically) bound protein (e.g. transcription factor or RNA polymerase) along a nucleic acid chain. It causes enhanced target location.

Sliding clamp Protein structure which ensures highly processive elongation reactions (transcription or replication) owing to topological ring closure around the nucleic acid strands.

snoRNAs Family of *small nucleolar RNAs* found in the nucleolus of eukaryotic cells. snoRNAs are involved in the processing, assembly and maturation of eukaryotic ribosomes. They also direct specific methylation and pseudouracil substitution into eukaryotic rRNAs.

snRNAs Family of *small nuclear RNAs* found in the nucleus of eukaryotic cells. snRNAs are involved in the process of splicing.

SOS response Regulation of a set of (mainly DNA repair) genes which are coordinately repressed by the regulator LexA. The SOS response is induced by DNA damage via proteolytic inactivation of the LexA repressor through modified RecA proteins.

Spacer region Sequence separating the promoter recognition elements (e.g. the −10 and −35 regions). For *E. coli* σ^{70} promoters the spacer sequence has a length of 17 ± 1 base pairs.

Specifier sequence Involved in the antitermination mechanism of *B. subtilis* aminoacyl tRNA synthetase genes. The specifier sequence, together with the T-box present in the leader region of aminoacyl tRNA synthetase genes, helps to bind the anticodon of cognat uncharged tRNA molecules thereby enabling the formation of an antitermination structure which causes transcription to continue into the structural genes.

Splicing Removal of intron sequences from a continuous pre-mRNA transcript; the exon boundaries are linked to a functional mRNA.

SpoT (ppGpp synthetase II, PSII) Enzyme responsible for the degradation of ppGpp. Contains a RelA-independent ppGpp synthetase activity which regulates the concentration of the effector nucleotide in response to the growth rate.

Stable RNAs Family of RNA molecules includes ribosomal RNAs, the RNA components of ribosomes, and transfer RNAs, which function as adapter molecules delivering amino acids to the ribosome. Stable RNAs are termed so because of their long half-life in the cell.

StpA Structural and partially functional homologue of H-NS.

Streptolydigin Antibiotic interacting with the β′ subunit of RNA polymerase. Streptolydigin inhibits the rate of transcription of bacterial RNA polymerases.

Stressed intermediate RNA polymerase initiation complex conformation between the open complex and the early elongating complex resulting from downstream end movement while the upstream end is still bound to the −35 promoter region.

Stringent response (stringent control) Pleiotropic response to amino acid starvation involving the activity of the effector nucleotide ppGpp.

Supercoiling Supercoiling or superhelical DNA results from the overwinding or underwinding of covalently closed circular DNA molecules or molecules fixed at both ends. Superhelical windings are characterized by the writhing number W_r. Supercoils can be constrained when the DNA is not wound around itself but wrapped around DNA-binding proteins. Note that winding around a protein core in left-handed turns, as it occurs in nucleosomes, corresponds to negative supercoils (right-handed superhelical windings) when the DNA is subsequently closed and the protein is removed.

Superhelical density The superhelical density of DNA is defined as T_w/W_r, the ratio of the number of superhelical windings writhe (W_r) to the twist (T_w). See also supercoiling.

Tac promoter Synthetic promoter with the perfect *E. coli* promoter core consensus sequence: TTGACA, 17 base pair spacer, TATAAT.

Tau protein Specific termination factor which affects early termination at the phage T3 or T7 terminators T3Te and T7Te.

T-box See specifier sequence.

TBP TATA box binding protein. Eukaryotic transcription factor that binds to the minor groove creating a sharp DNA kink.

Template strand DNA strand which is used as matrix for the transcription reaction. Reading of the template strand occurs in a 3′ to 5′ direction. Because of the incorporation of complementary nucleotides in a 5′ to 3′ direction the resulting transcript sequence is the reverse complement of the template strand.

Termination factor See Rho factor.

Ternary complexes Transcription complexes consisting of RNA polymerase, the template DNA and either the initiating NTP or a nascent transcript. Binary open complexes which consist of RNA polymerase and the promoter DNA are converted to ternary initiation complexes after binding of the first substrate NTP. Ternary elongation complexes consist of RNA polymerase, the template DNA and the nascent RNA chain.

Tetraloop Stable RNA secondary structure consisting of a loop with four nucleotides (UUCG or the GNRA family with N, any nucleotide and R, any purine). Tetraloops are stabilized by many unconventional stacking and H-bond interactions.

Tilt The rotation around the short axis of two neighbouring base pairs within a helical structure.

Topoisomerases Enzymes that keep homeostatic control of DNA supercoiling within a cell. Topoisomerases can either introduce negative superhelical turns into DNA under consumption of ATP (topoisomerase II or gyrase) or they relax negative supercoils (topoisomerase I).

τ-plot A kinetic analysis of transcription initiation from which isomerization rate constants (k_2)and primary binding constants (K_B) can be derived. In a τ-plot the times required for isomerization (τ) are plotted as a function of the reciprocal RNA polymerase concentration. τ-plots are useful to define promoter strength.

Transcription Synthesis of RNA according to a DNA template; it is an enzymatic reaction performed by RNA polymerase in the 5′ to 3′ direction.

Transcription bubble DNA region within the transcription complex where the template and non-template strands are not base paired to allow continuous recognition of the incoming substrate NTPs according to the template strand sequence. The unpaired DNA region of the transcription bubble is about 12–15 bases in length and moves with the elongating RNA polymerase complex in the downstream direction of transcription.

Transcription factor A protein that activates or represses transcription usually, but not exclusively, at the stage of initiation. Transcription factors bind to either DNA, RNA polymerase or both. They are of modular composition and their activity can often be changed by ligand binding or covalent modification (e.g.

phosphorylation). Some transcription factors are inactivated by proteolysis (e.g. LexA or the lambda repressor cI).

Transcript slippage Reiterative incorporation of the same nucleotide during transcription without movement of the RNA polymerase active centre relative to the template DNA. This may occur if more than three identical nucleotides are encoded in a row. Transcript slippage or reiterative transcription is observed as a regulatory method, e.g. during transcription of pyrimidine biosynthesis operons where transcript slippage depends on the cellular concentration of pyrimidine nucleotides.

Translation Synthesis of proteins encoded by the triplet sequences of an mRNA; it is an enzymatic reaction performed by ribosomes.

Turnstile attenuation (promoter occlusion) Physical occlusion of an upstream promoter by pausing at a downstream site close to the promoter.

Twist Rotation of a base pair around the helix axis. For B-DNA the average twist between two adjacent base pairs is approximately 35°.

Two-component regulatory systems Signal transduction system in bacteria involving two regulatory proteins, a sensor kinase and a response regulator.

Two hybrid system Method to detect protein–protein interactions *in vivo* taking advantage of the functional assembly of transcription factors from separated domains fused to the putative interacting components.

TyrR box Palindromic operator sequence to which regulators of the TyrR family bind.

UAS region (upstream activating sequence) Sequence upstream of strong promoters (around –150 to –40) which activates transcription. UAS regions are often associated with DNA curvature (AT-rich) and contain transcription factor binding sites.

UP element AT-rich DNA sequence at many strong promoters located at position –40 to –60 relative to the transcription start. The UP element functions in recruiting RNA polymerase through binding to the C-terminal domain (αCTD) of the RNA polymerase α subunit.

Writhe (Wr) The path of a helical DNA axis in space giving the number of superhelical turns. It is used to describe the superhelical status of circular DNA according to $L_k = T_w + W_r$ (see linking number and twist).

ω subunit A 10-kDa protein of unclear function associated with RNA polymerase through contacts with the β′ subunit.

Index